Anthropometry, Apparel Sizing and Design

D1700060

The Textile Institute Book Series

Incorporated by Royal Charter in 1925, The Textile Institute was established as the professional body for the textile industry to provide support to businesses, practitioners and academics involved with textiles and to provide routes to professional qualifications through which Institute Members can demonstrate their professional competence. The Institute's aim is to encourage learning, recognise achievement, reward excellence and disseminate information about the textiles, clothing and footwear industries and the associated science, design and technology; it has a global reach with individual and corporate members in over 80 countries.

The Textile Institute Book Series supersedes the former 'Woodhead Publishing Series in Textiles' and represents a collaboration between The Textile Institute and Elsevier aimed at ensuring that Institute Members and the textile industry continue to have access to high calibre titles on textile science and technology.

Books published in The Textile Institute Book Series are offered on the Elsevier web site at: store.elsevier.com and are available to Textile Institute Members at a substantial discount. Textile Institute books still in print are also available directly from the Institute's web site at: www.textileinstitute.org

To place an order, or if you are interested in writing a book for this series, please contact Matthew Deans, Senior Publisher: m.deans@elsevier.com

Recently Published and Upcoming Titles
in The Textile Institute Book Series:

New Trends in Natural Dyes for Textiles, Padma Vankar Dhara Shukla, 978-0-08-102686-1

Smart Textile Coatings and Laminates, William C. Smith, 2nd Edition, 978-0-08-102428-7

Advanced Textiles for Wound Care, 2nd Edition, S. Rajendran, 978-0-08-102192-7

Manikins for Textile Evaluation, Rajkishore Nayak Rajiv Padhye, 978-0-08-100909-3

Automation in Garment Manufacturing, Rajkishore Nayak and Rajiv Padhye, 978-0-08-101211-6

Sustainable Fibres and Textiles, Subramanian Senthilkannan Muthu, 978-0-08-102041-8

Sustainability in Denim, Subramanian Senthilkannan Muthu, 978-0-08-102043-2

Circular Economy in Textiles and Apparel, Subramanian Senthilkannan Muthu, 978-0-08-102630-4

Nanofinishing of Textile Materials, Majid Montazer Tina Harifi, 978-0-08-101214-7

Nanotechnology in Textiles, Rajesh Mishra Jiri Militky, 978-0-08-102609-0

Inorganic and Composite Fibers, Boris Mahltig Yordan Kyosev, 978-0-08-102228-3

Smart Textiles for In Situ Monitoring of Composites, Vladan Koncar, 978-0-08-102308-2

Handbook of Properties of Textile and Technical Fibres, 2nd Edition, A. R. Bunsell, 978-0-08-101272-7

Silk, 2nd Edition, K. Murugesh Babu, 978-0-08-102540-6

The Textile Institute Book Series

Anthropometry, Apparel Sizing and Design

Second Edition

Edited by

Norsaadah Zakaria

Deepti Gupta

The Textile Institute

WOODHEAD
PUBLISHING
An imprint of Elsevier

Woodhead Publishing is an imprint of Elsevier
The Officers' Mess Business Centre, Royston Road, Duxford, CB22 4QH, United Kingdom
50 Hampshire Street, 5th Floor, Cambridge, MA 02139, United States
The Boulevard, Langford Lane, Kidlington, OX5 1GB, United Kingdom

Library of Congress Cataloging-in-Publication Data
A catalog record for this book is available from the Library of Congress

British Library Cataloguing-in-Publication Data
A catalogue record for this book is available from the British Library

ISBN: (print) 978-0-08-102604-5
ISBN: (online) 978-0-08-102605-2

For information on all Woodhead publications
visit our website at https://www.elsevier.com/books-and-journals

Publisher: Candice Janco
Acquisition Editor: Brian Guerin
Editorial Project Manager: John Leonard
Production Project Manager: Debasish Ghosh
Cover Designer: Matthew Limbert

Typeset by SPi Global, India

Working together
to grow libraries in
developing countries

www.elsevier.com • www.bookaid.org

Contents

Contributors

Susan P. Ashdown Department of Fiber Science and Apparel Design, College of Human Ecology, University of Cornell, Ithaca, NY, United States

Jennifer Bougourd Consultant, London, United Kingdom

Zhe Cheng Wuhan Textile University, Wuhan, PR China

Inga Dābolina Riga Technical University, Institute of Design Technologies, Riga, Latvia

M.-E. Faust Philadelphia University, Philadelphia, PA, United States; Université du Québec à Montréal (Uqàm), Montreal, QC, Canada

Jelka Geršak Faculty of Mechanical Engineering, Institute for Textile and Garment Manufacture Processes, University of Maribor, Maribor, Slovenia

Deepti Gupta Department of Textile Technology, Indian Institute of Technology Delhi, New Delhi, India

Anke Klepser Hohenstein Institute for Textilinnovation gGmbH, Boennigheim, Germany

M. Kouchi National Institute of Advanced Industrial Science and Technology, Tokyo, Japan

V.E. Kuzmichev Ivanovo State Polytechnic University, Ivanovo, Russian Federation

Eva Lapkovska Riga Technical University, Institute of Design Technologies, Riga, Latvia

Christine Loercher Hohenstein Institute for Textilinnovation gGmbH, Boennigheim, Germany

Amir Feisal Merican Centre of Research for Computational Sciences and Informatics in Biology, Bioindustry Environment, Agriculture and Healthcare (CRYSTAL); Centre for Foundation Studies in Science, University of Malaya, Kuala Lumpur; Fashion Department, Faculty of Art and Design, University Technology MARA, Selangor, Malaysia

Simone Morlock Hohenstein Institute for Textilinnovation gGmbH, Boennigheim, Germany

Maja Mahnić Naglić Faculty of Textile Technology, University of Zagreb, Zagreb, Croatia

Slavenka Petrak Department of Clothing Technology, Faculty of Textile Technology, University of Zagreb, Zagreb, Croatia

Wan Syazehan Ruznan Textile and Clothing Technology Department, Universiti Teknologi MARA (UiTM), Shah Alam, Malaysia

Andreas Schenk Hohenstein Institute for Textilinnovation gGmbH, Boennigheim, Germany

Shaliza Mohd Shariff Centre of Research for Computational Sciences and Informatics in Biology, Bioindustry Environment, Agriculture and Healthcare (CRYSTAL); Institute Graduate Studies; Centre for Foundation Studies in Science, University of Malaya, Kuala Lumpur; Fashion Department, Faculty of Art and Design, University Technology MARA, Selangor, Malaysia

Asma Ahmad Shariff Centre of Research for Computational Sciences and Informatics in Biology, Bioindustry Environment, Agriculture and Healthcare (CRYSTAL); Centre for Foundation Studies in Science, University of Malaya, Kuala Lumpur; Fashion Department, Faculty of Art and Design, University Technology MARA, Selangor, Malaysia

Rosita Mohd Tajuddin Faculty of Art & Design, Jalan Kreatif, University Technology Mara (UiTM), Shah Alam, Selangor, Malaysia

Philip Treleaven University College London (UCL), London, United Kingdom

Aisyah Mohd Yasim Faculty of Art & Design, Jalan Kreatif, University Technology Mara (UiTM), Shah Alam, Selangor, Malaysia

Norsaadah Zakaria Centre of Clothing Technology and Fashion, IBE, Universiti Teknologi MARA (UiTM), Shah Alam; Malaysian Textile Manufacturing Association (MTMA) and Malaysian Textile and Apparel Centre (MATAC), Kuala Lumpur, Malaysia

Part One

Anthropometry and garment sizing systems

New directions in the field of anthropometry, sizing and clothing fit

1

Deepti Gupta
Department of Textile Technology, Indian Institute of Technology Delhi, New Delhi, India

List of Abbreviations

1-D	one-dimensional
2-D	two-dimensional
3-D	three-dimensional
4-D	four-dimensional
RTW	ready to wear
CAESAR	Civilian American and European Surface Anthropometry Resource project
MTM	made to measure
CAD	computer-aided design
AI	artificial intelligence
WEAR	World Engineering Anthropometry Resource
SMPL	skinned multiperson linear model
XR	extended reality

1.1 Introduction

The apparel manufacturing industry is concerned with the process of producing garments that fit the customers for whom they are intended. The method comprises several steps such as anthropometry or measuring the body, patternmaking or translating the measures into 2-D patterns, assembling or joining the patterns to make a 3-D shell, and fit testing or draping the shell on the 3-D form to assess how the garment looks and interacts with the body (Efrat, 1982). Rapid technological advances have taken place in each of these fields, and almost all operations have been digitalized.

Manual anthropometric tools yielding linear dimensions have been replaced by noncontact, full-body 3-D scanners that record hundreds of dimensions in a matter of seconds. Body scanners, in turn, are becoming smaller, more mobile, and smarter by the day. Scan data have, for the first time allowed designers and patternmakers to see what real bodies look like in 3-D and how they differ from the idealized models used in the fitting room. In addition to 3-D data, 4-D body movement data are also becoming available. Though studies with 4-D data are still limited, the technology has immense applications in the field of functional clothing.

Powerful software that can analyze and process scan data to offer deep insights regarding the diversity of human body shapes is available. Data repositories have been created to store, segregate, and process 1-D, 3-D, and 4-D anthropometric data. These

Anthropometry, Apparel Sizing and Design. https://doi.org/10.1016/B978-0-08-102604-5.00001-9

repositories are, as of now, small, local, or national but will grow to include international data and become more intelligent as they accumulate data. Advanced data mining techniques can process and retract data of specific populations for customized applications.

The process of patternmaking, considered so far to be a mix of art and science, is now wholly digitalized. Complete automation wherein patterns can be generated directly from body scans has not been accomplished yet, but it is possible to sketch garments directly on a 3-D avatar. The virtual garment thus created can be flattened out into 2-D pattern pieces through a series of operations, and the pattern pieces can be draped back on the virtual form to test the fit.

Drawing patterns on real bodies allows the designer to see and understand the peculiarities in body shape of specific population groups such as plus-size individuals or senior women. Accordingly, they can modify the patterns to suit the requirements of each body type. Inclusion of body movement data and availability of animated, real body parametric avatars into the patternmaking process will bring about a paradigm shift in how garments are designed. Designers are using these tools to study the effect of body posture and movement on the shape and body dimensions of swimmers, wheelchair-bound individuals, military personnel, fire fighters, or skiiers. Marginalized population groups who are unable to fit currently into clothes produced for idealized bodies will be able to get clothes specially designed for them. While on one hand their needs would be met, on the other hand, it will open up new market segments for the garment industry.

Traditional flat patterns that are the backbone of RTW industry are produced from linear measures using simple rules that are known to all pattern makers. However, the form and number of patterns created from 3-D avatars are varied and significantly different from the conventional blocks used by patternmakers. The way these pattern pieces are joined to create a complete piece of clothing determines the final look and fit of the garment.

Fit, defined as the relationship between an individual and their clothing, affects the comfort, appearance, performance, and self-esteem of the user. A well-fitted garment is produced from a combination of precise measurement and good pattern designing with proper consideration of the physical and mechanical properties of materials. In conventional systems, fit is assessed and approved by fit experts who work with in-house fit models using a combination of subjective assessment and objective evaluation (Yu, 2004). However, there is a disconnect between the experts' assessment and consumers' perception of fit leading to consumer dissatisfaction. Virtual methods of fit assessment based on body scanning are becoming popular.

A variety of digital fit-testing systems are becoming available to the individual customer, in the form of virtual try-on apps or tools on retail platforms. This is a revolutionary development as it establishes, for the first time, a two-way communication between the customer and manufacturer. Social media platforms built around clothing fit bring in the concept of peer review and create strong social connect with clothing. However, correlation between scan-based fit and virtual try-on fit is yet to be established (Gill, 2015).

To summarize, it can be said that apparel supply chain is witnessing dramatic changes due to technological developments in all fields of production. Apparel industry is moving from a low-tech, labor-intensive industry into a software-driven industry. Repercussions of this technological upheaval on jobs and livelihoods of labor employed in this sector will need due consideration. Experts from software technology field would have to be integrated into the apparel manufacturing process, while patternmakers and designers will have to scale up their digital skills and work closely with IT experts. Upskilling of labor force across the entire value chain would be required. While some changes would be inevitable, it is important to strike a balance between technology, skills, and livelihoods. Judicious selection and implementation of technologies that benefit the customer and the manufacturer appears to be the way forward. The following paragraphs discuss these developments in detail.

1.2 Anthropometry

Anthropometry is defined as the science of "measurement of the human body." The term includes the complete process of data collection, summarization, documentation, analysis, and communication. Over the last 100 years, vast amounts of anthropometric data were collected by tailors and garment manufacturers across the world. Based on these data, apparel companies built their own closely guarded database of anthropometric measures, patterns, and size charts. As a result, there was no standardization of data or sizes across companies or countries (Olds, 2004). However, in the last 40 years, several countries have conducted large-scale, standardized size surveys—some exclusively for clothing sizing and others for multiple applications in the design of automobiles, products, and clothing.

Methods and technologies for capturing of anthropometric data have evolved with advances in science and technology. In medieval times, body measurements were taken either with a measurement taker that was a long and narrow strip of parchment or paper on which the measurements were recorded or by draping method wherein fabric was wrapped around the body and thus attained its shape and size directly. Eventually, tape measures, scales, and calipers were used to collect large amounts of data. All these methods yielded data in the form of linear measures. Over a period of time, need was felt to include data on body shapes into anthropometric measures, and technologies were developed to capture three-dimensional shape of the body. Detailed discussion on traditional methods as well as digital 3-D anthropometry and methods of quality control of data are available in Chapter 2 of this book. Digital methods of scanning the body in 3-D are now the mainstay of any large-scale anthropometric survey across the world, even though manual surveys continue to be conducted. Details of digital anthropometric sizing surveys conducted across the world and the considerations in planning and conducting a large-scale survey are available in Chapter 3 of this book.

1.2.1 Development in body scanning devices

A comprehensive write-up on 3-D scanners appears in Chapter 6 of the book. However, an account of how the 3-D measurement systems have evolved and affected the industry during the last few years is summarized here.

The earliest scientifically planned survey using 3-D scanners was the Civilian American and European Surface Anthropometry Resource (CAESAR) project conducted in 1998–2000. One of the two scanners used in the study was the Cyberware WB4 scanner, which is composed of 16 cameras housed in a booth measuring approximately 3mx3mx3m, weighing about 450kg, and costing US$350,000 (Fig. 1.1A)

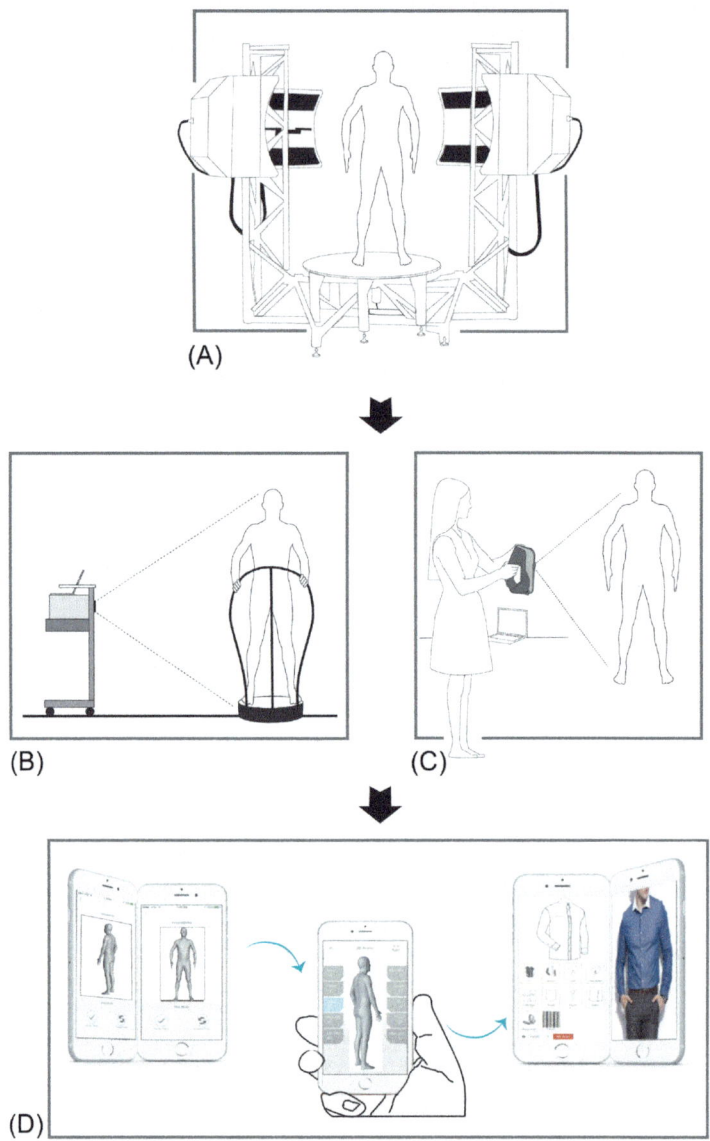

Fig. 1.1 Evolution of 3D body scanning systems. (A) Booth type body scanner. (B) Portable body scanner. (C) Hand held scanner. (D) Mobile scanner.

(Simmons, 2001). The total cost of the survey was 6 million dollars, and more than 4400 people were scanned over a period of 3 years (Robinette et al., 1999). With a view to demonstrate the complexity of the system, it might be mentioned that before use, the scanner had to be calibrated for the lighting scheme, camera scheme, landmark detection scheme, scanning garment colors, landmark marking sticker colors or shapes, luminance versus color for landmark detection and image stitching accuracy calibration, etc. (Robinette and Daanen, 2003). It took weeks to stitch the scans together to make the 3-D form and get the final output. A recent study by Lee et al. (2018) conducted on the head scans from CAESAR survey shows that scans have holes in occluded areas of the body and require tremendous amount of manual effort and processing to edit and landmark the scan data before the results are ready for use of product designers. Though the survey was conducted in 2000, the expertise to analyze and apply the data obtained from this survey became available only recently. This indicates that there is a huge phase lag between the time a technology such as this is developed and the time it takes for practitioners to adopt it. A lot of developments can happen during this large window of time, and a technology may actually be rendered redundant by the time people accept and understand it completely. This is what seems to have happened in the case of 3-D body scanners. While companies were working to make the booth-type scanners more compact, efficient, and user-friendly, the market has been flooded with cheaper and more user-friendly options.

A variety of portable, easy-to-use 3-D scanners have come into the market since the CAESAR survey. One example is the BodyLux scanner by ViALUX, Germany (www.vialux.de). The company, set up in the year 2000, provides scanners that measure 3-D shape based on a combination of micromirror projection and phase-encoded photogrammetry. The scanner has two parts—a cart with the integrated sensor unit and a turntable with handhold for the subject to stand on. It requires no special clothing and scans the body for about 50 s. A coordinate triplet, independent of the neighborhood, is calculated by means of projected pattern sequences for each camera pixel. The 3-D model is generated in real time, and body measurements are automatically calculated in accordance with specific standards. The lower-body scanner, BodyLux classic, is configured for customizing the sizing of compression wear. It calculates all required circumference and length measures from the scan and compares the dimensions of the customer with those in the size tables of compression garment manufacturers. The operator helps the customer to select the best product match from the enlisted suppliers of compression stockings, and the order is generated directly completing the circle from measurement to product ordering. The system costs about US$ 22,000 and weighs about 35 kg. While the system is cheaper and easier to use, the scans suffer from similar issues of occlusion and data loss in some parts of the body as with the booth scanners.

Further development in the field of scanning is that of high-resolution handheld scanners such as Artec Eva, developed by Artec3D, Luxembourg. The scanner weighs less than 1 kg, costs about US$20,000, and scans and processes data in real time (Fig. 1.1B). Because of the portability and ease of use, handheld scanners can be used for special applications that require customization, for example, to measure the leg, arm, or face for the design of pressure garments. As they can be taken close to the body, they can be used along with the static scanners to scan the body areas such as top of the head, groin, underarm, and side seam area, which are difficult to scan with static scanners.

Handheld scanners are relatively new in the market, and their applications are slowly being explored. Conkle et al. (2017) conducted an anthropometric study on more than 400 children (0–5 years) to define the malnutrition status and report promising results with a handheld scanner. Salleh et al. (2018) demonstrate the possibility of developing a body measurement system using a handheld scanner. Data acquisition was done using a handheld scanner, and a computer program developed using MATLAB was used to transform the data into a 3-D body model.

While the booth-type and handheld scanners can only be used by professionals under controlled conditions, a range of low-cost consumer devices like webcams, smartphones, or Kinect can be used directly by customers to scan themselves. However, the quality of such scans remains doubtful. A game-changing development in the field of anthropometry for garment industry is reported by the Indian arm of a United States-based company Mirrorsize (mirrorsize.com). Their free to download mobile app allows a customer to scan her fully clothed body in a matter of seconds, anywhere, anytime (Fig. 1.1C). The app uses advanced computer vision, deep learning, 3-D, and mesh processing technologies, to generate a body measurement chart with up to 95% accuracy in real time. The app is device-agnostic and can be run on any mobile phone or tablet. Syncing the app with websites of online clothing retail companies will allow the customer to pick a garment from a collection and drape it on their 3-D avatar for virtual try-on and virtual fit testing. Mirrorsize is being tested by companies in the United States and EU for possible integration. The company is in talks with major offline and online retailing brands in India to launch its product in the market in 2019.

Attempts have also been made to generate 3-D scans without the use of any scanning device. In a recent paper, Molyboga and Makeev (2018) employed powerful algorithms to obtain 2-D contour of the human body from a single image of a clothed person. Various filters and detectors are used to extract the parameters of 2-D slices of selected body areas. Three-dimensional model is subsequently generated from the contour and body measurements.

It may thus be concluded that exciting developments have been driving the field of whole-body scanning. With rapid developments in mobile and computing technologies, body measurement technology has gone from being an expensive, time-consuming, cumbersome, and centralized activity controlled by garment manufacturer to an easy, free, private, and completely individualized process controlled by the user. Future research will be targeted on the integration of this large volume of scan data into online and offline retail operations to offer value-added customized services to customers.

1.2.2 Applications of 3-D body scanning

In addition to acquiring data for the development of body size charts, visualization of 3-D scan data can be used for further understanding of the complex geometry and diversity of human forms. Three-dimensional data from real bodies allow designers to see the peculiarity of each body type and frame customized patternmaking rules

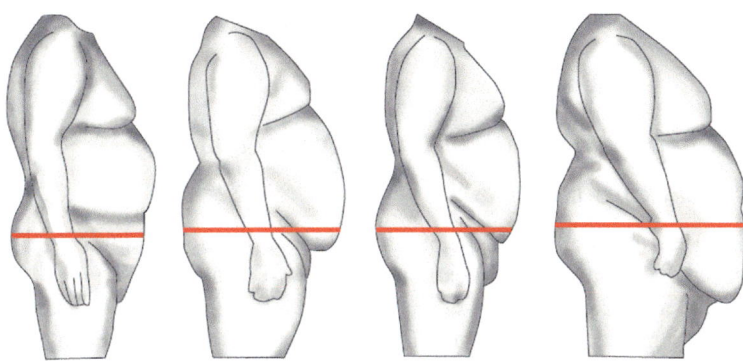

Fig. 1.2 Body shapes of Plus size men.

for each body shape. For example, while designing clothes for senior citizens, a designer need not use the same block pattern as is used to design for younger women. The increase in curvature of back and change in bilateral symmetry of the body of women with age can be quantified and visualized from the scans. A deeper understanding of variations in shoulder shapes and angles can help solve fit problems that are common in the shoulder area of women (Ashdown and Na, 2008).

A scanning survey of plus-size men in Germany conducted by Morlock et al. (2016) provides an insight into the high variability existing in the shape of bodies of plus-size men. It is clear from the scans that several body shapes exist within the plus-size group and it cannot be treated as a single homogeneous group. As seen in Fig. 1.2, the shape and curvature of chest, upper back, hip, and abdomen area is different for each size group. The abdomen starts to hang after a certain size, and this changes the body morphology altogether. It is obvious that each one of these body shapes requires a different pattern shape for the garment to give a good fit. Detail discussion on size charts for plus-size men and women can be found in Chapter 16 of this book. Such shape data of real bodies can be used to redefine body morphotypes and redesign patterns to accommodate the varied body shapes found in each group.

Another application of 3-D scans has been demonstrated for the study of body-shaping garments. While these garments are being used extensively by customers, little is known about the actual sculpting effect these garments have on the body. In an interesting study, Klepser et al. (2018) used 3-D scans to measure changes in body geometry brought about by shaping garments. Cross sections of scans of users with and without the shaping garments show the bridging, compression, and silhouette smoothing effect of shaping products. Results underline the importance of material properties, body geometry, and body tissue on shaping effects. Crucial information about which areas of the body are shapeable, and quantitative data regarding the shapeability of various types of garments can provide useful inputs to designers of shapewear.

Applications of 3-D shape data in the field of medical devices and prosthetics as well as assistive devices have also been explored. Veitch et al. (2012) used body scanning to study the complex 3-D geometry of the female breast in preoperative breast cancer patients. Results of the study have applications in breast reconstructive surgery and in design of custom-fit mastectomy bras.

1.2.3 Kinanthropometry or body motion analysis

1.2.3.1 Background

Protective wear such as firefighter clothing, military clothing, and sportswear are used under extreme conditions and in extreme postures such as shown in Fig. 1.3 (Song, 2011; Baytar et al., 2012; Aldrich et al., 1998). If clothing does not fit the wearer perfectly during actual conditions of use, then it may compromise the health, safety, and performance of the user. Criticality of fit in special missions and activities was recently demonstrated, when a historic event featuring all-woman spacewalk planned by NASA in March 2019 was canceled at the last moment as one of the female astronauts could not find a spacesuit that fit perfectly (New York Times, 2019).

Because of these considerations, along with shape and size data, designers also need to integrate body movement data into patterns while designing performance garments. Change in body dimensions brought about by movement needs to be quantified and taken into consideration to minimize the restraining effect of clothing on the movements of the user. Currently, there is no measurement standard that considers size reference and function-oriented motion of the body in the design of products.

1.2.3.2 Definition

Kinanthropometry is defined as the process of measuring a body in motion. Measurements that capture the change in dimensions of body parts and the range of movement of body joints are termed as "dynamic," "functional," "ergonomic," or "4-D" measures. These measures yield data that can help, for example, to locate the axes of rotation of joints (to design shoulder area of a garment), map the change in body lengths and girths with movement (to design sports and protective clothing), and show compression and elongation of tissues (for body-shaping and medical garments).

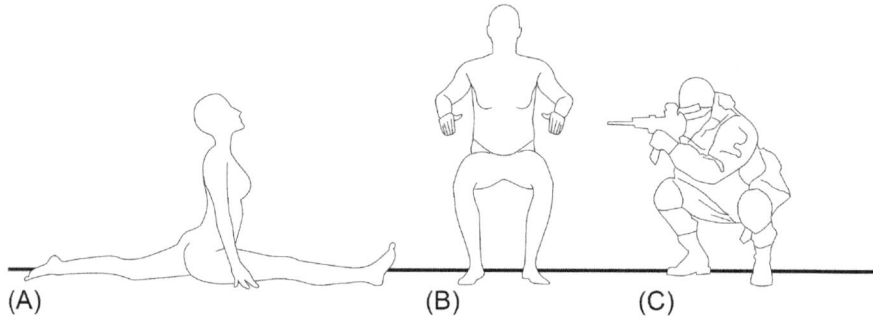

(A)   (B)   (C)

Fig. 1.3 Extreme postures adopted during specific activities. (A) Gymnast. (B) Motorcycle rider. (C) Soldier.

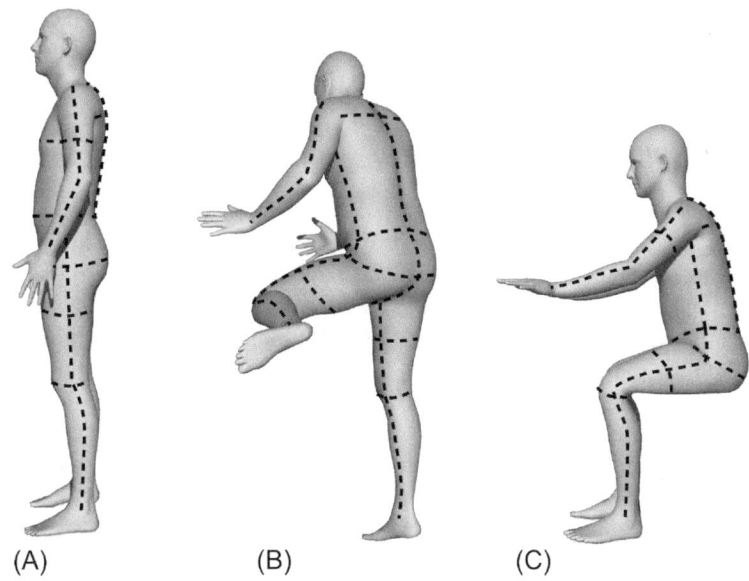

(A)   (B)   (C)

Fig. 1.4 Key body postures (A–C) of a motorcycle rider.

Motion data for a body can be captured by scanning the body in two ways:

(a) By using a 3-D scanner to scan the body in static poses typical for a particular activity. These measurements are called as functional measures.
(b) By using a 3-D motion scanner to capture the measurements during movement. These are called as dynamic measures.

Steps involved in capturing static measures in animated mode include (a) identifying work-related postures and motion sequences, (b) capturing each posture one step at a time, (c) measuring the body in movement poses, and (d) developing a motion-related sizing system (Loercher et al., 2018). Fig. 1.4 shows the typical poses identified for a motorcycle rider for capturing the functional measures in static mode. There are several limitations of this method since it requires the subject to hold each pose for a few seconds. Holding a static pose does not simulate true movement; reproducibility of such measurements is poor as it may not be possible for all subjects to hold a position to the required extent; positions may vary from person to person; props may be required to hold the pose that leads to shadowing of body areas.

True motion (4-D) anthropometry is carried out with the help of 3-D motion scanners, also known as 4-D scanners, which measure the body while it is in motion. Motion scanners are now available in companies such as 3dMD, ViALUX, and SinfoMed. They scan the moving body and automatically generate a continuous 3-D polygon surface mesh from all synchronized stereo pairs (http://www.3dmd.com/). These systems require a large number of sensors and are therefore sensitive and expensive at the moment. As technology advances, cost will come down. They are being used extensively in medical applications (http://www.3dmd.com/3dmd-customer-research/). Some work that reported the use of motion scanning in the field of clothing is discussed in the succeeding text.

1.2.3.3   Applications of 4-D anthropometry in garment industry

Effect of movement on body measurements

Loercher et al. (2018) of the Hohenstein Institute, Germany, collected motion data of several subjects and calculated the difference in measurement between the static pose and movement poses. The smallest and the largest difference in a body part caused by different poses were estimated. Designing the clothing to cover this entire range of measurements around that body part can ensure ergonomic clothing comfort. Motion-related studies can be particularly useful in studying the bodies of people with disabilities. As each body varies in shape and size, peculiarities of each body type can be observed and clothing designed accordingly. Information about the deformation in soft tissues and permanent change in body shape can also be collected. Interaction of body with clothing and how it changes with different body poses can also be studied.

1.2.3.4   Estimation of air volume and ease

Estimation of air volume between the body and clothing is a measure of clothing ease and determines the ease of movement of the user. At the same time the volume of air trapped between the body and clothing and within the multilayers of clothing determines the thermal insulation of the body. It is known that insulation increases linearly with increase in air gap, as long as no convection is present. If the gap is too much, convection causes a bellow effect, and insulation is reduced. Therefore a single large air space is less insulating than a number of smaller spaces trapped between layers. In order to apply these principles of insulation to clothing design, an accurate estimation of the air volume inside clothing is required. It is a complex parameter to measure on a clothed and moving body. The 3-D scanning technique has been shown to be an accurate and reproducible method for quantification of air volume under clothing (Daanen et al., 2002).

Petrak et al. (2018) used a 3-D body scanner to quantify the overall and microclimatic change in the volume of an outerwear jacket brought about by upper-limb positions simulating functional reach movements of aircrew personnel. The overall volume of a clothed versus unclothed body was determined and calculation of air layer between the second- and third-layer garment also estimated. Data from such studies can help quantify the exact amount of ease required in the garment at specific areas such as the chest and waist so as to achieve optimum wearing comfort in performance wear.

Mert et al. (2018,b) used motion scanning to quantify the change in air gap thickness and contact area between a moving body and clothing during Alpine skiing. A 3-D avatar wearing a loose- and tight-fit coverall was simulated in alpine skiing poses, and the dynamic change of air gap thickness due to these positions was estimated. Results are useful in quantifying heat and mass transfer from the moving body during activity. Garment designers can optimize thermal comfort and protective performance of garments using these data.

Liu et al. (2019) used a motion scanner to estimate how the distribution of static and dynamic ease of clothes changes at bust and waistlines with body movement. Static ease distributions showed different patterns at different body landmarks. The frequency and magnitude of ease variations increased with walking speed, although the concentration areas of ease remained unchanged. This understanding can help optimize patterns for high-performance sportswear and improve the dynamic fitting process of virtual clothes. Effects of fabric properties, clothing sizes, and body shapes on ease distribution can further enhance knowledge in this field.

1.2.4 Challenges

Even though a large volume of 3-D scan data are available now, most patternmakers use only the linear measures extracted from the scans and disregard the shape data. Some of the issues that need to be resolved before designers can make full use of these tools are discussed in the succeeding text.

1.2.4.1 Skill gap

Designers and patternmakers trained in traditional 2-D methods are not able to integrate vast amounts of 3-D data into the patternmaking process. Only a few people have had experience of using and operating a 3-D scanner during their employment or educational programs. Strengthening the human interface between designers and technology will be a critical area of focus in the future.

1.2.4.2 Location of body landmarks

Landmarks define the relationship between anthropometry and body morphology. They are used to measure and map the body using the principles of triangulation. A standardized method for locating, marking, and tracking landmarks on the body is still not available. Correlation between manual and scanned landmarks has not been established. There is no clarity on the exact point from which a measurement is to be taken to which point. For motion tracking, body marks are used to identify the landmarks, and the displacement of the same in x, y, and z planes is recorded. The pose or posture of the body at the time of measurement can shift the landmark. Collating the body landmarks with those on the pattern pieces is an issue still to be tackled. Errors in landmark location can lead to problems in pattern design, sizing, and fitting.

1.2.4.3 Standardized data repositories

Currently, it is difficult for a designer to collate and reconcile anthropometric data sets generated at different places and different times using different technologies. This poses a major hurdle in the application of data for commercial purposes. Freely available standardized data repositories of measurements of international populations are not available. Data sets of different populations with data tagging are a requirement.

1.2.4.4   3-D avatars

For designing of garments that fit, designing needs to be done directly on the 3-D form. Raw body scans obtained from various scanning sources are not ready for use and have to undergo several steps of cleaning and manipulation before they are ready for use. Current CAD software for patternmaking cannot automatically extract and use data from 3-D measurement acquisition systems (Dāboliņa et al., 2018). Animated avatars that can simulate various poses will help the designers to visualize the stresses and strains in the skin and body. Avatars that can simulate body softness and compressibility and can be morphed to simulate real bodies as they grow, age, and move are required (Ballester et al., 2014).

1.2.5   Future outlook

Three-dimensional body scanners have been used successfully by companies and institutions across the world to carry out a large-scale anthropometric survey of their target populations. Companies such as Levi's and Brooks Brothers tried to use them at their retail outlets to offer customized clothing services to clients. However, none of the attempts have been successful commercially. Given the high cost and lack of success at the retail end, it appears that large booth-type scanners may no longer remain viable in the long term. Recent developments in mobile scanning technologies, coupled with powerful computing and graphics capabilities, have a tremendous potential to shape the future of the apparel industry. Scanning systems that are precise and yield reproducible scans not affected by the clothing or pose of the user will become popular. Individual scan data will grow, but issues of comparability of data between scanners, accuracy of data processing and output, reproducibility of scans, standardization, and marking of landmarks would have to be resolved. Small companies and large online companies with no physical presence will be able to use scan data uploaded on the cloud by individuals through scanning apps. This would allow companies access to a truly global, updated database, without ever having to own a scanner or scan anyone physically. This vast database could be mined using deep learning and AI tools for MTM and RTW apparel manufacturing.

1.3   Anthropometric data analysis and output

A key process in anthropometry is data analysis and the form in which the output is made available to practitioners. Till the end of 20th century, good-quality anthropometric data were rare and precious resource, and only linear measures were available. The output was published in the form of size charts derived from clustering and averaging the body measurements of a population. Based on this approach, size charts for men, women, and kids were developed for US, UK, French, Chinese, Korean, and other populations across the world.

Development of 3-D body scanners, along with customer requirements for better fitting products led to a race for the collection of anthropometric data by private enterprises and governments. Individual data collected by customers have further added to this pool. During 2000–10, when most large-scale 3-D anthropometric surveys were conducted, data sources were proprietary and/or costly due to the sheer cost involved in conducting a survey. Participants of the survey got the first right to use the data, and subsequently, they became available to other users. While companies continue to use the linear data available with them, raw and processed data from 3-D scanners are now mostly available in 3-D form. Anthropometric data can be represented in the form of 3-D body scan, multiview images, one image, body contour, etc. (Molyboga and Makeev, 2018). As a result the analysis, management, mining, and disbursing of such diverse data have become a humongous task. Demand for algorithms that can extract precise information about the human body from digital data is growing. Accordingly, agencies that specialize in data management and analysis are now a part of the data collection team.

Anthropometric data can be purchased online from various sources such as CAESAR store (Store.sae.org, 2019) for data of CAESAR survey and DINED (Data.4tu.nl, 2019) for data of the Netherlands. World Engineering Anthropometry Resource (WEAR) (https://www.bodysizeshape.com/page-1855750) is a consortium that shares nearly 200 anthropometric data sets comprising 2-D and 3-D data from 20 countries on a single platform. Fee-based bundled services are offered by companies such as Anthrotech (Anthrotech, 2019) that scan and analyze specific population data provided by clients or provide solutions from their own data sets. Human Solutions GmbH through their portal iSize provides 3-D processed data of French and German populations (Anon, 2019). Global databases that cover all populations of the world are not yet available.

Avatars created from statistically derived body scan data of a large number of subjects of a particular population group are available publicly. These templates represent typical body shapes and sizes extracted from a large database of scans and bring 3-D and 4-D data directly to the designer (Ballester et al., 2016). Grouped and parametrized avatars of adult males and females obtained by principal component analysis and partial least squares regression method are available (www.inkreate.eu) (www.i-size.net). These are statistically derived sets of high-resolution 3-D templates. In some cases, body movement data are incorporated into the forms through static poses (Inkreate.com).

Prof Michael Reed of the University of Michigan has made several 2-D data sets available on his website (Mreed.umtri.umich.edu, 2019) in public interest. Downloadable, 3-D parametric forms of adults and children in different poses are also available on their site http://humanshape.org/. It is also possible to individualize the process of avatar making at the level of the designer. Tools are available for converting any raw body scan from any scanning device into a simulation-ready individual avatar (https://www.inkreate.eu/) for use of the designer.

In all examples discussed earlier, the avatars are available in fixed static poses, and data about movement are not available. The latest and most exciting advancement in this field aims to fix this gap. Max Planck Institute, Germany (http://smpl.is.tue.mpg.de/), has developed a fully animated parametric skinned multiperson linear model (SMPL) (Fig. 1.5). SMPL is a realistic learned model of human body shape and pose that is compatible with existing rendering engines, allows animator control, and is

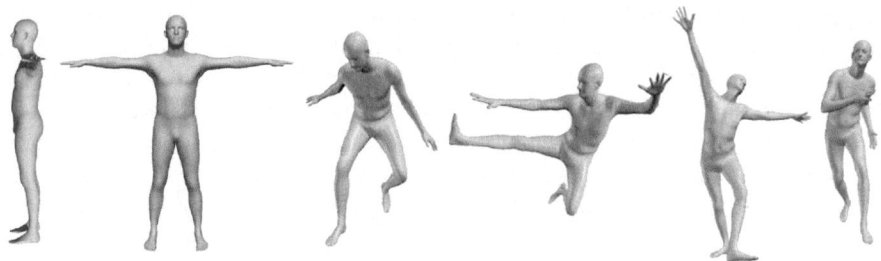

Fig. 1.5 SMPL model developed by Max Planck Institute Germany.

available for research purposes. The designer can animate the model in real working postures and see the natural pose-dependent deformations and soft-tissue dynamics in the body. This can be applied to the design of ergonomic clothing as the model is fast to render, easy to deploy, and compatible with existing rendering engines such as Maya. Such a tool is of immense use to the designers of functional clothing as it can be used to realistically represent a wide range of human body shapes in extreme postures.

1.4 Patternmaking

History of garment making is as old as civilization, and the process has evolved and changed with the changing needs of the times and with the developments in science and technology in allied fields. A key requirement for producing well-fitted garments is to use patterns that are based on accurate anthropometric measures. Patterns form the basis of all garment production, and the fit of a garment is directly dependent on the accuracy of the manufacturer's block patterns. In made-to-measure settings, a designer strives to provide a perfect fit for the individual customer. In the RTW scenario the aim is to produce a garment that will be a good fit for a large number of people in that size range. It requires a mix of art and science to develop patterns that produce a well-fitted garment (Efrat, 1982).

A pattern represents a deconstructed 3-D garment in 2-D form. Several methods are used to create patterns:

Modeling or draping method is used to drape fabric directly on the body. After a satisfactory fit is obtained, draped fabric is cut out and flattened to develop patterns for the garment. This method is suitable for designing bespoke garments having complex designs as it is time-, skill-, and material-intensive. In a modified version, Hutchinson (1977) "molded" a warm sheet of thermocol on the body of the subject and removed it when cool. Thermocol retained the 3-D form effectively and was cut and flattened into a 2-D pattern that was then used as a block pattern for individual subjects. This technique can be considered to be a physical forerunner of the modern-day practice of designing digital patterns in 3-D.

Drafting method is used by tailors to produce made-to-measure clothing. Traditional tailors were successful and continue to be so because while they measure the body with a tape measure and record linear measurements, they note the details and the peculiarities of the 3-D form and factor these "shape data" into the patterns. The process is skill-based and suitable for bespoke tailoring. With the growth of ready-to-wear industry, faster and simpler methods of making patterns were needed.

Flat-pattern method has been the mainstay of the RTW industry for the past nearly 100 years. In this method, basic blocks are developed using empirical rules and averaged body measurement data or "size charts" as a guideline. The basic block is graded up and down to cover the target sizes in the population. Over the years, companies have optimized their own set of patterns, based on the preferences of their target consumers.

1.4.1 Problems with conventional flat-pattern method

Even though the industry started using CAD systems for making patterns a long time ago, no scientific investigations were conducted to study the theory of patternmaking. As a result, simple approximations of body morphology, which were the basis of manual practices, were imported into the digital domain and continue to be used for the drafting of blocks in digital systems. In these systems the patternmaker is totally detached from the body shape and size of the real customer and works only with averaged linear measures, with the assumption that actual fit would be achieved by "nipping and tucking" the garment during fit trials (Scott and Sayem, 2018). The method is inadequate in that it does not numerically isolate ease from the body measurements and as such prevents a comparison between 3-D body and 2-D pattern dimensions. When industry shifted from manual patternmaking to CAD, this same approach was transferred from the physical to the virtual world (Scott and Sayem, 2018), and so were the problems of fit. Even though 3-D shape data are widely available now, they cannot be integrated into the patternmaking process as the existing patternmaking systems are programmed to use only 1-D measurements and designers are also trained to work with 1-D measures only.

1.4.2 Technological advances in the field of patternmaking

Due to rapid developments in the field of digital technologies and extended reality (XR) in particular, several tools are now available to the patternmaker. These tools if deployed effectively can be used to solve the problems of fit, inadequate size charts, the lack of data, and the lack of consumer confidence that have been plaguing the retail and fashion industry. Some of these are summarized in the succeeding text:

(a) Three-dimensional avatars that simulate real body shapes and sizes of all segments of the population (Ballester et al., 2014)
(b) Animated avatars that allow the designer to change the poses of the avatar and see how the garment would interact with the body during use
(c) Tools to sketch garments directly on the avatar (https://www.inkreate.eu/)
(d) Tools to generate 2-D patterns directly from the sketch
(e) Tools to design in 2-D and visualize the garment in 3-D in real time

(f) Digital library of fabrics, styles, accessories, closures, and garment sewing specifications that can be cut and pasted to design garments, clo3D
(g) Tools to make technical drawings and do a tentative costing of the garment directly from a sketch
(h) Visualize the fit of the garment in 3-D at the time of sketching (clo3D)
(i) Exceptional, easy-to-use interfaces for better communication between the client and the designer

1.4.3 Applications of advanced patternmaking technologies

Shortcomings and inconsistencies in conventional patternmaking methods can be addressed with the help of data provided by advanced technologies. A major application area for the technological tools discussed earlier is in clothing used in protective, sports, or medical fields (Ashdown et al., 2005; Daanen and Sung-Ae Hong, 2008; Hlaing et al., 2011; Baytar et al., 2012; Maghrabi et al., 2015; Aluculese et al., 2016; Petrak et al., 2016; Naglic et al., 2017; Bogović et al., 2018; Traumann et al., 2019). Some cases studies are presented in Chapters 12, 13, 15, and 16 of the book.

The process of making patterns on the avatar, known as 3-D/2-D/3-D, is demonstrated in Fig. 1.6 for the design of motorcycle riders' clothing. Garment is sketched directly on the avatar (a); a mesh is generated (b), patterns are flattened into 2-D patterns (c); 2-D patterns are assembled and sewn together and draped on the avatar for virtual fit testing (d). If the avatar is animated, fit of the garment in various poses can also be tested. The process can be used for MTM and RTW applications. Though the patterns obtained by this method yield good fit, the pattern shapes obtained from flattened mesh may not always conform to the accepted principles of fabric grain and dart manipulation followed by patternmakers. This issue may have to be addressed by the patternmaker manually (Mahnic et al., 2016).

Patterns generated directly from the body scan capture the peculiarities of shape or posture of the customer. Aluculese et al. (2016) captured 3-D data of wheelchair-borne women in different postures. The pose data were mapped onto human templates and

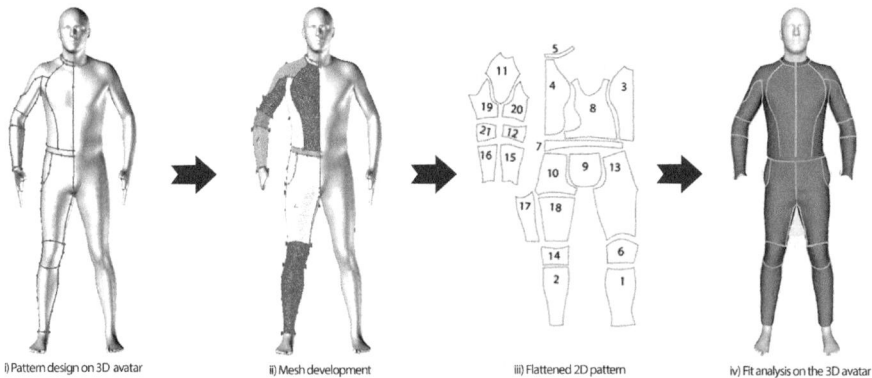

i) Pattern design on 3D avatar   ii) Mesh development   iii) Flattened 2D pattern   iv) Fit analysis on the 3D avatar

Fig. 1.6 Zoning for motorcycle rider.

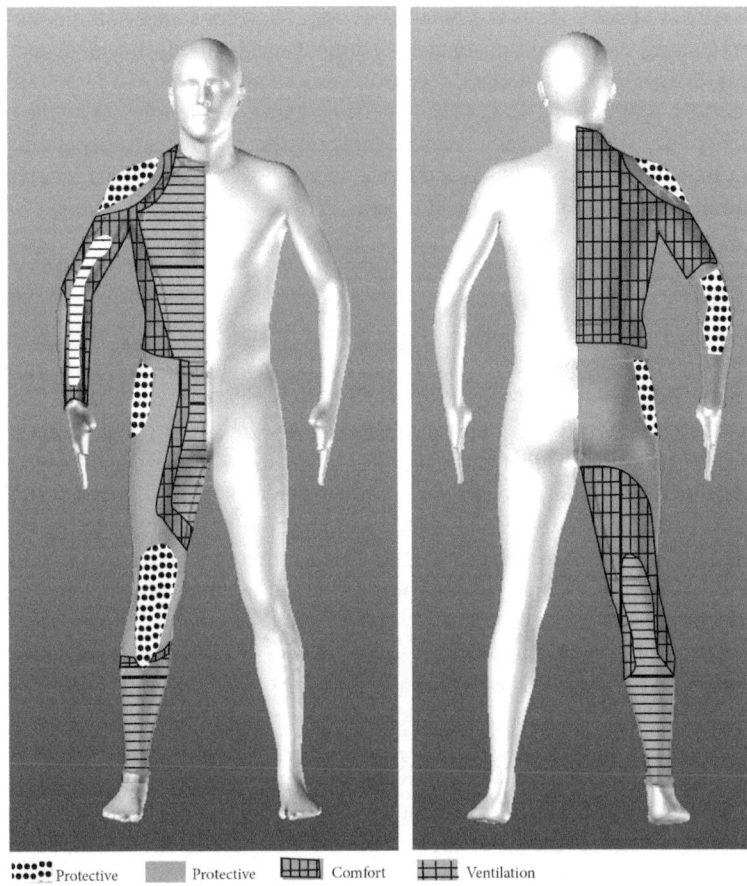

 Protective   Protective   Comfort   Ventilation

Fig. 1.7 3D-2D-3D method of pattern making.

exported into a CAD program for patternmaking. Patterns for trousers were designed directly on the scans as per personalized needs of comfort and performance of the users.

Design of functional clothing poses several challenges to a patternmaker. Protective clothing, for example, requires the designer to use several fabrics with heterogeneous properties to create a zoning effect. Design of a motorcycle riding suit with zoning effect is shown in Fig. 1.7. The suit is divided into a protective zone for high abrasion resistance, a comfort zone to allow for the body to be comfortable while riding the bike, and a ventilation zone for breathability. Availability of an avatar for such applications can greatly facilitate a patternmaker by helping her design as per the human anatomy, balancing the requirements of protection and comfort. Pattern shapes obtained for the suit, shown in Fig. 1.6, are complex, and the number of pattern pieces is large. Each pattern piece can be mapped with fabric having required properties, and its performance in different poses can be studied.

The aforementioned processes though automated are still based on the skill of the pattern designer.

Methods have been proposed wherein a designer or a consumer with no knowledge of patternmaking can design customized garments with highly detailed geometries. Li and Lu (2014) propose a method for creating garment models in 3-D on individualized avatars by combining 3-D parts from existing garment examples. With the plethora of technologies available for capturing and analyzing body scan data, all patterns should fit. If they don't fit, then it should be interpreted either as a problem of data analysis or its application by the designer (Scott and Sayem, 2018).

1.5 Fit

Even though most consumers consider fit to be the single most important aspect of a garment that influences final purchase decision, it is the least documented and least understood element of the garment supply chain. As a result, RTW garment industry is plagued by fit problems that cause dissatisfaction to consumers and losses to manufacturers due to lost sales and clothing returns. The lack of visualization and understanding about the diversity in body shapes, changes in body shape with size or age, use of models with idealized body shapes in the fitting room, and the lack of factoring of movement data into fit analysis are some of the factors that contribute to poor fit of garments.

Fit is defined as the relationship between garment size and body size. The understanding of fit, however, is different for academicians who study fit, practitioners who finalize fit, and the consumers' experience and expectations of fit. Hardly any systematic studies have been conducted to study the fit of actual mass-produced garments on a varied population (Gill (2015)). Experts agree that a good fit can be achieved through a combination of five garment factors, namely, ease, line, grain, balance, and set (Erwin et al., 1979). When all these five factors are in alignment, the garment conforms closely to body shape, lies smoothly without any wrinkles, allows adequate ease for movement, and appears to be a part of the body. According to Hutchinson (1977) the critical area of the body from the fit point of view is the shoulder. A garment that fits the figure perfectly between the neck and the horizontal line of the figure at the level of the armhole would fit the whole body well. It is therefore clear that the shoulder not only is important from aesthetic point of view but also is crucial for ease of movement of the upper body. But a scientific method to achieve this perfect fit is not known.

In traditional RTW systems, fit of a garment is determined, assessed, and controlled by fit experts at the production site. Though a customer can complain about the poor fit and return garments that do not fit, a two-way communication between the manufacturer and the customer is not there. New digital technologies are changing this. The process of fit assessment is undergoing rapid and dramatic changes, fuelled by advances in digital technologies, deep learning, data analysis tools, and AI. For a detail review on developments in the field of fit, the reader is referred to the works of Gill (2015) and Miell (2018).

A variety of digital interfaces and platforms are now available directly to the consumer to evaluate garment fit in online clothing purchase space. Websites providing reality and augmented reality fit programs allow the customer to experience a 360^0 retail experience within the privacy of their home, allowing them to make data-supported intelligent clothing choices. Some of the platforms are data-driven, while others are strong in visualization effects and tech-intensive depending on the composition of the team Gill (2015). These interfaces can be classified into those that offer fit recommendation (Fitbay, Fits.me) or size recommendation (Dressipi and True fit) or fit visualization (Metail, My Virtual Model) or a combination of all three. These vary widely in how they communicate with the consumer and how they convey fit and drape (Miell, 2018).

The basis of all virtual trial rooms are the 3-D avatars. The avatar can be a generic body or a scan of the customer or a parametric avatar that can be modified to simulate the body of the customer. Animated avatars that simulate postures can also be provided for a more realistic simulation of garment fit during actual use. Most of the currently available avatars are rigid and do not realistically portray the deformations of a soft body when a garment is worn, for example, the compression of the soft breast tissue by a bra or bikini top. In a recent development, Harrison et al. (2018) reported the development of soft body avatars that simulate the compressibility of body tissues. Any rigid 3-D scan can be converted into a soft avatar. Using finite-element methods the behavior of a body-fitting garment on a deformable body can be simulated in a realistic manner. Stresses and strains in the garment and the body are simulated together, with two-way coupling of forces and displacements.

A number of companies are now proposing solutions to size and fit issues Gill (2015). These solutions aim to engage the customer in the garment purchase process. The customer is asked to input details and some body measurements, the system makes some recommendations, and the customer can visualize the options in a virtual trial room before making the final purchase decision. Initial results from these systems indicate that several consumer groups such as the senior women buyers have low fashion confidence. For such customers, visualization of fit alone is not enough. They prefer the fit and size recommendation tools that offer expert styling advice on what would look good on them or which current fashion trends would suit them (Miell, 2018).

Data scientists and styling experts are working together to build highly intelligent recommendation tools to engage and guide the customer to make the right choices. Customers input height, weight, and bra size data into the system (Limited, 2019), which are used to compute waist and hip dimensions. The tool makes a size recommendation based on these inputs. Recommendations regarding what style to buy are made by algorithms based on the details provided by the customer regarding their body shape, coloring, life style, age, and previous buys and returns (Dressipi.com, 2019). Data analytics are used to analyze what is returned most and relate it to specific body shapes and sizes, and algorithms are modified so as not to make those recommendations (Miell, 2018).

Some systems take the consumer experience further by making recommendations about the total look. For example, while trying on a party jacket in the virtual trial room, recommendations are made regarding the bags and shoes that will go with the outfit. These recommendations are based on previous choices made by customers

of similar demographics. This experience is similar to the traditional customized tailoring methods, where the customer discussed their fit preferences with an experienced tailor who in turn gave feedback based on experience with many clients
(Gill, 2015). Augmented reality tools are employed to help the customer imagine
themselves in the selected outfit to create an aspirational effect. Customers on such
platforms may buy up to four more items than those who shop without these tools.

As data accrue, these deep learning tools will become more and more intelligent.
Stronger visualization tools along with style descriptions are being developed to edu-
cate the customer about the style of the garment and how it is to be worn (Miell, 2018).
These technological developments have placed the onus of garment fit determination
and assessment quite literally into the hands of the customer. More importantly, they
have bridged the gap between the customer and manufacturer by facilitating real-time
two-way communication. Incoming data will provide key insights into purchase preferences of consumers. Data about body type and fit preferences as well as peer group
recommendations can be incorporated into the garment design process. Ultimately the
designer has to make the transition from real fit evaluation to the virtual environment.
While consumer interfaces for self-fit evaluation have been developed, the knowledge
to use them may not be there. All customers do not choose to engage with these interfaces. While solutions are available for testing and visualization of fit, the understanding of the theory of garment fit is still limited. In order for the field to grow, the
theoretical principles need to be studied and applied systematically to draw up the
tools, techniques, and standards for assessing and achieving a good fit in garments
(Gill, 2015).

1.6   Conclusions

It can be concluded from the discussion earlier that automation and customization is
the key to garment manufacturing of the future. Technological developments in the
field of measuring and classifying the body, patternmaking, and fit testing are
expected to change the way in which garments are designed, produced, tested, and
sold in the future. Companies can make significant savings, and supply-chain sustainability can be enhanced by shifting a majority of operations from physical to virtual
environments.

While these technologies are here to stay, more work is required to integrate them
into mainstream commercial systems. The methods need to be modified and adapted
to make them compatible with current systems of garment production. Increasing use
of digital technologies and the availability of all data in digital format imply that the
process of patternmaking and garment production, in general, will become open to
new skill sets. Computer programmers, 3-D designers, and engineers will become
an essential part of the garment industry. Designers would have to work closely with
engineers who can carry out all the computerized operations required to "engineer the
garment." Garment designers will have to be skilled in software handling, and software engineers would need to understand the principles of garment design and production. This transformation of a garment designer from a creative person into a

tech-savvy person is a disruptive intervention in the industry. The effect this automation activity will have on the creativity of designers is something that needs to be considered. The reskilling of manpower would be a major challenge in near future.

Conventional garment manufacturing hubs, especially those located in developing countries with abundant labor supply, will find it difficult to accommodate these changes. The effect this has on livelihoods and economies will only be known over a period of time.

1.7 Future directions

Theoretical understanding of shape, size, and fit and their incorporation into patternmaking processes needs to be established. Development and incorporation of the mathematical basis for darting and shaping patterns so that they conform to real and not idealized body morphology needs to be done.

Three-dimensional scans will be used to carry out fundamental studies on body shapes, relation between clothing and the body as it moves, quantification of microclimatic air volume, and the overall area and volume of clothing. The knowledge will be used to design patterns with improved fit, performance, insulation, and comfort for users. Development of new rules for garment design and production has to be established and incorporated into the garment technology curriculum at all levels of education.

Acknowledgments

The author is highly thankful to Kanika Jolly for preparing the images for the chapter.

References

Aldrich, W., Smith, B., Dong, F., 1998. Obtaining repeatability of natural extended upper body positions: its use in comparisons of the functional comfort of garments. J. Fash. Mark. Manag. 2, 329–351.

Aluculese, B., Krzywinski, S., Curteza, A., Meixber, C., 2016. Animation of 3D human scanning data—extracting the sitting posture for a better understanding of the garment development for wheelchair users. In: 16th Romanian Textiles and Leather Conference— CORTEP 2016 Iasi, 27–29 October 2016.

Anon, 2019. [online] Available at: https://portal.i-size.net/SizeWeb/pages/home.seam. Accessed 9 April 2019.

Anthrotech, 2019. Home | Anthrotech. [online] Available at: https://anthrotech.net/. Accessed 9 April 2019.

Ashdown, S.P., Na, H.-S., 2008. Comparison of 3-D body scan data to quantify upper-body postural variation in older and younger women. Cloth. Text. Res. J. 26, 292–307. https://doi.org/10.1177/0887302X07309131.

Ashdown, S.P., Solcum, A.C., Lee, Y.A., 2005. The third dimension for apparel designers: visual assessment of hat designs for sun protection using 3-D body scanning. Cloth. Text. Res. J. 23 (3), 151–164.

Ballester, A., Parrilla, E., Piérola, A., Uriel, J., Pérez, C., Piqueras, P., Alemany, S., 2016. Data-driven three-dimensional reconstruction of human bodies using a mobile phone app. Int. J. Digit. Hum. 1 (4), 361–388.

Ballester, A., Parrilla, E., Uriel, J., Pierola, A., Alemany, S., Nacher, B., Gonzalez, J., Gonzalez, J.C., 2014. 3d-based resources fostering the analysis, use, and exploitation of available body anthropometric data. In: Proceedings of 5th Int. Conf. on 3D Body Scanning Technologies, Lugano, Switzerland, 2014, pp. 237–247. https://doi.org/10.15221/14.237.

Baytar, F., Aultman, J., Han, J., 2012. 3D body scanning for examining active body positions: an exploratory study for re-designing scrubs. In: 3rd International Conference on 3D Body Scanning Technologies, Lugano, Switzerland.

Bogović, S., Stjepanovič, Z., Cupar, A., Jevšnik, S., Rogina-Car, B., Rudolf, A., 2018. The use of new technologies for the development of protective clothing: comparative analysis of body dimensions of static and Dynamic postures and its application. Autex Res. J. https://doi.org/10.1515/aut-2018-0059.

Conkle, J., Ramakrishnan, U., Flores-Ayala, R., Suchdev, P.S., Martorell, R., 2017. Improving the quality of child anthropometry: manual anthropometry in the body imaging for nutritional assessment study (BINA). PloS One 12 (12), e0189332.

Daanen, H., Hatcher, K., Havenith, G., 2002. Determination of clothing microclimate volume. In: Tochihara, Y., Ohnaka, T. (Eds.), Proceedings of the 10th International Conference on Environmental Ergonomics. Kyushy Institute of Design, Fukuoka, Japan, pp. 665–668.

Daanen, H., Sung-Ae Hong, S.A., 2008. Made-to-measure pattern development based on 3D whole body scans. Int. J. Cloth. Sci. Technol. 20 (1), 15–25. https://doi.org/10.1108/09556220810843502.

Dāboliņa, I., Viļumsone, A., Dāboliņš, J., Strazdiene, E., Lapkovska, E., 2018. Usability of 3D anthropometrical data in CAD/CAM patterns. Int. J. Fash. Des. Technol. Educ. 11 (1), 41–52. https://doi.org/10.1080/17543266.2017.1298848.

Data.4tu.nl, 2019. 4TU—DATASETS. [online] Available at: https://data.4tu.nl. Accessed 9 April 2019.

Dressipi.com, 2019. Home. [online] Available at: https://dressipi.com/. Accessed 14 April 2019.

Efrat, S., 1982. The Development of a Method for Generating Patterns for Garment That Conform to the Shape of the Human Body. Ph.D. Thesis, Leicester Polytechnic. https://www.dora.dmu.ac.uk/handle/2086/5846. Accessed 5 April 2019.

Erwin, M.D., Kinchen, L.A., Peters, K.A., 1979. Clothing for Moderns, sixth ed. Macmillan publishing Co., Inc., New York.

Gill, S., 2015. A review of research and innovation in garment sizing, prototyping and fitting. Text. Prog. 47 (1), 1–85. https://doi.org/10.1080/00405167.2015.1023512.

Harrison, D., Fan, Y., Larionov, E., Pai, D.K., 2018. Fitting close-to-body garments with 3D soft body avatars. In: Proceedings of 3DBODY.TECH 2018—9th Int. Conf. and Exh. on 3D Body Scanning and Processing Technologies, Lugano, Switzerland, 16–17 October 2018, pp. 184–189. https://doi.org/10.15221/18.184.

Hlaing, E.C., Krzywinski, S., Roedel, H., 2011. Development of 3D virtual models and 3D construction methods for garments. In: 2nd International Conference on 3D Body Scanning Technologies, Lugano, Switzerland, 25–26 October 2011.

Hutchinson, R., 1977. The Geometrical Requirements of Patterns for Women's Garments to Achieve Satisfactory Fit. Leeds University. M. Sc. Thesis.

Klepser, A., Stephanie Hiss, S., Mahr-Erhardt, A., Morlock, S., 2018. Three-dimensional quantification of foundation garments shaping effects. In: Proceedings of 3DBODY.TECH 2018—9th Int. Conf. and Exh. on 3D Body Scanning and Processing Technologies, Lugano, Switzerland, 16–17 October 2018, pp. 92–104. https://doi.org/10.15221/18.092.

Lee, W., Lee, B., Yang, X., Jung, H., Bok, I., Kim, C., Kwon, O., You, H., 2018. A 3D anthropometric sizing analysis system based on North American CAESAR 3D scan data for design of head wearable products. Comput. Ind. Eng. 117, 121–130.

Li, J., Lu, G., 2014. Modeling 3D garments by examples. Comput. Aided Des. 49, 28.

Limited, M., 2019. Home—Metail. [online] Metail. Available at: https://metail.com/. Accessed 14 April 2019.

Liu, Z., He, Q., Zou, F., Ding, Y., Xu, B., 2019. Apparel ease distribution analysis using three-dimensional motion capture. Text. Res. J. https://doi.org/10.1177/0040517519832842.

Loercher, C., Morlock, S., Schenk, A., 2018. Design of a motion-oriented size system for optimizing professional clothing and personal protective equipment. J. Fash. Technol. Textile Eng. S4. https://doi.org/10.4172/2329-9568.S4-014.

Maghrabi, H., Vijayan, A., Wang, L., Deb, P., 2015. Design of seamless knitted radiation shielding garments with 3D body scanning technology. In: The International Design Technology Conference, DesTech2015, 29th of June—1st of July 2015, Geelong, Australia.

Mahnic, M., Petrak, S., Stjepanovic, Z., 2016. Analysis of tight fit clothing 3D construction based on parametric and scanned body models. In: Proceedings of the 7th International Conference on 3D Body Scanning Technologies, Lugano, 2016.

Mert, E., Psikuta, A., Arévalo, M., Charbonnier, C., Luible-Bär, C., Bueno, M.A., Rossi, R.M., 2018. A validation methodology and application of 3D garment simulation software to determine the distribution of air layers in garments during walking. Measurement 117, 153–164.

Mert, E., Psikuta, A., Joshi, A., Marlene, A., Charbonnier, C., Luible-Bär, C., Annaheim, S., Rossi, R., 2018. The distribution of air gap thickness and the contact area during alpine skiing. In: Proc. of 3DBODY.TECH 2018—9th Int. Conf. and Exh. on 3D Body Scanning and Processing Technologies, Lugano, Switzerland, 16–17 October 2018p. 105.

Molyboga, G., Makeev, I., 2018. Statistical model for human measurements. In: Proceedings of 3DBODY.TECH 2018—9th Int. Conf. and Exh. on 3D Body Scanning and Processing Technologies, Lugano, Switzerland, 16–17 October 2018p. 125.

Morlock, S., Schenk, A., Klepser, A., Schmidt, A., 2016. XL Plus mMen—new data on garment sizes. In: Proceedings of 7th Int. Conf. on 3D Body Scanning Technologies, Lugano, Switzerland, 2016, pp. 255–264. https://doi.org/10.15221/16.255.

Mreed.umtri.umich.edu, 2019. Downloads. [online] Available at: http://mreed.umtri.umich.edu/mreed/downloads.html. Accessed 9 April 2019.

Naglic, M.M., Petrak, S., Gersak, J., Rolich, T., 2017. Analysis of dynamics and fit of diving suits. In: 17th World Textile Conference AUTEX 2017- Textiles—Shaping the Future.

New York Times, 2019. First All-Female Spacewalk Canceled Because NASA Doesn't Have Two Suits That Fit. https://www.nytimes.com/2019/03/25/science/female-spacewalk-canceled.html. Accessed 4 April 2019.

Olds, T., 2004. The rise and fall of anthropometry. J. Sports Sci. 22 (4), 319–320. https://doi.org/10.1080/02640410310001641593.

Petrak, S., Marina, Š., Naglić, M., Maja, 2016. Design and 3D simulation of unique women's pants collection. In: Glogar, M.I., Grilec, A. (Eds.), Proceedings of the 9th Scientific and Professional Consultancy Textile Science and Economy—Creative Mixer. In: 2016, University of Zagreb Faculty of Textile Technology, Zagreb, pp. 118–121.

Petrak, S., Spelic, I., Mahnic Naglic, M., 2018. The volumetric analysis of the human body as starting point for clothing pattern design. In: Proc. of 3DBODY.TECH 2018—9th Int. Conf. and Exh. on 3D Body Scanning and Processing Technologies, Lugano, Switzerland, 16–17 October 2018, pp. 83–91. https://doi.org/10.15221/18.083.

Robinette, K.M., Daanen, H., 2003. Lessons learned from CAESAR: a 3-D anthropometric survey. In: Proceedings of the XVth Triennial Congress of the International Ergonomics Association, Ergonomics in the Digital Age, August 24–29, 2003, p. 00730.

Robinette, K.M., Daanen, H., Paquet, E., 1999. The CAESAR project: a 3-D surface anthropometry survey. In: The Second International Conference on 3D Imaging and Modelling: 4–8th October 1999; Ottawa, pp. 380–386. https://doi.org/10.1109/IM.1999.805368.

Salleh, M., Lazim, H., Lamsali, H., 2018. Body measurement using 3D handheld scanner. Mov. Health & Exerc. 7 (1), 179–187. https://doi.org/10.15282/mohe.v7i1.213.

Scott, E., Sayem, A.S.M., 2018. Landmarking and Measuring for Critical Body Shape Analysis Targeting Garment Fit. In: Proc. of 3DBODY.TECH 2018 – 9th Int. Conf. and Exh. on 3D Body Scanning and Processing Technologies, Lugano, Switzerland, 16–17 October 2018, pp. 222–235. https://doi.org/10.15221/18.222.

Simmons, K., 2001. Body Measurement Techniques: A Comparison of Three-Dimensional Body Scanning and Physical Anthropometric Methods. PhD thesis.

Song, G., 2011. Improving Comfort in Clothing, first ed. Woodhead Publishing Limited, Sawston, United Kingdom.

Store.sae.org, 2019. CAESAR: Civilian American and European Surface Anthropometry Resource Project. [online] Available at: http://store.sae.org/caesar/. Accessed April 2009.

Traumann, A., Peets, T., Dabolina, I., Lapkovska, E., 2019. Analysis of 3-D body measurements to determine trousers sizes of military combat clothing. Text. Leath. Rev. 2(1). https://doi.org/10.31881/TLR.2019.2.

Veitch, D., Burford, K., Dench, P., Dean, N., Griffin, P., 2012. Measurement of breast volume using body scan technology (computer-aided anthropometry). Work (Reading, MA) 41 (Suppl 1), 4038–4045. https://doi.org/10.3233/WOR-2012-0068-4038.

Yu, W., 2004. Clothing Appearance and Fit: Science and Technology. Woodhead Publishing, p. 31.

Further reading

3DMD, 2019. 3D Scanning Technologies. http://www.3dmd.com/. Accessed 4 April 2019.

Ashdown, S.P., 2007. Sizing in clothing. In: A volume in Woodhead Publishing Series in Textiles. Elsevier.

Ashdown, S.P., Loker, S., Schoenfelder, K., Lyman-Clarke, L., 2004. Using 3D scans for fit analysis. J. Text. Appar.Technol. Manag. 4, 1–12.

Biomecánica, 2019. Preliminary Results of the InKreate Project. http://www.biomecanicamente.org/component/k2/item/1123-rb65-inkreate-english. Accessed 8 April 2019.

BodyLux®, 2019. 3D Body Measurement. Vialux GmbH.https://www.vialux.de/en/bodylux.html. Accessed 4 April 2019.

Civilian American and European Surface Anthropometry Resource Project (CAESAR), 2019. http://store.sae.org/caesar/. Accessed 4 April 2019.

CLO, 2019. 3D- Fashion Design Software. https://www.clo3d.com. Accessed 4 April 2019.

Human Shapes, 2019. Realistic Human Body Shape Modeller Based on Real Data. http://humanshape.org/. Accessed 5 April 2019.

Inkreate, 2019. www.inkreate.eu. Accessed 5 April 2019.

iSize, 2019. Human Solutions GmbH. https://portal.i-size.net/SizeWeb/pages/home.seam. Accessed 5 April 2019.

Krzywinski, S., Siegmund, J., 2017. 3D product development for loose-fitting garments based on parametric human models. J. Fash. Technol. Text. Eng. S3, 005. https://doi.org/10.4172/2329-9568.S3-005.

Mahnic, M., Petrak, S., Gersak, J., Rolich, T., 2017. Analysis of dynamics and fit of diving suits. In: 17th World Textile Conference AUTEX 2017- Textiles—Shaping the Future.

Michael Reed, 2019. Adult Anthropometric Data. University of Michigan. http://mreed.umtri.umich.edu/mreed/downloads.html. Accessed 5 April 2019.

Miell, S.L., 2018. Enabling the Digital Fashion Consumer Through Gamified Fit and Sizing Experience Technologies. University of Manchester Ph.D. thesis.

Mirrorsize, 2019. Simplifying the World of Fashion. https://www.mirrorsize.com. Accessed 4 April 2019.

MPI SMPL, Loper, M., Mahmood, N., Romero, J., Pons-Moll, G., Black, M.J., 2015. A Skinned Multi-Person Linear Model (SMPL). ACM Trans. Graph. (Proc. SIGGRAPH Asia) 6 (35), 248:1–248:16.

Vialux GmbH, 2019. 3D Body Scanners. https://www.vialux.de/en/index.html. Accessed 4 April 2019.

WEAR, 2019. World Engineering Anthropometric Resource. https://www.bodysizeshape.com/page-1855750. Accessed 4 April 2019.

Yip, J.M., Mouratova, N., Jeffery, R.M., Veitch, D.E., Woodman, R.J., Dean, N.R., 2012. Accurate assessment of breast volume: a study comparing the volumetric gold standard with a 3D laser scanning technique. Ann. Plast. Surg. 68 (2), 135–141.

Zhang, L., Shin, E., Hwang, C., Baytar, F., 2017. The use of 3D body scanning technology to assess the effectiveness of shapewear: changes in body shape and attractiveness. Int. J. Fash. Des. Technol. Educ. 10 (2), 190–199.

Anthropometric methods for apparel design: Body measurement devices and techniques

2

M. Kouchi
National Institute of Advanced Industrial Science and Technology, Tokyo, Japan

2.1   Introduction

Anthropometric methods are intended to ensure the comparability of measurements obtained by different measurers and repeated measurements by the same measurer. For this purpose, postures for measurement, points on the body used to define measurements (known as landmarks), instruments, measurement procedure, and measurement attire are standardized. Anthropometry started in the field of anthropology and dates back >100 years. In traditional anthropometric methods, human measurers decide the landmark locations and take measurements manually using traditional tools such as calipers and a tape measure. These instruments are not very expensive, but such traditional methods require time to complete the measurements for each person and are prone to error.

In recent decades, noncontact human body measuring systems (hereafter, three-dimensional (3-D) body scanners) have been available and used in sizing surveys. Using 3-D body scanners, the 3-D body surface shape and landmark locations can be obtained, and one-dimensional (1-D) measurements can be calculated from these data. Body scanners are much more expensive than traditional tools, but more people can be measured in a limited time compared with the traditional methods. It may require time after the scan, and scan-derived 1-D measurements are not always comparable with those obtained by traditional methods.

Fig. 2.1 shows a flow diagram of anthropometry. In the traditional methods, landmarking and measurement are conducted by measurer(s). In 3-D anthropometry, landmarking must still be done by a measurer (3-D anthropometry 1), or landmark locations can be calculated from body surface data (3-D anthropometry 2). Landmarking is the most important process for ensuring that anatomical locations correspond between subjects and defining body dimensions and homologous body models.

The anthropometry method used depends on the purpose of the measurement. Specifically the chosen method will depend on which type(s) of data are required (1-D measurements, 3-D landmark locations, or 3-D surface shape) and how these obtained

Anthropometry, Apparel Sizing and Design. https://doi.org/10.1016/B978-0-08-102604-5.00002-0

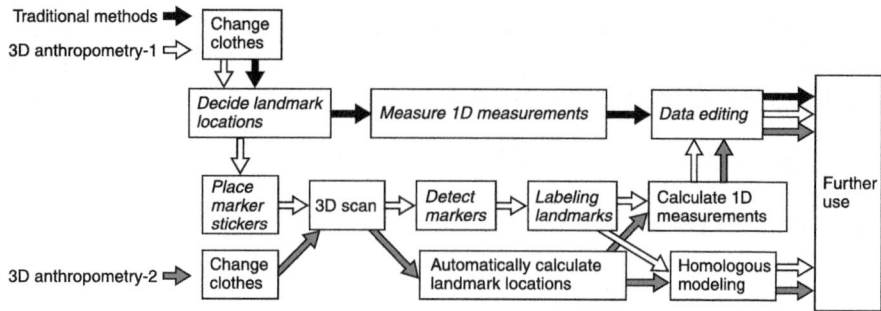

Fig. 2.1 Flow diagram of anthropometry. Manual procedures are shown in *italics*.

data will be applied. The required accuracy of measurements and measuring attire also depend on the purpose of the anthropometry. Measurements for a sizing survey for establishing national standards should be more accurate than measurements for selecting the proper size of clothes for individual customers. The proper measurement attire may be different for designing underwear and designing outerwear. If measurements are compared with existing data obtained by traditional methods, scan-derived 1-D measurements need to be comparable with those obtained by traditional methods. If measurements are used only within a company, they need not be comparable with measurements obtained by other methods.

Users of anthropometric data expect three different types of quality: validity of data, comparability of measurement items, and accuracy and precision of measurements. The validity of data means that the subject population of an anthropometric survey meets the target population of the user. Comparability of measurement items means that 1-D measurements with the same name are measured using exactly the same method. In this chapter, anthropometric methods are described with special attention on the quality of obtained data. Individual measurement items are not described. Please see anthropometry textbooks introduced in Section 2.7 for descriptions of individual measurement items.

2.2 Traditional anthropometric methods

2.2.1 Basic postures

In the basic standing posture, the subject stands erect with feet together (see Fig. 2.7A). The shoulders are relaxed, and the arms are hanging down naturally. The head is oriented in the Frankfurt plane; that is, the Frankfurt plane of the subject is horizontal. The Frankfurt plane is defined using three landmarks of the head, the right tragion, left tragion, and left orbitale (Fig. 2.2). The orientation of the head affects the accuracy of measurements such as the height and neck girth. The tragion is the notch just above the tragus. The orbitale is the lowest point on the lower edge of the orbit (eye socket), which can only be located by palpation. The three landmarks should be marked in advance, and the measurer should confirm that the line

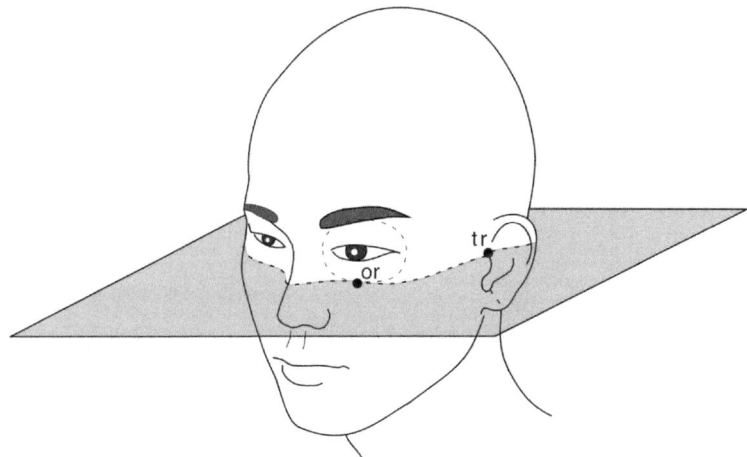

Fig. 2.2 Frankfurt plane (or, orbitale; tr, tragion).

connecting the left tragion and the left orbitale is horizontal just before taking a measurement. In the basic sitting posture, the subject sits erect with thighs fully supported by a hard horizontal plane. The head is oriented in the Frankfurt plane. Proper instructions are necessary to ensure that the subject maintains the proper posture during the measurement.

2.2.2 Measuring instruments

The instruments that are used in traditional anthropometry are calibrated in millimeters. Measurements are read to the nearest millimeter except small measurements of the head, hand, or foot. Fig. 2.3 shows the main instruments used in traditional methods.

The anthropometer and tape measure are usually used in anthropometry for garment design. The anthropometer is used to measure a vertical distance from the floor to a specific landmark. Four rods are put together to make an anthropometer (Fig. 2.3A), and a straight arm is inserted into the cursor. The measurer holds the rod of the anthropometer vertical, slides the cursor, and places the tip of the arm on the target landmark. The measurer should keep the tip of the arm away from the eyes of the subject.

A large sliding caliper consists of one or two rod(s) of the anthropometer and two arms (Fig. 2.3B, left). It is used for measuring large distances between two landmarks and for breadth and depth measurements. Curved arms are used instead of straight arms when necessary (e.g., measuring the chest depth in the midsagittal plane). The length of the two arms must be the same except when the projected distance between two landmarks is measured.

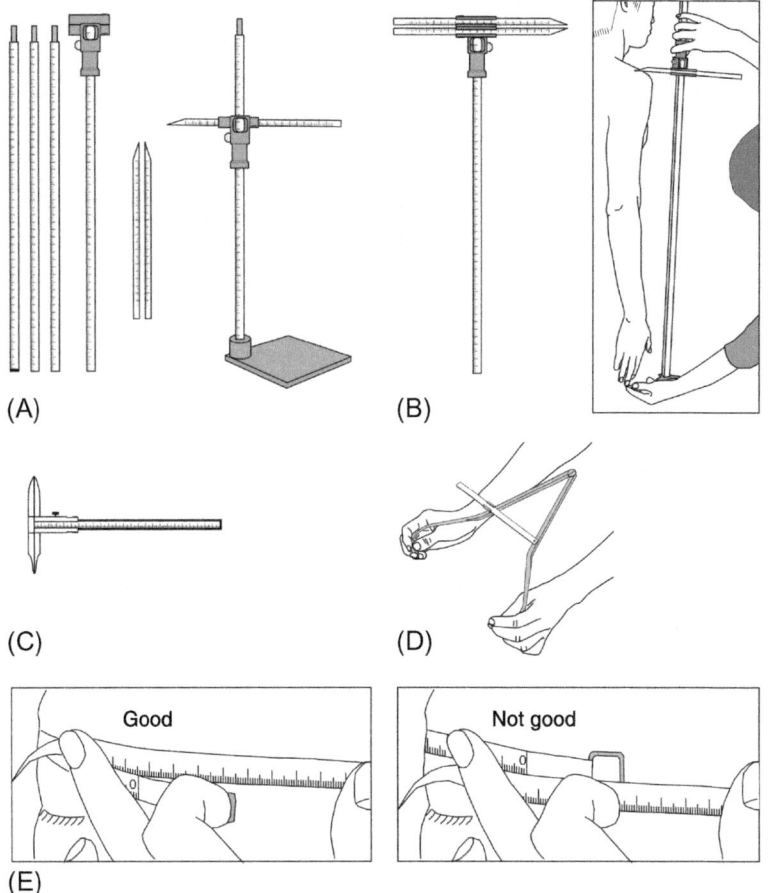

Fig. 2.3 Traditional instruments: (A) anthropometer, (B) large sliding caliper, (C) sliding caliper, (D) spreading caliper, and (E) tape measure.
(A) and (B) are from Mochimaru, M., Kouchi, M., 2006. Biomechanism Library. Measurement of Man: Size, Shape and Motion. Tokyo Denki University Press, Tokyo (in Japanese).

A sliding caliper is used to measure small breadth measurements and the distance between two landmarks (Fig. 2.3C). Pointed-tip jaws should not be used to measure living people.

A spreading caliper is used to measure the distance between two landmarks when two tips of a sliding caliper cannot touch the landmarks because a part of the body is in the way, such as chest depth in the midsagittal plane and head length (Fig. 2.3D). The large sliding caliper with curved arms can be used for the same purpose.

A tape measure is used for measuring the girth and surface distance. The material of a tape measure should not stretch by tension or by wetting. The tape measure should be cleaned with alcohol as necessary. When a tape measure is wrapped around a subject, the zero point of the tape measure should overlap the scale on the tape measure as shown in the left image in Fig. 2.3E.

An inclinometer is used to measure the shoulder slope. Place the inclinometer on the shoulder line (see "Armscye line" in Section 2.5.5) with an end of it at the side neck point, and read the angle. Prepare two inclinometers, and measure the right and left sides simultaneously when both sides are measured.

2.2.3   Role of the measurer

In traditional methods a measurer and an assistant work together to take measurements. The measurer and assistant must be female when the subject is a female. When the subject is a male, a measurer of the same sex is preferable. Measurers should be properly trained before starting a survey to obtain accurate measurements in minimal time. The training includes lectures on basic anatomy for understanding landmark locations, definitions of landmarks and measurement items, measurement errors, and physical training of landmarking and measurements with several subjects with different body shapes.

The measurer is in charge of the measurement process. She/he must decide and mark the locations of landmarks, give proper instructions to the subject for maintaining the correct posture, give proper instructions to the assistant, and take actual measurements.

2.2.4   Role of the assistant

The assistant records the measured value in the correct cell of a data sheet. She/he should repeat the value aloud before writing it down in order to avoid mistakes. If the assistant realizes that the value is unusual, she/he must ask the measurer to take another measurement. The assistant helps the measurer by checking the posture of the subject that is not visible to the measurer (orientation of the head, rotation of the torso, etc.), checking the orientation of the anthropometer (the anthropometer should be vertical), holding the tape measure at the back of the subject, and passing the small articles necessary for landmarking or measurement to the measurer (Fig. 2.4). To minimize the time for measurement, the assistant should be aware of what she/he should do without instruction from the measurer.

2.2.5   Measurement errors

The accuracy of measurements is affected by factors related to the instrument, the measurer, and the subject. In traditional methods the instruments are simple and easy to calibrate. The accuracy in landmarking and measurements depends on the skill of the measurer. A proper measuring posture and its repeatability are subject-related factors. However, posture is a part of the definition of a measurement item and can be controlled by proper instruction from the measurer. Since measurements are taken quickly, the effect of body sway is negligible. Therefore the skill of the measurer is the main cause of errors in traditional methods.

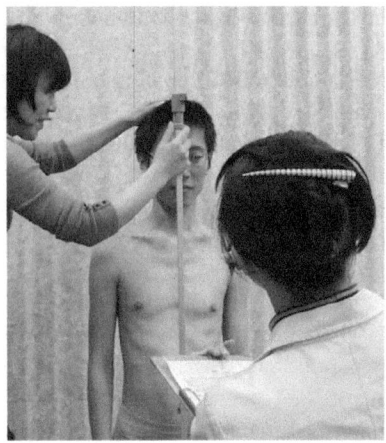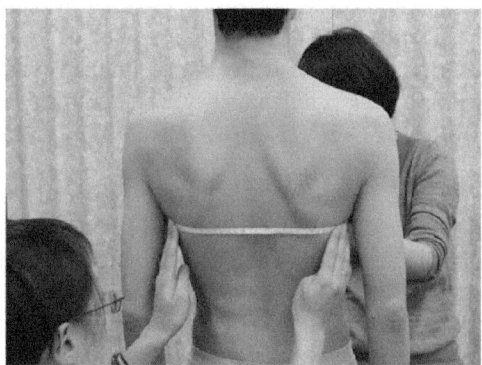

Fig. 2.4 Possible roles of the assistant: check the posture of the subject and orientation of the instrument, record measurements, and hold the tape measure.

Suppose a measurer performed a 1-D measurement x of N subjects. The measurement of the ith subject, x_i, is described as Eq. (2.1), where M is the mean of N subjects, s_i is the characteristic of the ith subject, o is the effect of the measurer, and e_i is the random error.

$$x_i = M + s_i + o + e_i \tag{2.1}$$

Since the mean of the random error is 0, the mean of the subject population will be $M + o$; that is, the effects of random errors are canceled out, but the effect of the observer remains as a bias. On the other hand the variance of random errors is not 0. Therefore the variance of x is larger than the between-subject variance. One of the purposes of training measurers is to reduce the random error of each measurer. The variance of random errors by a specific measurer can be calculated from two repeated measurements of N subjects using Eq. (2.2), where x_1 and x_2 are the first and the second measurement of each subject, respectively:

$$V[e] = \sum (x_1 - x_2)^2 / 2N \tag{2.2}$$

The square root of the random error variance is called the technical error of measurement and is one of the indicators of the degree of repeatability of measurements by a measurer.

When multiple measurers participate in an anthropometric survey, systematic differences (or biases) between measurers also increase the variance. Interobserver error, the difference between measurements taken by two different measurers, depends on the magnitude of the bias and the magnitude of the random error by each measurer. The interobserver error is larger than the intraobserver error due to the systematic bias between the two measurers. Another purpose of training is to reduce the bias between measurers to an allowable range when plural measurers participate in an anthropometric survey. A common understanding of locations of landmarks on the body between measurers is most important for reducing the bias. A practical protocol for locating a landmark helps to reduce both bias between measurers and random error within a measurer. A protocol established in ISO 20685-1 (see Section 2.5) can be used for evaluating if the differences between measurements taken by two measurers are within an acceptable range.

2.3 Three-dimensional anthropometry

2.3.1 Basics of 3-D measurement

Triangulation is used for obtaining the depth information in 3-D body scanners (Fig. 2.5). In one type of body scanner, a single slit light is projected on the body surface, which is observed by a camera from a different angle. The length between the light source and the camera (L in Fig. 2.5) and angles between the line connecting the light source and camera and the line connecting a point on the body surface and the light source or the camera (α and β in Fig. 2.5) are known. Therefore the depth (D in Fig. 2.5) can be calculated. The time required for scanning the entire body by slit light ranges from several seconds to over 10 s according to the system used.

In another type of body scanner, structured patterns are projected on the body surface, which are observed by a camera from a different angle. In this method the body can be measured within 1 s, in principle. However, plural cameras and plural

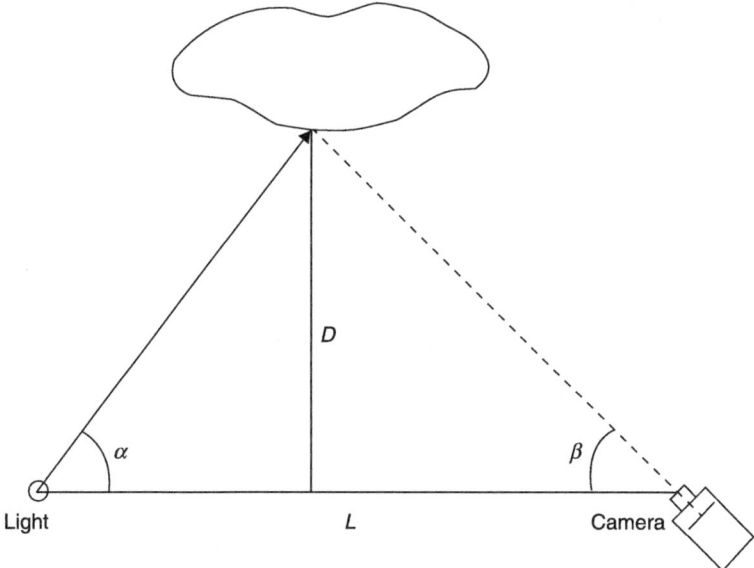

Fig. 2.5 Principle of 3-D measurement.

Table 2.1 Different methods to obtain landmark coordinates in 3-D anthropometry

Decision of landmark location	Detection of marker location	Labeling
Manual	Manual	Manual
	Automatic	Automatic
Automatic	Automatic	Automatic

projectors are often used to minimize the occluded area, and the time for a scan is several seconds to nearly 10 s according to the number and the arrangement of cameras and projectors for avoiding the interference of patterns from different projectors.

There are several different methods for obtaining 3-D coordinates of landmark locations (Table 2.1). Important points that affect the throughput are:

(1) the method to decide landmark locations on the body: a measurer decides landmark locations manually, or a system calculates landmark locations automatically;

(2) the method to obtain 3-D coordinates of marker stickers: an operator manually picks the centers of marker stickers, or a system automatically recognizes marker stickers and calculates 3-D coordinates; and

(3) the method to name landmarks (labeling): an operator or a system automatically names each marker.

Deciding landmark locations on the human body and placing marker stickers on them are time-consuming. However, picking marker centers and naming markers are also time-consuming. Automatic calculations of landmark locations save time, but the calculated landmark locations do not always match with landmark locations chosen by skilled anthropometrists. The difference can actually be very large.

There are systems in which all three processes are manual, or process (1) is manual but processes (2) and (3) are automatic, or all three processes are automatic (Table 2.1). It is very important to confirm the time necessary for scanning and for obtaining the 3-D coordinates of landmarks and 1-D measurements before purchasing a 3-D body scanner.

2.3.2 Scanning posture

A basic standing posture used in traditional methods may not be suitable for body scanning due to larger occluded areas at the axillae and the crotch. Occluded areas are smaller when arms and legs are abducted, but the shape of the shoulders and body dimensions around the shoulders and hips change (Kouchi and Mochimaru, 2005). Fig. 2.6 shows the relationships between the arm abduction angle and acromial height (shoulder height) or shoulder (biacromial) breadth measured for two subjects using a motion capture system (Vicon MX) when their arms are slowly abducted from a basic standing posture. The acromial height (Fig. 2.6A) remains stable as long as the abduction angle is smaller than approximately 20 degrees, but the biacromial breadth (Fig. 2.6B) becomes smaller when the abduction angle is over approximately 5

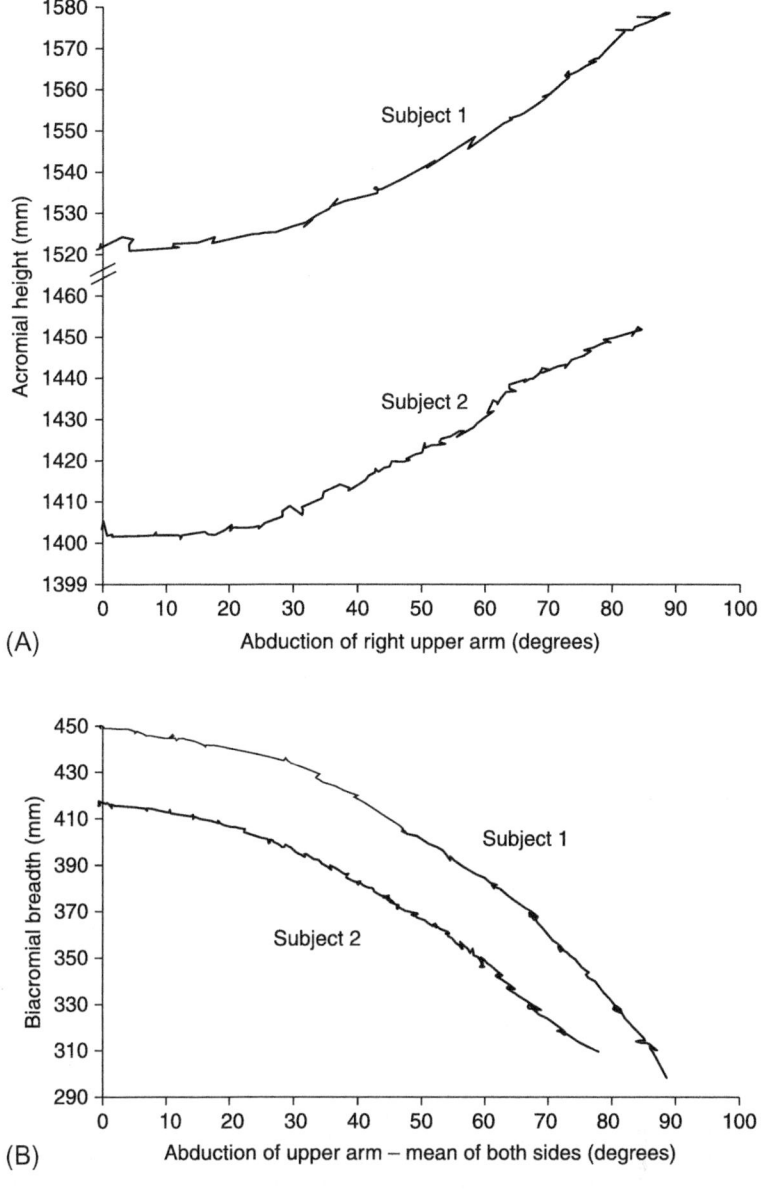

Fig. 2.6 Relationships between the arm abduction angle and right acromial height (A) or biacromial breadth (B). Stature is 168 cm for subject 1 and 181 cm for subject 2. The abduction angle in the basic standing posture is 0 degrees.

From Kouchi, M., Mochimaru, M., 2005. Causes of the measurement errors in body dimensions derived from 3D body scanners: differences in measurement posture. Anthropol. Sci. (Jpn Ser.) 113, 63–75 (in Japanese with English abstract).

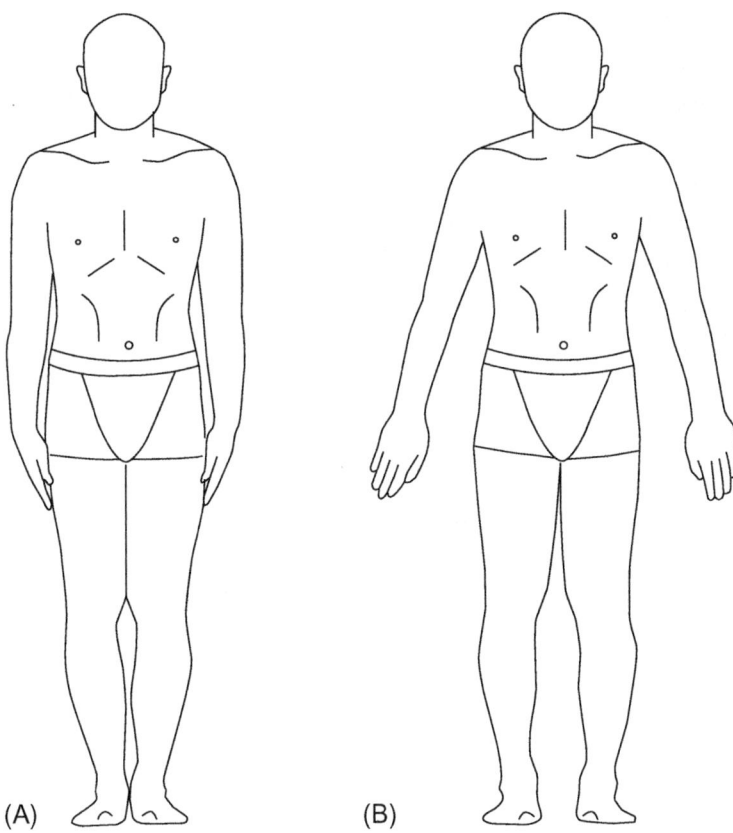

Fig. 2.7 Scanning postures recommended in ISO 20685-1. (A) Posture for height measurements. (B) Posture for girth measurements.

degrees. Kouchi and Mochimaru (2005) compared body dimensions of 40 subjects in several different postures using the traditional methods. They found significant differences in measurements defined using the acromion (shoulder point in ISO 8559-1) and armpit front fold point as well as the hip breadth and hip girth between a basic standing posture and a posture recommended in ISO 20685-1 in which the arms are abducted 20 degrees and the distance between the foot axes of both feet is 20 cm (Fig. 2.7B).

The scanning posture is a compromise between the occluded area and shape deformation. It should be noted that many scan-derived 1-D measurements obtained from a scan with a posture with arms and legs abducted are not comparable with traditional measurements because the postures are different. ISO 20685-1 recommends two standing postures for scanning: One is the basic standing posture for height measurements (Fig. 2.7A), and the other is a posture with arms and legs abducted for girth measurements (Fig. 2.7B). In the latter posture the head is oriented in the Frankfurt plane, the long axes of the feet are parallel to one another and 20 cm apart, the upper arms are abducted to form a 20° angle with the side of the torso, the elbows are straight, the palms face backward, and the subject is breathing quietly. Lu et al. (2010) examined the relationships between the arm posture and differences between scan-derived and traditional measurements. They found that when the palms faced backward rather than forward or to the side of the body, the space between the arm and body was larger, and the differences between the scan-derived and traditional measurements were smaller.

2.3.3 Scanning attire

Scanning attire should be neither too loose nor too tight and should be appropriate for the purpose of the measurement. Fig. 2.8 shows the scan results of the same person wearing three different garments. It is clear that the shape changes according to the garment. ISO 20685-1 recommends that the subject being measured wears garments that expose landmarks and a pattern that results in no side seams on the thigh. We recommend a seam on the center back of the lower garment.

In most 3-D body scanners, dark colors cannot be captured. Long hair should be pulled up using a rubber band and/or cap so that it does not hide the neck and shoulders of the subject.

2.3.4 Comparability of scan-derived and traditional 1-D measurements

The quality of scan-derived measurements depends on the accuracy of the machine and the performance of software for detecting and calculating the coordinates of the center of marker stickers and for calculating body dimensions using landmark locations. It also depends on the skill of the measurer to decide landmarks if the system uses manually chosen landmark locations and on the skill of the operator to pick the center of marker stickers using a mouse if the system uses manually chosen marker locations. The quality of scan also depends on the body sway of the subject during the scan because a human cannot stand completely still for 10 s. The effects of body sway are smaller when the scan time is shorter and the scan direction is from the top down rather than from side to side. Many more factors affect the accuracy of scan-derived 1-D measurements compared with traditional 1-D measurements in which the skill of measure(s) is the dominant factor.

Scan-derived and traditional measurements have been compared in several studies (e.g., Bougourd et al., 2000; Paquette et al., 2000; Han et al., 2010), but the criterion of judging comparability is not always clear. ISO 20685-1:2018, first published in 2006, establishes a protocol for evaluating the comparability between the scan-derived and traditional 1-D measurements. In ISO 20685-1 the criterion of judgment is based on interobserver differences of experienced anthropometrists in traditional 1-D measurements. The procedure for evaluation is described in the succeeding text. See the original document for details:

1. Measure N subjects ($N \geq 40$) by traditional methods (measurer should be a skilled anthropometrist) and by a 3-D body scanner.

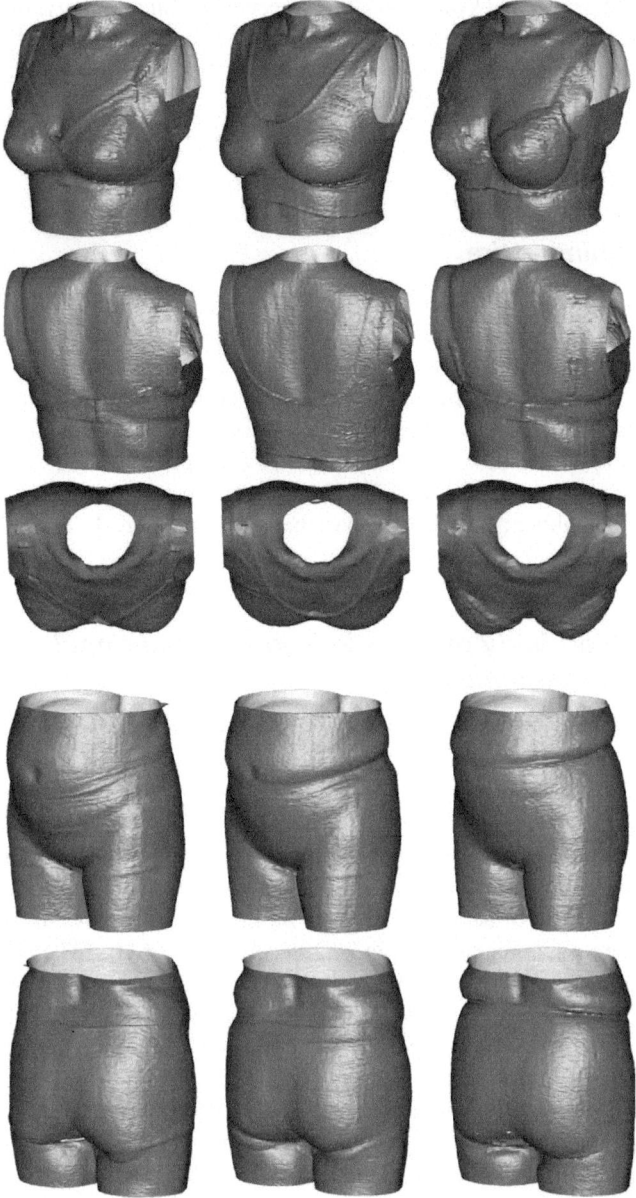

Fig. 2.8 Shapes of the same person in different garments. *Left*, garment used in Digital Human Research Center, National Institute of Advanced Industrial Science and Technology, Japan; *middle*, garment used in Size *JPN* project; *right*, commercially available lift-up bras and soft girdle.

2. Calculate the error for all subjects. The error is defined as the difference between the scan-derived measurement and the measurement by the skilled anthropometrist (scan-derived measurement minus manual measurement).
3. Calculate the 95% confidence interval of the mean error. The lower and upper limits of the 95% confidence interval are calculated as the mean error $\pm 1.96 \times$ the standard error. The standard error is calculated as the standard deviation divided by the square root of N.
4. When the following equations are satisfied, the two types of measurements are considered sufficiently similar: $-$the maximum allowable error $<$ lower limit of the 95% confidence interval of the mean error, and the upper limit of the 95% confidence interval of the mean error $< +$the maximum allowable error. The maximum allowable error is 4 mm for height, small girth, and body breadth measurements; 5 mm for segment length and body depth measurements; 9 mm for large girth measurements; 2 mm for foot measurements and head measurements including the hair; and 1 mm for hand measurements and head measurements not including the hair.

The procedure is explained using an example of back neck height (cervical height). Back neck height was measured for 74 subjects (39 females and 35 males) by a bodyline scanner (Hamamatsu Photonics K.K., Hamamatsu, Japan) and by a skilled anthropometrist. The error was calculated as the difference between the scan-derived measurement and the traditional measurement for the 74 subjects. Means and standard deviations are shown in Table 2.2. The lower limit of the 95% confidence interval is calculated as mean error $- 1.96\,\text{S.E.} =$ mean error $- 1.96\,\text{S.D.}/\sqrt{N} = 5.6 - 1.96 \times 4.9/\sqrt{74} = 4.48$ [mm]. The upper limit of the 95% confidence interval is calculated as mean error $+ 1.96\,\text{S.E.} = 5.6 + 1.96 \times 4.9/\sqrt{74} = 6.72$ [mm]. The maximum allowable error of height measurements is 4 mm. As the upper limit of the 95% confidence interval is larger than the maximum allowable error, the results do not satisfy the condition, and we conclude that the scan-derived and traditional measurements are not comparable in this example; this bodyline scanner systematically gives larger values. Fig. 2.9 shows the relationship between the 95% confidence interval of the mean error and the maximum allowable error in this example.

2.3.5 Accuracy of scan-derived 3-D measurements

ISO 20685-2:2015 establishes a protocol for evaluating the performance of 3-D body scanners. The basic idea is to measure an object with known shape and size and compare the measured results with the actual values. An artifact (ball) calibrated using a coordinate measuring machine (CMM) that is traceable to the international standard

Table 2.2 Comparison of scan-derived and traditional measurements of back neck height (cervical height) ($N = 74$) (unit: mm)

	Scan-derived measurement	Traditional measurement	Error
Mean	1388.2	1382.6	5.6
S.D.	76.9	76.1	4.9

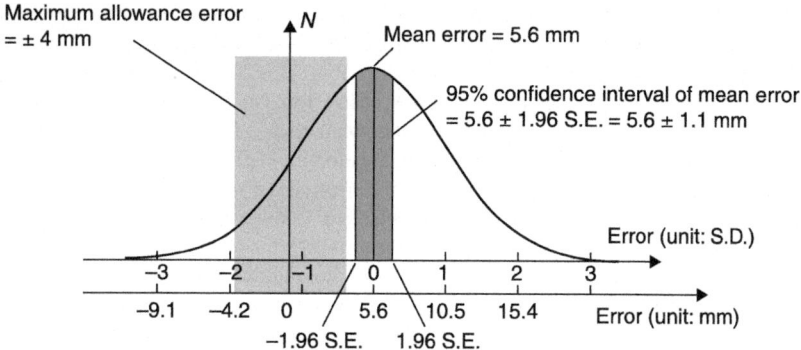

Fig. 2.9 Distribution of the error and the relationship between the 95% confidence interval of the mean error and the maximum allowable error using the data of Table 2.2. *S.D.*, standard deviation; *S.E.*, standard error.

of the length is used as a test object. An artifact is used because only by using a test object with actual values we can evaluate the accuracy of the measured data. This procedure may provide a basis for the agreement on the quality of scan-derived surface shape between the data providers and data users.

In ISO 20685-2 a ball is measured at nine different positions within the scanning volume of a whole-body scanner to be evaluated. A function of the sphere is fitted to the scanned point cloud, and the diameter is calculated. The error of the diameter measurement is calculated as the difference between the measured diameter ($d_{calculated}$ in Fig. 2.10, left) and the actual diameter. The distance between the center of the best-fit sphere and each of the measured data points is calculated (r_i in Fig. 2.10, right). The error of the spherical form measurement is evaluated using spherical form dispersion value (90%), the radial thickness of the spherical shell (shaded area in Fig. 2.10,

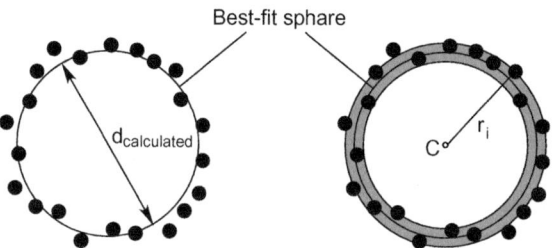

Fig. 2.10 Evaluation of 3-D body scanners. *Left*: the error of the diameter measurement is calculated as the difference between the diameter of the best-fit sphere ($d_{calculated}$) and the actual diameter given by a coordinate measuring machine. *Right*: the error of the surface shape measurement is evaluated using the spherical form dispersion value (90%), the radial thickness of the shell *(shaded area)* that includes 90% of the measured data points, and the standard deviation of the distances (r_i) from the center of the best-fit sphere (C) to each of the measured data points.

right) that includes 90% of all the measured data points. The standard deviation of r_i is also used to evaluate the error of the spherical form measurement.

ISO 20685-2 also establishes a protocol for evaluating the repeatability of measured landmark locations using an anthropomorphic dummy representing the size and shape of a natural human. Repeatability rather than the accuracy of landmark location is evaluated because shape of anthropomorphic dummies is too complicated to give the actual values using a CMM. Marker stickers are placed on landmarks premarked on the dummy. The dummy is scanned 10 times at slightly different positions. Landmark coordinates are calculated for the 10 scans. Landmark location data from 10 scans are superimposed simultaneously. The error is calculated as the distance between corresponding landmark locations for all possible pairs of landmark locations data ($N = {}_{10}C_2 = 45$). Mean error and the maximum error are reported. Examples of evaluation results using a ball with the diameter of 120.01593 mm and an anthropomorphic dummy are published in Kouchi et al. (2012).

Occluded areas due to the complicated body shape and quality of scan-derived landmark locations are issues specific to anthropometry. Occluded areas cannot be evaluated using a test object. ISO 20685-2 provides a protocol for evaluating the occluded area using human subjects.

It may not be easy to access an artifact calibrated using a CMM. A hard and nondeformable ball of appropriate surface treatment and size can be used for daily verification. This can be done by measuring the test object using a 3-D body scanner to be evaluated and comparing the measured diameter with the diameter measured with a calibrated vernier caliper.

The following information should be considered when selecting a 3-D body scanner: scan volume, time necessary for one scan, scan direction, resolution, accuracy, function to capture the texture, function to capture premarked landmark locations and its method (manual or automatic), number of cameras arranged in different directions, function to merge data obtained by different cameras, function to calculate 1-D measurements, and function to generate a homologous model. The final decision depends on the required type(s) of data (1-D measurements, landmark location, and surface shape), accuracy of data, and funds (time and space allowed to obtain required data, price of a 3-D body scanner, and running cost).

2.4   International standards related to anthropometric methods

It is implicitly assumed that measurements from different surveys are taken using exactly the same method when the measurement name is identical. Unfortunately, this is not always true. Measurement items named identically but defined differently or named differently but defined identically cause confusion. To avoid such unnecessary confusion, textbooks and standards are used as references. The most frequently used reference is an anthropometry textbook, Martin's textbook of anthropology (Martin and Knußmann, 1988), but this textbook does not focus on anthropometry for garment design. There are several international standards related to anthropometry, ISO 7250 series, ISO 8559-1, ISO 15535, and ISO 20685 series. ISO 20685-1 and ISO 20685-2 were already introduced in Section 2.3.4. ISO 7250-1, ISO 8559-1, and ISO 15535 are introduced in the succeeding text. ISO/TR 7250-2:2010 is a data book of body dimensions measured according to ISO 7250-1. Standards, however, are not intended to be a manual for anthropometry. For details of measurement items and measurement procedures, please refer to the textbooks and manuals described in Section 2.7.

2.4.1   ISO 7250-1 and ISO 8599-1: Definitions of measurement items

ISO 7250-1:2017 and ISO 8559-1:2017 describe measurement items for technological design and garment design, respectively. Both standards have been recently revised.

Sixty-two measurement items for technological design are described in ISO 7250-1. Six measurement items were added in the 2017 revision. Many landmark names are based on anatomical terms. There are only 11 measurement items measured using a tape measure, and most are length, breadth, and depth measurements measured using an anthropometer or calipers.

Ninety-three measurement items for garment design, 26 landmarks and levels, three lines, and two planes are described in ISO 8559-1. Thirty-eight measurement items were added in the 2017 revision. ISO TC 133/WG 1 decided to use landmark names and terms familiar to the apparel field. Therefore landmark names are not the same with those in ISO 7250-1 even if the definitions are the same. Among the 93 measurement items, 54 items are measured using a tape measure, and 18 items are measured using an anthropometer. One item is measured using a scale and one item by an inclinometer. Eight items are calculated from manually measured measurements.

Table 2.3A compares 15 pairs of landmarks from ISO 7250-1 and ISO 8559-1 with the same definitions or locations that are close. Twelve of the 15 pairs are identical or practically identical. However, three pairs are not identical.

Table 2.3B compares 19 pairs of measurement items from ISO 7250-1 and ISO 8559-1 with the same or similar definitions. Twelve pairs are identical and three pairs are practically identical. However, four pairs are not identical.

2.4.2   ISO 15535 and how to eliminate irregular values

ISO 15535:2012 specifies the general requirements for establishing anthropometric databases, such as the number of subjects to be measured and a protocol for eliminating irregular values from measured data.

In traditional methods, irregular values are inevitable due to mistakes. The effects of these irregular values on statistics such as the standard deviation, skewness, maximum, and minimum can be very large. ISO 15535 establishes a protocol for eliminating these irregular values.

In this protocol, measured values smaller than or larger than the range of the mean ± 3 standard deviations are reviewed individually. This way, very large or very

Table 2.3 Comparison of definitions of landmarks and measurement items in ISO 7250-1:2017 and ISO 8559-1:2017

ISO 8559-1	ISO 7250-1	Comparison
A. Landmarks		
3.1.1 Shoulder point	5.2 Acromion	Identical
3.1.2 Center point of brow ridge	5.6 Glabella	Identical
3.1.3 Tragion	5.20 Tragion	Identical
3.1.4 Orbitale	5.13 Orbitale–infraorbitale	Identical
3.1.5 Lowest point of chin	5.9 Menton	Identical
3.1.6 Back neck point	5.3 Cervicale	Identical
3.1.12 Center chest point	5.10 Mesosternale	Identical
3.1.15 Lowest rib point	5.8 Lowest point of rib cage	Identical
3.1.22 Waist level	Not defined as a level, but the same level is used for defining the waist circumference	Identical
3.1.11 Bust point	5.18 Thelion Bust point is not identical with thelion, but bust point is used for females wearing bras	Identical for females wearing bras
3.1.16 Highest point of the hip bone	Not defined as a landmark, but the highest point of the hip bone is used for defining the level of waist circumference	Identical
Not defined as a landmark, but the same point is used for measuring stature	5.22 Vertex (top of head)	Identical
3.1.10 Elbow point	5.12 Olecranon	Not identical: elbow point is defined with arm hanging freely downward, while olecranon is defined with the elbow flexed 90°
3.1.19 Wrist point	5.21 Ulnar stylion	Not identical: wrist point is the most prominent point of the bulge of the head of ulna. Ulnar stylion is the most distal point on the ulnar styloid

Continued

Table 2.3 Continued

ISO 8559-1	ISO 7250-1	Comparison
3.1.26 Inside leg level	5.4 Crotch level	Not identical: inside leg level is decided by visual inspection with subject standing feet apart. Crotch level is decided by using an anthropometer
B. Measurement items		
5.1.1 Stature	6.1.2 Stature (body height)	Identical
5.1.18 Back neck height (sitting)	6.2.3 Cervicale height, sitting	Practically identical: feet are not supported in ISO 8559-1, but are supported in ISO 7250-1
5.2.1 Hip breadth	6.1.12 Hip breadth, standing	Identical
5.2.5 Chest depth	6.1.9 Chest depth, standing	Identical
5.2.6 Bust depth	6.2.15 Thorax depth	Identical
5.3.1 Head girth	6.3.12 Head circumference	Identical
5.3.2 Neck girth	6.4.9 Neck circumference	Identical
5.3.4 Bust girth	6.4.10 Chest circumference	Identical
5.3.10 Waist girth	6.4.11 Waist circumference	Identical
5.3.20 Thigh girth	6.4.13 Thigh circumference	Identical
5.3.24 Calf girth	6.4.14 Calf circumference	Identical
5.5.4 Index finger length	6.3.4 Index finger length	Identical
5.5.5 Foot length	6.3.7 Foot length	Practically identical: longitudinal axis of the foot is not defined in ISO 8559-1
5.5.6 Foot width	6.3.8 Foot breadth	Practically identical: longitudinal axis of the foot is not defined in ISO 8559-1
5.6.1 Body mass	6.1.1 Body mass (weight)	Identical
5.1.15 Inside leg height	6.1.7 Crotch height	Not identical: inside leg height is the height of inside leg level, measured with the subject standing with feet shoulder width apart. Crotch height is the height of crotch level measured with the subject standing feet together

Table 2.3 Continued

ISO 8559-1	ISO 7250-1	Comparison
5.3.19 Wrist girth	6.4.12 Wrist circumference	Not identical: the tape measure passes the wrist point in ISO 8559-1, while the tape measure passes at the level of radial stylion and just distal to the ulnar stylion in ISO 7250-1
5.5.2 Hand length (wrist crease)	6.3.1 Hand length (stylion)	Not identical: hand length is measured from the distal wrist crease in ISO 8559-1, while it is measured from the line connecting the radial stylion and ulnar stylion in ISO 7250-1
5.5.3 Palm length perpendicular	6.3.2 Palm length	Not identical: palm length is measured from the distal wrist crease in ISO 8559-1, while it is measured from the line connecting the radial stylion and ulnar stylion in ISO 7250-1

small irregular values (caused by, e.g., the wrong measurement unit) can be identified. Most of the irregular values are, however, not extreme values. These irregular values can have effects on correlations rather than univariate statistics. To eliminate such irregular values, draw a scattergram using two measurement items highly correlated with each other, and review outliers identified by visual inspection. If the outliers are due to mistakes in data input, correct the values. If the cause is unknown, delete the values.

In the example shown in Fig. 2.11, 217 male subjects are plotted using body height (stature) and iliospinal height. It is easy to locate an outlier by visual inspection though

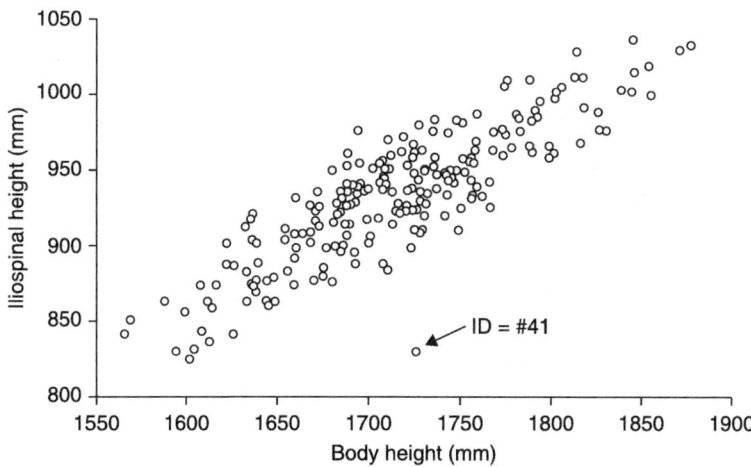

Fig. 2.11 Find outliers in a scattergram by visual inspection.

both measurements of the outlier are within the range of the mean ± 3 standard deviations. Unfortunately, it is unknown which of the two measurements is incorrect only from this scattergram. In this situation, choose another measurement item highly correlated with both iliospinal height and body height (e.g., acromial height). Draw a scattergram using the iliospinal height and acromial height and another scattergram using the body height and acromial height. The measurement item with the outlier in two scattergrams has the incorrect value.

2.5 Landmarking

Landmark locations decided by measurers are used for defining anatomical correspondence between individuals. Many of these landmarks are defined on specific locations of bones or easily defined features of soft tissues such as nipples and the navel. Landmarks and imaginary lines on the body specific to anthropometry for garment design are defined using anatomical features and small articles such as a neck chain (chainette). In the following sections, main lines and landmarks are described using several manuals and ISO 8559-1 as references (Clauser et al., 1988; JIS L0111:2006; National Institute of Bioscience and Human Technology, 1994; O'Brien and Shelton, 1941). Basically, landmark names in ISO 8559-1 are used except when it is preferable to use names in ISO 7250-1 to make the anatomical definition clear.

2.5.1 Small articles for landmarking

Landmark locations should be marked on the skin when one landmark is used for measuring several measurement items in traditional anthropometry, during the training of anthropometry, or for body scanning. An eyeliner pencil is useful for marking the skin. The mark can be easily removed by makeup remover, and it is easier to mark the skin compared with an eyebrow pencil.

A neck chain is used to define the neck base line and side neck point (Fig. 2.13). A small ruler is used to define the armpit back fold point (posterior axilla point) (Fig. 2.15). A waist belt is used to define the natural waistline (see Fig. 2.17).

2.5.2 Posture and measurement attire

When marking landmark locations the marking must always be done for a subject in the correct posture for the measurement. Otherwise the actual landmark location and the mark on the skin can be very different. For example, the tip of the spinous process of the seventh cervical vertebra can be easily palpated at the back of the base of the neck when the subject bends his/her neck forward (Fig. 2.12B). Suppose that the tip of the spinous process is marked in this posture. The mark on the skin slides away from the tip of the spinous process while the subject lifts his/her head to be oriented in the Frankfurt plane.

When a landmark is covered by a garment (e.g., trochanterion), it is desirable to mark the skin directly rather than on the garment to avoid effects of relative movement

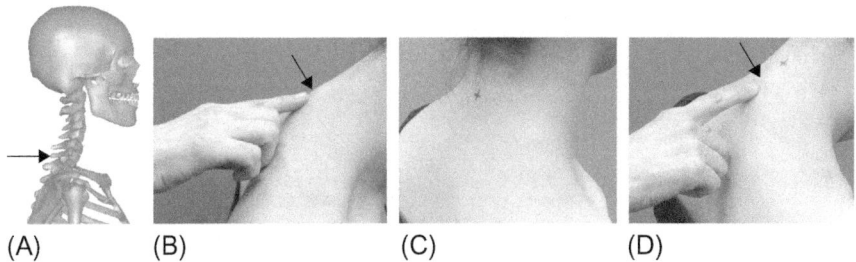

(A)   (B)   (C)   (D)

Fig. 2.12 Back neck point (cervicale). The *arrow* indicates the tip of the spinous process of the seventh cervical vertebra on the bone (A) or on the living human (B, D). The *cross mark* indicates the location of the back neck point marked on the subject in the basic standing posture (C, D). Note the difference between the location of the cross mark and the protrusion of the spinous process of the seventh cervical vertebra when the subject bends her neck forward (D).

between the skin and garment. When taking a measurement, use the mark on the skin. When the body is scanned, place a marker sticker on the garment immediately before scanning.

When a subject is wearing a brassiere, use the bust point rather than the thelion (the center of the nipple). The bust point is the most anterior point of the bust for subjects wearing bras.

2.5.3 Size and shape of marks

Marks on the skin should be clear and visibly large enough to avoid confusing them with moles. When a marker sticker attached on a mark for 3-D body scanning falls off, the paste removes the mark made with the eyeliner pencil. Putting another marker sticker at the same location is easy if the mark is a cross that is larger than the marker stickers.

2.5.4 Neck base line and related landmarks

2.5.4.1 Back neck point (Cervicale)

The back neck point is the tip of the spinous process of the seventh cervical vertebra (Fig. 2.12A). The tip of the spinous process of the seventh cervical vertebra is visible and easily palpated when the subject bends the head forward (Fig. 2.12B). The back neck point must be marked with the subject holding his/her head in the Frankfurt plane (Fig. 2.12C). The location of the marked back neck point while the head of the subject is oriented in the Frankfurt plane is considerably different from the position of the tip of the spinous process of the seventh cervical vertebra while the subject bends the head forward (Fig. 2.12D). In a few subjects the spinous processes of two or three vertebrae are equally prominent. In such cases, select the one that makes the most natural neck base line.

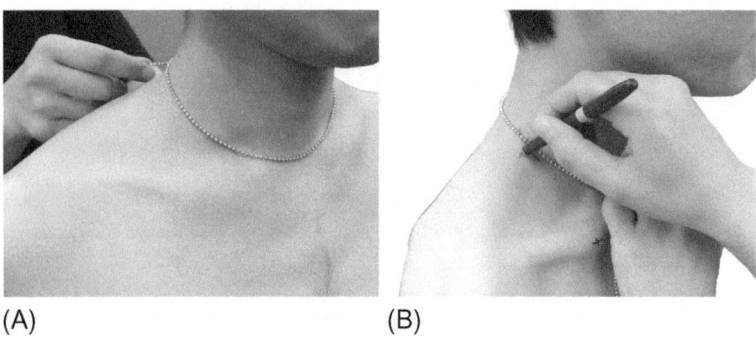

(A)   (B)

Fig. 2.13 Neck base line (A) and side neck point (B).

2.5.4.2   Front neck point

The front neck point is the crossing point of the median line and a line tangent to the highest points of the medial (or sternal) extremities of the right and left claviculae.

2.5.4.3   Side neck point and neck base line

The side neck point is the crossing point of the neck base line and the anterior border of the trapezius muscle. The neck base line is defined using a neck chain (Fig. 2.13A). The neck chain is placed around the neck so that it passes the back neck point and the front neck point. The anterior border of the trapezius muscle can be easier to palpate by asking the subject to place his/her hand on his/her opposite shoulder. Find the anterior border of the trapezius muscle, and draw a line along the anterior border with the subject hanging both arms naturally downward (Fig. 2.13B). Draw a line along the neck chain to make a cross at the side neck point.

2.5.5   Armscye line and related landmarks

2.5.5.1   Acromion, shoulder point, and shoulder line

ISO TC 133/WG 1 decided to use acromion as the shoulder point in ISO 8559-1:2017. The acromion is the most lateral point of the lateral edge of the acromial process of the scapula (Fig. 2.14A, point a). The measurer stands at the back of the subject and palpates the lateral borders of the right and left acromial processes with the balls of the index fingers (Fig. 2.14B). By moving the ball of the index finger in the mediolateral direction, locate the edge between the superior and lateral aspects of the acromial process. Draw a line along the edge. Draw a short line perpendicular to this line at the most lateral point of the lateral edge (Fig. 2.14C, point a). (See also Section 2.4.1.)

The shoulder point is used to define the shoulder line, a line connecting the shoulder point and side neck point. Usually, acromion is located posterior to the ridgeline of the shoulder, and this may not be desirable for defining the shoulder line. In this case

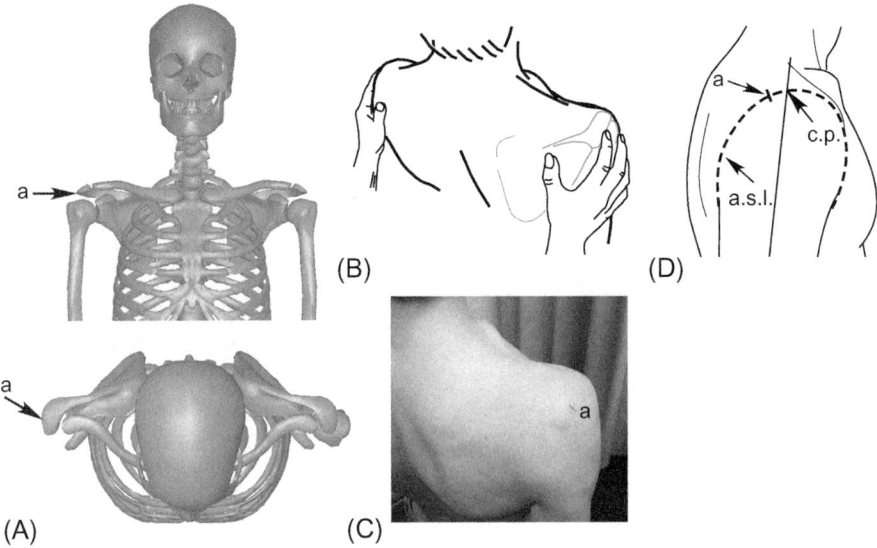

Fig. 2.14 Armscye line and related landmarks (a, acromion; a.s.l., armscye line; and c.p., crossing point of the armscye line and a line that bisects the anteroposterior diameter of the upper part of the upper arm). (A) Shoulder point (acromion) on the bone, (B) palpate the right and left acromial processes from the back of the subject, (C) shoulder point (acromion) on the skin, (D) armscye line.
Panel (B) is from *Mochimaru, M., Kouchi, M., 2006. Biomechanism Library. Measurement of Man: Size, Shape and Motion. Tokyo Denki University Press, Tokyo (in Japanese).*

the crossing point of the armscye line and a line that bisects the anteroposterior diameter of the upper part of the upper arm can be used (c.p. in Fig. 2.14D).

2.5.5.2 Armscye line

The armscye line is defined using a string or narrow tape or a neck chain. A string is placed under the arm of the subject abducting his/her arm approximately 30°. The subject hangs down his/her arms naturally. Both ends of the string are brought up and crossed over the acromion (Fig. 2.14D).

2.5.5.3 Armpit back fold point

The armpit back fold point is the highest point of the axilla (armpit) in the posterior aspect of the trunk. The armpit back fold point is defined using a ruler and a tape measure. The subject abducts the arm approximately 30 degrees. Place the edge of the ruler firmly into the axilla in a horizontal position and ask the subject to carefully lower the arm to the side. Make sure that the ruler is level. Draw a short horizontal line on the trunk at the top of the ruler on the posterior side (Fig. 2.15A). Remove the ruler, and place a tape measure along the armscye line to extend the armscye line below while the arm of the subject is hanging down naturally (Fig. 2.15B). The armpit

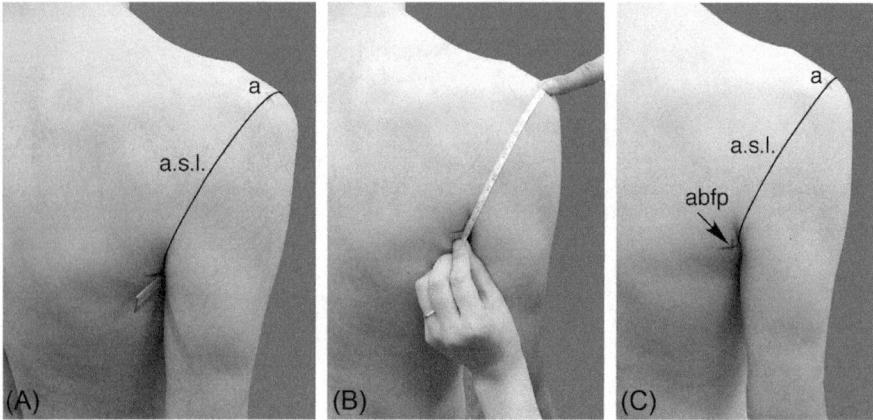

Fig. 2.15 Armpit back fold point (a, acromion; a.s.l., armscye line; abfp, armpit back fold point). (A) A ruler placed under the arm and the mark at the upper edge of the ruler, (B) place a tape measure to define the armscye line, and (C) armpit back fold point and armscye line.

back fold point is the crossing point of the tape measure and the horizontal line indicating the top edge of the ruler (Fig. 2.15C).

2.5.6   Waist girth and different definitions

2.5.6.1   Waist level in ISO 8559-1 and ISO 7250-1

The waist circumference in ISO 7250-1 and waist girth in ISO 8559-1 is a horizontal girth measured midway between the lowest rib point (inferior point of the bottom of the rib cage (10th rib) projected horizontally, 45 degrees from the midsagittal plane, to the surface of the skin) and the highest point of the hip bone (highest point at the side of the upper border of the iliac crest) (Fig. 2.16A). The lower part of the rib cage is visible when the subject inhales deeply (Fig. 2.16B). Palpate to find the lowest point of the rib cage. The mark should be made while the subject relaxes his/her muscles and breathes normally. Palpate to find the upper edge of the iliac crest (Fig. 2.16C). Locate the midpoint of the lowest rib point and the highest point of the hip bone (Fig. 2.16D) and mark (Fig. 2.16E).

2.5.6.2   Natural waist girth

The natural waistline may not be horizontal. The natural waistline is defined using a waist belt. Ask the subject to wrap a waist belt around his/her waist where it settles naturally (Fig. 2.17A). Draw a short horizontal line at the center front, center back, and right and left sides at the level of the midline of the waist belt (Fig. 2.17B). Remove the waist belt. The natural waist girth is measured by passing the tape measure along the four marks on the natural waistline. The four marks may not be on the same horizontal plane.

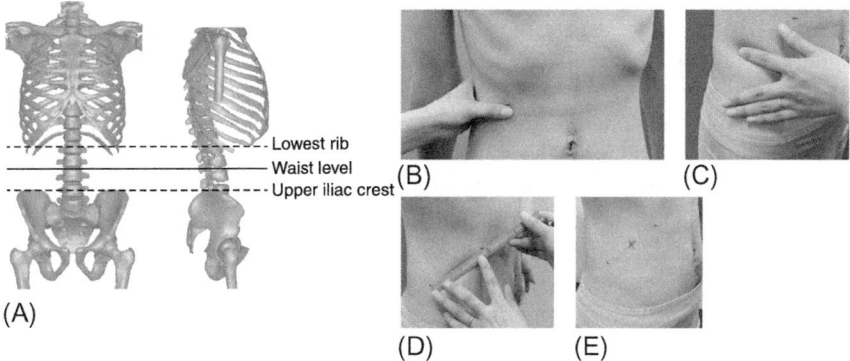

Fig. 2.16 Waist level in ISO 7250-1 and ISO 8559-1. (A) Waist level in relation to the bones, (B) rib cage is visible when the subject inhales deeply, (C) palpate to find the upper edge of the iliac crest (below the index finger in this example), (D) locate the midpoint of lowest rib point and the highest point of the hip bone, and (E) the *cross mark* indicates the waist level.

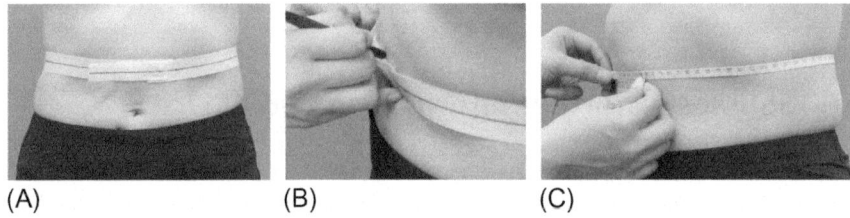

Fig. 2.17 Natural waist girth. (A) Waist belt on the natural waist, (B) mark at center front, center back, right and left sides at the level of the midline of the waist belt, (C) measure the natural waist girth.

2.6 Future trends

The main focus of this chapter is to introduce methods to obtain high-quality anthropometric data. When measuring the human body for designing garments that fit the shape of the body and conform to changes in shape due to body movements, anatomical landmarks are the basis for defining interindividual correspondence or homology. In this sense, traditional methods are still the basics of anthropometry, and at present, no software can replace a skilled anthropometrist. Unfortunately, there are not many educational institutes that provide training courses for anthropometry. The best and only way to learn anthropometric methods is through one-on-one training. It is especially important to teach the same methodologies at all institutes through national and international networks to harmonize the methods. Furthermore a protocol for evaluating the skills of measurers needs to be established. For example, a method developed by Kouchi and Mochimaru (2010) can be used to evaluate the skill of landmarking.

There are two trends in 3-D anthropometry. One is high-quality scanning for a national sizing survey using a high-end scanning system. The other is in-shop or in-house scanning for retail use using an inexpensive scanning system or a smartphone. Only a few high-end scanning systems are now available, because two manufacturers of high-quality whole-body scanners discontinued scanner production due to the small market size. Users (institutes) can select a suitable scanner for their sizing surveys based on a standard quality assessment report. Moreover, institutes should conduct quality control for their sizing surveys.

On the other hand, inexpensive whole-body scanners are available for use in shops. Recently released, very inexpensive sensors may have accelerated the general trend for a lower price. These scanners may have sufficient accuracy to make made-to-measure garments or select the best fitting size (Daanen and Haar, 2013). Distributed scanners can accumulate a large amount of scan data in a short time. In the future large-scale but not very accurate scan data and small-scale but very accurate data may be utilized for developing software utilized for customer services.

For the quality control of large-scale data, it may be necessary to establish a protocol for evaluating the comparability between the automatically calculated landmark locations and landmark locations decided by a skilled anthropometrist.

2.7   Sources of further information and advice

Definitions of landmarks and measurement items are described in textbooks of anthropometry. Several manuals of anthropometry are available through the Internet. Manuals for a specific survey provide practical information on procedures to decide landmark locations and to take measurements, though measurement items specific to garment design are not included in most of these textbooks and manuals.

Martin's textbook (Martin and Knußmann, 1988) is the most frequently used reference. Landmarks and measurement items for physical anthropology and ergonomics are briefly described. Sometimes, it is unclear how to handle the instrument to take a measurement as described. A book by Cameron (1984) describes measurement items used for the study of child growth using photos. A textbook by Norton and Olds (1996) describes measurement items used for sports and health sciences.

A reference manual by Lohman et al. (1988) describes a measurement technique and information on literature and the reliability for several measurement items. This book also includes chapters on measurement errors and which side to measure.

A measurer handbook used in the anthropometric survey by the US Army (Clauser et al., 1988) describes over 100 measurements including some for garment design. Each landmark and measurement item is described in detail in a practical way using a photo and a figure.

A color atlas by van Sint Jan (2007) describes bones and how to palpate bones to find landmarks. Though most of these landmarks are not used in anthropometry for garment design, it provides useful anatomical information. Errors in landmarking in traditional methods are quantified in Kouchi and Mochimaru (2010).

Though published in 1941 the report of an anthropometric research project by the US Department of Agriculture provides a good manual of anthropometry for garment design (O'Brien and Shelton, 1941).

For the quality control of 1-D measurements, ISO 15535 establishes a protocol for eliminating irregular values as described in Section 2.4.2. For the quality control of scan-derived 1-D measurements, ISO 20685-1 establishes a protocol for evaluating comparability with traditional measurements as described in Section 2.3.4. ISO 20685-2 establishes a protocol for evaluating accuracy of 3-D scanners as described in Section 2.3.5.

References

Bougourd, J.P., Dekker, L., Ross, P.G., Ward, J.P., 2000. A comparison of women's sizing by 3D electronic scanning and traditional anthropometry. J. Text. Inst. 91 (2), 163–173. Part 2.

Cameron, N., 1984. The Measurement of Human Growth. Croom Helm, London.

Clauser, C., Tebbetts, I., Bradtmiller, B., McConville, J., Gordon, C.C., 1988. Measurer's Handbook: US Army anthropometric survey 1987–1988. Technical Report Natick/TR-88/043, United States Army Natick Research, Development and Engineering Center, Natick, MA. 01760-5000.

Daanen, H.A.M., Ter Haar, F.B., 2013. 3D whole body scanners revisited. Displays 34, 270–275.

Han, H., Nam, Y., Choi, K., 2010. Comparative analysis of 3D body scan measurements and manual measurements of size Korea adult females. Int. J. Ind. Ergon. 40, 530–540.

ISO 15535:2012. General requirements for establishing anthropometric databases.

ISO 20685-1:2018. 3-D scanning methodologies for internationally compatible anthropometric databases—Part 1: Evaluation protocol for body dimensions extracted from 3-D body scans.

ISO 20685-2:2015. 3-D scanning methodologies for internationally compatible anthropometric databases—Part 2: Evaluation protocol of surface shape and repeatability of landmark positions.

ISO 7250-1:2017. Basic human body measurements for technological design—Part 1: Body measurement definitions and landmarks.

ISO 8559-1:2017. Size designation of clothes—Part 1: Anthropometric definitions of body measurement.

ISO/TR 7250-2:2010. Basic human body measurements for technological design—Part 2: Statistical summaries of body measurements from national populations.

JIS L 0111:2006. Glossary of terms used in body measurements for cloths.

Kouchi, M., Mochimaru, M., 2005. Causes of the measurement errors in body dimensions derived from 3D body scanners: differences in measurement posture. Anthropol. Sci. (Jpn Ser.) 113, 63–75. in Japanese with English abstract.

Kouchi, M., Mochimaru, M., 2010. Errors in landmarking and the evaluation of the accuracy of traditional and 3D anthropometry. Appl. Ergon. 42 (3), 518–527.

Kouchi, M., Mochimaru, M., Bradtmiller, B., Daanen, H., Li, P., Nacher, B., Nam, Y., 2012. A protocol for evaluating the accuracy of 3D body scanners. Work: A J. Prev. Assess. Rehabil. 41 (Suppl 1), 4010–4017.

Lohman, T.G., Roche, A.F., Martorell, R., 1988. Anthropometric Standardization Reference Manual. Human Kinetics Books, Champaign, IL.

Lu, J.M., Wang, M.J., Mollard, R., 2010. The effect of arm posture on the scan- derived measurements. Appl. Ergon. 41, 236–241.

Martin, R., Knußmann, R., 1988. Anthropologie. Band I, Wesen und Methoden der Anthropologie, 1. Teil Wissenshafstheorie, Geschichte, mophologische Methoden. Gustav Fischer, Stuttgart (in German).

National Institute of Bioscience and Human Technology ed, 1994. Reference Manual of Anthropometry in Ergonomic Designing. Japan Publication Service in Japanese.

Norton, K., Olds, T. (Eds.), 1996. Anthropometrica. A Textbook of Body Measurement for Sports and Health Courses. UNSW Press, Sydney.

O'Brien, R., Shelton, W.C., 1941. Women's Measurements for Garment and Pattern Construction. United States Department of Agriculture, Washington, DC.

Paquette, S., Brantley, J.D., Corner, B.D., Li, P., Oliver, T., 2000. Automated extraction of anthropometric data from 3D images. In: Proceedings of the IEA 2000/HFES 2000 Congress. vol. 6, pp. 727–730.

Van Sint Jan, S., 2007. Color Atlas of Skeletal Landmark Definitions. Guidelines for Reproducible Manual and Virtual Palpations. Elsevier, Philadelphia, PA.

Further reading

Mochimaru, M., Kouchi, M., 2006. Biomechanism Library. Measurement of Man: Size, Shape and Motion. Tokyo Denki University Press, Tokyo (in Japanese).

National size and shape surveys for apparel design

Jennifer Bougourd[a], Philip Treleaven[b]
[a]Consultant, London, United Kingdom, [b]University College London (UCL), London, United Kingdom

3.1 Introduction

The clothing and textile industry is a global economic success, worth in excess of $3 trillion (Global Fashion Statistics, 2016), but its supply chain has a significant environmental and social impact, with the consequence that it is one of the world's most polluting industries. Millions of tons of clothing are being committed annually to landfill, some of which is the result of poorly fitting garments that could, in some cases, be traced to a lack of accurate anthropometric data. The success of the clothing industry comes at huge environmental cost, and part of this problem has arisen because many countries have seen major demographic changes—life expectancy beyond 80 years, a broadening ethnic mix, and increasingly sedentary life styles—all of which have affected body shapes and sizes. These changes have led to growing complaints from consumers about the fit of clothes, while retailers perceived them as presenting commercial opportunities.

The key to meeting some of these challenges has been the advent of reliable three-dimensional (3-D) whole-body scanning systems, with accurate measurement extraction capability. Such systems have enabled many size and shape surveys to be undertaken over the past two decades.

This chapter outlines the global context of the apparel industry and how size and shape data can help to provide a more sustainable supply chain; introduces the concept of inclusive design; and looks at ways in which surveys can be designed and processed.

3.2 A global context

Concern with the environment is not new as can be seen in the publications by Carson (1962) and Meadows et al. (1972). The first UN international conference on the environment was held in the same year as the Meadows publication, and it was not until this event that the concept of sustainable development was introduced to the international community as a new paradigm for economic growth, social equality, and environmental sustainability. Later, in 1987, that broad definition of environmental sustainability was conceived as "development that meets the needs of the present without compromising the ability of future generations to meet their own needs,"

Anthropometry, Apparel Sizing and Design. https://doi.org/10.1016/B978-0-08-102604-5.00003-2

the central idea being one of intergenerational equity. The report of Our Common Future (UN, 1987) led to earth summits in 1992 and 2002 and, in 2012, to Rio +20. Following the 2012 conference, it was suggested that sustainable development is not a destination but a dynamic process of adaptation, learning, and action. It was described as being about recognizing, understanding, and acting upon interconnections between the economy, society, and the natural environment (UN, 2012a).

However, despite earlier concerns that the world was not yet on this path, a framework to meet the impact of some of the drivers of current change (production and consumption patterns, innovation, demographics, political dynamics, and changes in the global economy) was designed. This framework has since become the 17 Sustainable Development Goals adopted by the General Assembly in 2015 and published as "Transforming our world: The 2030 Agenda for Sustainable Development" (UN, 2015).

During the period, these goals were being developed; there were different responses from member states. Some regions and governments drew up their own early strategies, with significant international and societal dimensions. A UK publication of 2005 incorporated five principles with an explicit focus on the environment:

- helping people make better choices
- one planet economy, sustainable consumption, and production
- confronting the greatest threat—climate change and energy
- a future without regrets: protecting our natural resources and enhancing the environment
- from local to global: creating sustainable communities and a fairer world

(HM Government, 2005)

The second of these principles, "One planet economy, sustainable consumption, and production," became the umbrella under which a sustainable clothing UK "roadmap" was launched in 2007 (Defra, 2011). It aimed to improve the sustainability of clothing across the life cycle, from crops that are grown to make fabric to the design and manufacture of garment, to retail, to use, and to end of life (Ibid).

The clothing industry is, as indicated earlier, a global economic success, and while the roadmap is a UK initiative, it is linked to Asia, the EU, and the United States, as most clothes consumed in the United Kingdom have a global supply chain (ibid). Many of those supply chains have, however, a significant environmental and social impact (exacerbated by continuing high consumption levels). This position was made clear as early as the Rio+20 conference (where the fashion industry was described as being not only one of the world's most polluting industries but also one that exploited labor across the globe) (UN, 2012b). Aspects of which continue to be challenged (e.g., Greenpeace, 2017) and investigated by the recently launched UK parliamentary inquiry into the impact of fast fashion on the environment (UK Environmental Audit Committee, 2018).

Notwithstanding these recent affirmations of the adverse effects of the apparel industry, considerable progress continues to be made to reverse these global impacts. Life cycle assessment of product, process, and services is being widely applied for sustainability accounting (e.g., ISO 14000:2009 series). Clothing and textile organizations are actively engaged in promoting and supporting sustainability programs and providing certification of achievement (e.g., Sustainable Apparel Coalition and the Higg Index, 2018), while individual companies who are at the forefront of life cycle changes in their supply chains include Nike (2010), Patagonia (1980), and Marks and Spencer (M&S) (2018). A further, more ambitious solution suggested is the first Planetary Boundaries Assessment report of collaborative research between Houdini and Albaeco. They looked to improve materials, processes, and operations as a whole, not only to remove negative impacts but also to exercise a long-term positive influence on the planet (Sustainable Brands, 2018).

An ideal outcome of these activities is for companies to have a closed-loop strategy, where products are returned to the company for refurbishment, repair, resale, and recycling or to nature at the end of product life. That is a circular rather than a linear economy (take, make, dispose) adopted by, for example, Nike, Patagonia, and Paramo, and promoted by the Ellen MacArthur Foundation (2017a). Although not without its' critics (Greenpeace International, 2017) further support for this circular strategy is being acknowledged through recent EU legislation (European Legislation 2018a), the EU-China "global" agreement (European Commission, 2018b), and "Circular Economy Multinational Awards" given in 2017 to Patagonia at the Economic Forum annual meeting in Davos (Bryers, 2017). However, while this strategy is easier for sportswear or outerwear companies, it is considerably more complex for fast fashion. It has been reported that the fashion industry is still wasteful and polluting:

- Every second the equivalent of one garbage truck of textiles is land filled or burned (e.g., Burberry, 2018).
- An estimated USD 500 billion value is lost every year due to clothing barely worn and rarely recycled.
- Clothes release half a million tonnes of microfibers into the ocean every year equivalent to more than 50 million plastic bottles.

(Ellen MacArthur Foundation, 2017b)

It is because of a growing awareness of these "unsustainable challenges" that many organizations, both national (e.g., M&S and Tesco) and international (e.g., Nike and Adidas), that joined the original UK roadmap initiative are continuing to resolve their ongoing Sustainable Clothing Action Plan (SCAP) that is to target areas in the clothing product life cycle.

- Improving environmental performance across the supply chain, including the following:
  - sustainable design
  - fibers and fabrics
  - reuse, recycling, and end-of-life management
  - clothes cleaning
- Consumption trends and behavior;
- Awareness, media, education, and network;
- Creating market drivers for sustainable clothing;
- Instruments for improving traceability along the supply chain (environment, ethics, and chain).

(Defra, 2010)

Such targets have helped companies to engage with these actions (WRAP, 2017) and to fuel a cultural shift in the clothing industry. It is suggested that "sustainability will be the centre of innovation in the fashion industry in 2018" (BOF/McKInsey, 2018) with key drivers being the adoption of principles of a circular economy, design for longevity, use of sustainably materials, and more importantly commitment to change (Op de Beeck, 2018).

However, despite these extensive, national and international activities and calls for personalized clothing, little information has been found that directly addresses the problem of poorly sized and shaped, mass-produced clothing. According to a study conducted by Body Lab, 64% of purchases returned were due to sizing issues in 2016 (Cilley, 2016). This high percentage is due, in part, to the rise of online purchases and business practices such as try-before-you-buy (TBYB) and free-returns strategy described as a "returns tsunami" (Brightpearl, 2018). In addition, large volumes of markdowns, unsold clothing, and the subsequent volume of landfill are all unsustainable practices, and while not all these issues can be directly attributed to garment size and shape, it is the capture and application of contemporary and accurate anthropometric data that has the potential to help reverse these trends.

3.3   Importance and significance of national size and shape surveys

The value of size and shape surveys for clothing is considerable. Nevertheless, if we recognize the need to confront urgent sustainability issues (the growth of and changes within populations and variable life styles), then we shall need to undertake anthropometric studies at regular intervals, to help accurately reflect such developments.

Countries comprise people of differing cultures, ethnicities, life styles, and ages, which in turn influence the size and shape of people within populations. Data collected during anthropometric surveys (scientific study of the physical dimensions shapes and sizes) can be used for a variety of applications requiring different ranges of static and dynamic dimensions. For this reason the type of anthropometric data collected during a sizing survey needs to be appropriate for the people, the products they use, and their purposes.

Interest in collecting body data from large groups of people for clothing applications has been well practiced for the military, but little was achieved for civilian groups until the mid-20th century and then primarily for women. These early studies, together with an increase in mass-produced clothing, continue to prompt many countries to undertake national and clothing-specific anthropometric studies, and although manual studies are still being executed (see Table 3.1), the advent of three-dimensional body scanning systems has accelerated those interests, particularly during the last 20 years (see Table 3.2).

The importance of these 3-D technologies lies in their capacity not only to extract large amounts of accurate data very quickly and without subject contact but also to do so in a way that enhances and extends traditional, one-dimensional measurement. Anthropometric data collection can now capture information for one, two, three, and four dimensions (e.g., 3dMD, 2016). Three-dimensional and 4-D scanning and data extraction systems offer opportunities to increase our understanding of the static

Table 3.1 Manual sizing: a compilation of recently reported civilian surveys

Location	Date	Comments
Australia	2001	Study of the elderly
Korea	2003–04	Major survey
Turkey	2007	Medium-scale masters project in 18 cities across Turkey
India	2008–10	Small study of younger people
Croatia	2009	Major survey
Malaysia	2009	Small study
Nigeria	2009	Small study of the hand, foot, and ear among young people
Korea	2015	Major survey

and dynamic shapes of a population rather than an aggregate of its one-dimensional sizes.

Funding for anthropometric studies can vary. They can be purely commercial endeavors (e.g., Size North America), solely government sponsored (e.g., Size Korea), or part government with contributions from industry (e.g., SizeUK). The following discussion sets out advantages of undertaking studies, which can be accrued for some or all of the funding streams.

Data collected through surveys using 3-D and 4-D technologies (whether for whole bodies or body parts—the head, hand, or foot) can provide a wealth of benefits for the government; academia; industry; and, ultimately, consumer and help make a collective contribution to meet established global, economic, and social sustainability demands.

3.3.1 Government

In addition to giving governments an opportunity to foster innovation and to support the use of new technology, national body shape, and size surveys can generate tremendous interest in both its scientific applications and in the provision of highly detailed data on a population, supplying direct commercial benefit to a nation.

3.3.2 Academia

There are educational opportunities for university students and staff. Research teams can gain experience of the initial organization and implementation of surveys, enabling them to explore technologies and, when data sets are available, to continue to develop new applications. For example,

- to advance knowledge of shape analysis and classifications (Tahan et al., 2003; Simmons and Istook, 2003; Ball et al., 2012; Morlock et al., 2016),
- to aid understanding of the size and shape of older populations (NDA, 2010),
- to propose methods for capturing and/or animating virtual body models (Ruto, 2009; Lane, 2017).

Table 3.2 Digital anthropometric sizing surveys

Country, survey	Date	Ages	No: male and/or female	Scanner	Postures/measurements
Japan: HQL	1992–94	7–90	34,000, M and F	Voxelan, Hamano	Not known/178
Netherlands: CAESAR	1999–2000	18–65	1255, M and F	Vitronic	(As US)
Germany: intimate apparel	1999–2000	14–80	1500, F	Vitronic	3/84
United States: CAESAR	1998–2000	18–65	2375, M and F	WB4, Cyberware	3/57+40 manual)
United Kingdom: SizeUK	1999–2002	16–90+	11,000, M and F	TC2	2/130+10 manual
Italy: CAESAR	2000–01	18–65	801, M and F	WB4, Cyberware	(As US)
Germany: E-Tailor	2001	16–70	500, F	Vitronic	2 / 10
Germany: elderly women	2002	50–80	1300, F	Vitronic	2/84+1 manual
United States: SizeUSA	2002–03	18–65	10,500, M and F	TC2	2/130+10 manual
Korea	2003–04	8–75	5000, M and F	Hamamatsu	Not known/359
Japan: Size-JPN	2004–07	18–89+	6700, M and F	Not known	Not known/217
France	2005–06	5–70	11,562, M and F	Vitronic	2/55+10 manual
China: (Heads) SizeChina	2006	18–71+	2500, M and F	Cyberware (Head scanner)	1/3+5 manual
Thailand	2006–08	16–60+	13,442, M and F	TC2 NX16	1/140
France: Seniors	2007	70–100	400, M and F	Vitronic	2/"Varied"
Germany: SizeGermany	2007–08	6–87	12,132, M and F	Vitronic	4/43
Spain	2007–08	12–70	9159, F	Vitronic	3/95
France (children)	2009	0–15 days; 72–78 mths	2177, M and F	Vitronic	38 (over 3 yrs. electronic/under 3 yrs. manual)

Germany (Plus Size)	2009	18–87	2265F	Vitronic	1/143 + 1 manual
Sweden	2005–10	18–65	367, M and F	Vitronic	4/55 + 10 manual
Mexico: SizeMexico	2010	18–65+	17,364, M and F	TC2 NX16	1/200+
United Kingdom: NDA pilot[a]	2010	60–75	30, M and F	TC2 NX16	2/130 + 10 manual
Korea	2010–13	7–69	14,012	Hamamatsu	177 + 139 manual
Italy: SizeItaly	2012–13	18–75	5873M and F	Vitronic	4/44 + 5 manual
Spain	2013–14	13–65	1400M	Vitronic	2/50
Germany (Plus Size)	2013–15	18–81	814M	Vitronic	1/123 + 1 manual
Spain (Children)	2014–15	0–12	1000C	Vitronic	1/50 (under 3 yrs. manual)
Size North America (in progress)	2017–18	6 to upper age not known	17.000 M, W, C	Vitronic	4 /100
India (in progress)	2018–21	15–65	25.000	Size Stream	Not known / 120
Malaysia (in progress)	2018–21	18–65	11.000	TC2	Not known / 130

[a]The New Dynamics of Ageing (NDA, 2010): Newsletter for Summer 2010 see also McCann and Bryson (2015).

Furthermore, such pioneering research offers confirmation of "the continuation of world-class, dynamic and responsive research …" and contributes to national assessments and ratings to ensure accountability of public funding, establish reputational yardsticks, and support allocation of future funding (Research Excellence Framework [REF], (2018) the United Kingdom—e.g., UCL's contribution to SizeUK).

3.3.3   Apparel industry: Size and shape

The main driving force for the majority of national surveys has been better body data for the apparel industry. Initial interest in surveys on the part of industry was limited to the use of one-dimensional data to update, augment, or create new body sizing systems. However, provision of new shape and size data is enabling a much wider range of products for clothing design and development, and as tools for interrogating survey shape data continue to develop (Ruto, 2009), these shape data help to structure the basis for:

- new body shape and sizing systems
- understanding the impact of body shape through the ageing process
- identifying body shapes within a specified market
- confirming the body shape and the body size of physical fit models within selected shape and size ranges
- accessing a 3-D avatar library relative to market requirements
- three-dimensional digital design
- automating block and style pattern generation
- morphological classifications and grading
- creation of physical fit mannequins relative to identified body shapes
- digital in-store and online fitting and purchase

The availability of these products and of market-specific applications created from 3-D anthropometric data (e.g., Sizemic Ltd) has the potential to meet some of the environmental, social, and economic challenges discussed earlier and, in doing so, help to streamline the fit of products across the supply chain, by:

- enabling apparel designers and technologists to understand demand in specific shape and size categories and encourage expansion of shape and size ranges to meet the needs of a whole population;
- improving garment fit and consistency of fit tests across shapes, sizes, and products, bringing a positive impact to sustainable development and on customer loyalty and sales;
- allowing both physical and digital fitting to take place on all the same body shapes and sizes within a target market that can increase the efficiency of the product development process and establish a fit standard for all products across supply chains;
- reducing the number of iterations of samples with shorter lead times, with a positive impact on sales and longevity of garment wear;
- achieving efficiencies in the product development cycle with consequent environmental and cost benefits (fewer samples, fewer fit sessions, lower fit model costs, and a reduction in staff time for processing and managing sample approval);

- increasing suppliers' sample approval rates with lower sample making and delivery costs and considerable time saving;
- enhancing the overall level of quality control (due to physical mannequin provision), which provides an excellent QA communication tool;
- improving margins due to reduced returns and better sell-through rates/lower markdowns;
- facilitating virtual garment visualization, digital design and range development, and approval.

(Sizemic, 2012)

Further sustainable practice can be achieved by using representative virtual shapes from a survey. These scan shapes can be uploaded into virtual design systems (such as Optitex and Assyst), where both contour and some free form clothing designs can be created and patterns unwrapped and flattened (e.g., Morlock et al., 2016; Kung, 2012; Kirchdoerfer et al., 2011; Krzywinski et al., 2005) from which patterns can be morphologically graded or resized (Kung, 2012; Sayem et al., 2014a,b). See images in Fig. 3.1.

In addition, two-dimensional cross-sectional data are being used to help evaluate and select digital systems and identify air gaps between body surface and garment to

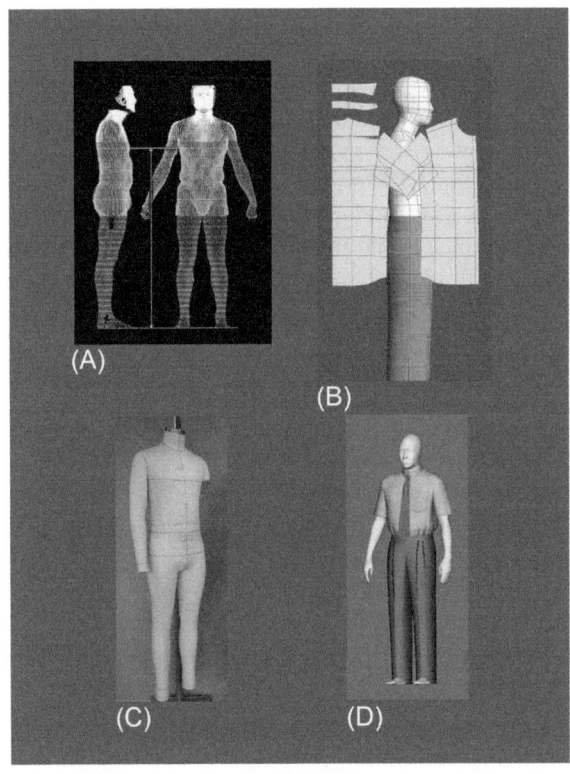

Fig. 3.1 Streamlining the fit of products across the supply chain using 3D data: (A) scan and automatic measurement extraction; (B) automatic pattern and morphological grading; and (C and D) physical and virtual fit mannequins. *Sources:* TPC Hong Kong Ltd.; Sizemic Ltd.

help establish fit, while apparel design teams and retail buyers are using 3-D virtual body shapes and two-dimensional pattern wrapping techniques to remotely assess fit and approve garment designs. And looking at emerging applications, 4-D temporal techniques are being used to capture people in clothing for realistic virtual try-on (3dMD).

3.3.4 Consumers

The aim of all regularly conducted clothing-specific size and shape surveys is to provide anthropometric data that can be used by designers and technologists to create well-fitting, ready-to-wear (RTW) garments for all consumers in a population. Evidence discussed (see Section 3.3.3) suggests that if all design and product development tools (e.g., 3-D scans, sizing systems, patterns, avatars, and mannequins) across the supply chain are created from the same segmented data, then consumers of identified market segments could expect to find clothes that have an appropriate fit. This is the case for purchasing RTW garments, in-store or online, and, in some cases, mass customization where size and shape data are drawn from national surveys.

Consumers will only be satisfied with garment fit and subsequently maximize a garment's life use if the data produced for a survey accurately reflect the entirety of a population that it is designed to serve, for example, for all ages, sizes, and shapes, and if designers and retailers chose to use those data to embrace the concept of inclusive design—a process that results in inclusive products or environments that can be used by everyone, regardless of age, gender, or disability (CEBE, 2002)—a definition that can be extended to address not only the criteria earlier but also, in the context of sustainable practice, race, income, education, culture, etc. (Ibid). Addressing size and shape in the context of inclusive design and sustainability has the potential to help design out waste and pollution, keep products and materials in use, and regenerate natural systems, principles of which point to the rise of the "conscious consumer."

3.4 Planning a national anthropometric survey for clothing

Running a national anthropometric survey is akin to planning and executing a military operation. Even with the most thorough planning and preparation, some issues will only appear during the survey and will need to be resolved.

There are several reports setting out the process of planning, executing, applying, and reviewing anthropometric surveys. They describe similar stages, but the aims and data to be collected vary. Studies conducted during the previous two decades include

- clothing-specific studies (e.g., SizeUK, 2018 and SizeUSA, 2018),
- clothing with automotive and health studies (e.g., Size Thailand, 2008),
- clothing with technological design (e.g., Size Germany, 2018),
- clothing with technological and automotive design (e.g., Size North America, 2018).

The priority is therefore to set out the focus and objective for the survey, the population, and the clothing products for which the data are to be used at the beginning of the survey.

The following sections—infrastructure (3.4.1), preparation (3.4.2), implementation (3.4.3), and data storage (3.4.4)—will be confined to clothing-specific surveys, though it is recognized that data collected may have secondary application in such areas as health studies (Wells et al., 2008).

3.4.1 Infrastructure

A number of fundamental decisions need to be taken when planning a survey:

- identification of the partners who will collaborate to promote and conduct the survey
- source of funding for the survey: whether the data can and will be sold to third parties to help to recoup survey costs, legal ownership of the results of the survey
- if and when participating clothing textile companies receive data in return for funding;
- whether the whole or any part of the processed data will be made publicly available;
- if and when press and publicity material needs to be prepared

3.4.1.1 Organizing committee

If the organizing committee and partners are to maximize the impact of a national study, it is important to involve the government, leading clothing organizations, research centers, and technology companies. A lead organization is required to manage the survey, the finance, press, and publicity and to negotiate all legal requirements. Normally a committee will comprise representatives of government, clothing and textile organizations, major clothing companies, and universities.

3.4.1.2 Management of a survey

In addition to planning and organizing meetings, the management committee and team handles issues involving the scheduling of the time, the distribution of the work, and the completion of outcomes that are vital to the success of the survey.

3.4.1.3 Survey funding

As indicated earlier in Section 3.3, funding for surveys can vary, but financial support can typically come from government, from clothing companies, or a combination of both. Cash from government or from sponsor companies may support a whole study, for example, India Size (2018) and Size North America (2018), respectively. In a study where support comes from both sources, cash can cover equipment and staff, and contributions from companies may be made in cash or in-kind shopping vouchers for participants, the donation of staff time, equipment, and consumable materials such as underwear. Underwear may be needed, which meets style and color requirements for a scanner, local culture requirements, commercial partners, or those specified in ISO 20685. In-kind items can be given a monetary value and offset against a notional purchase price for the data.

3.4.1.4   Legal considerations

Legal requirements will vary from nation to nation or region to region, but the following issues may need to be considered:

- Collaboration agreements. These would normally be negotiated when a government funds a survey.
- Intellectual property rights (IPR). Legal rights of access to data and results can be protected and controlled through licensing.
- Ethical considerations. Approval for scanning or manual measurement should be sought with reference to relevant data protection legislation (e.g., Regulation (EU) 2016/679, which came into law in May 2018 and specifies "sensitive personal data"); such requirements would usually be processed by the lead organization. Ethics approval documentation can comprise a letter explaining the aims and nature of the project, formal participation, and confidentiality agreements that enable the survey to collect anonymized personal data, scan data, and measurement data from subjects in each of the local data collection venues.
- Protection. The interests of children and vulnerable adults must be protected during the conduct of a sizing survey. For example, in Britain, legislation is in place requiring those responsible for the conduct of a sizing survey to be certified as having been subject to a check against a record of those with relevant criminal convictions (Disclosure and Barring Service (DBS), 2018).
- Confidentiality. All members of the data collection team should be required to complete an agreement to protect the confidentiality of subject data. Survey data need to be analyzed, applied, and stored to maintain the anonymity of the subjects who have been scanned and measured (see reference earlier to current EU law).
- License agreements. It may be that following a sizing survey processed data are to be made available to companies that did not participate in the initial arrangements; the sale of such data should be licensed on the basis that confidentiality can be maintained.

3.4.1.5   Press and publicity

As indicated earlier, a national sizing survey can generate considerable public interest and publicity for participating organizations. It is important that the lead organization manage the interaction with the media by, for example, establishing a press office to prepare press packs and make arrangements for the release of news and statements. Typically, press events can be the means by which subjects can be encouraged to participate in the study, both local and national launch events, and announcement of headline results.

3.4.2   Preparation

The key issues to be dealt with may include the preparation of logistics software, subject registration, a life style questionnaire, an anthropometric measurement set, benchmarking and the purchase of equipment, statistics, a model for a data collection center, a proofing study, and the selection of survey staff and their training.

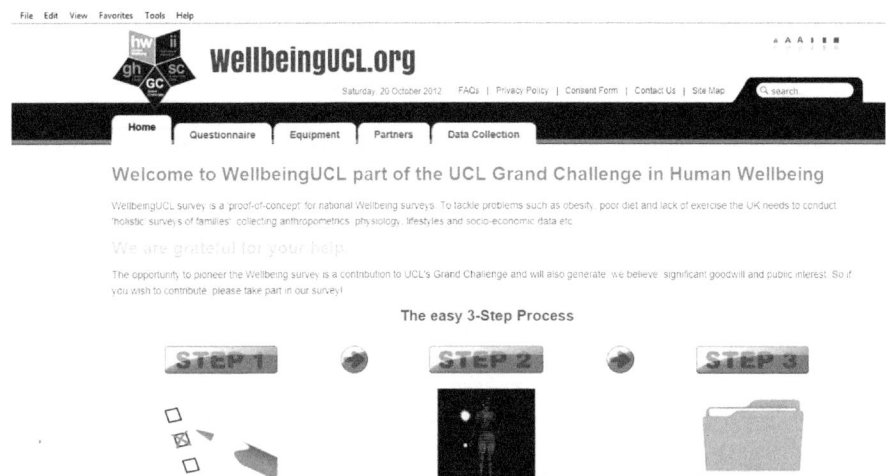

Fig. 3.2 Screenshot showing a recent example of a logistics software package. It shows a "Home/Welcome" page.
Source: University College London.

3.4.2.1 Logistics software

A comprehensive set of software is needed for subject registration, subject selection, the life style questionnaire, and uploading of the body scan and other data collected and tracking statistics (see Fig. 3.2).

3.4.2.2 Subject registration

One of the aims of encouraging online registration is that subject details can be processed and their details checked against the recruitment frame. This check enables the acceptance or rejection of a subject according to the numbers required for each age/gender/region (see example in Table 3.3). If a subject is accepted, a bar code could be issued and appointments for visits to venues agreed well in advance of the data collection program (i.e., details needed for recruitment selection—age; gender; ethnicity; key measurements—height, weight, chest/bust, waist, and hip; location; and contact details). In addition, information for subjects related to the project (ethical papers, confidentiality agreement, a comprehensive description of 3-D body scanning, data collection process, and preparation for scanning) could all be accessed and, where necessary, processed prior to appointment.

3.4.2.3 Life style questionnaire

Life style questionnaires would normally be designed and tested in conjunction with the marketing departments of the commercial partners. They also tend to be gender-specific (and, in the case of children's study, age-specific). A questionnaire would normally include subject clothing preferences; size and fit issues; shopping habits (both in-store and online); and personal details related to body measurements, health and income, etc (see Fig. 3.3).

3.4.2.4 Anthropometric measurement set and postures

The selection of measurements and postures will depend on the objectives of the survey. If the data required are to be intergenerational (i.e., all adults and/or all children) to include a full range of shapes, sizes, and postures and the majority of clothing product types, then it will be necessary to

- identify the body postures to be scanned (e.g., ISO 8559-1:2017);
- assess the measurements listed in ISO 8559-1:2017 and, should the ISO measurements not be sufficient to meet the requirements of inclusive design and the development of the majority of clothing product types, then there may be a need to include additional measurements;
- create a survey-specific measurement set.

(The latter two proposals would need to include relevant body landmarks, body locations, images, and an estimation of allowable errors.)

Such an assessment would be required in order to evaluate available measurement extraction software offered during scanner benchmarking. The majority of scanner providers currently offer measurements listed in ISO 8559 and ISO 7250 (recently updated) for clothing and technological design. However, if there is incompatibility (between the survey-specific set and software offered), then it may be possible for a scanner manufacturer to extend their current measurement provision, for example, the development of TC2 software for SizeUK measurement set. In addition, if

Fig. 3.3 Screenshot of a recent lifestyle questionnaire incorporated into logistics software, discussed above.
Source: University College London.

intergenerational adult studies (18–100 years) were undertaken, a means of addressing life-span morphological changes would be required (e.g., spinal posture—see Ashdown and Na, 2008).

It is important to note that both of these standards (ISO 8559 and ISO 7250) were revised in 2017, and although there are a few basic measurements, common to both standards, there is now a much clearer distinction between the amount and type of landmarks, measurements, and postures required for clothing as opposed to techno-logical design.

Table 3.3 The recruitment of subjects participating in the SizeUK survey. Figures for seven age bands recruited for each of three regional centers (total populations in each region were matched for accurate representation)

Age group	16–25	26–35	36–45	46–55	56–65	66–75	76+	Sum
Region 1	188	188	188	188	188	188	188	1316
Region 2	188	188	188	188	188	188	188	1316
Region 3	188	188	188	188	188	188	188	1316
Totals	564	564	564	564	564	564	564	3948

Allen et al. (2003).

3.4.2.5 Equipment benchmarking, selection, and purchase

All subjects need to register with the survey and would ideally, complete an online life style questionnaire before selection (see discussion earlier). However, in order not to exclude any potential volunteers, an additional set of PCs and paper copies of questionnaires would need to be made available at each data collection center.

To maximize national benefit, it may be necessary to supplement 3-D body scanners with head or foot scanners and include a height gauge and a body composition monitor to automatically record subject height and weight. Depending on the objectives of the study and the 3-D scanner selected, it may also be necessary to include traditional anthropometric equipment, underwear/scan wear, head covers, and materials for the hygienic maintenance of equipment.

It is advisable to benchmark all equipment to be used during the survey. This includes scanners, height, weight, and any other manual measuring equipment (see 8559-1 2017). Types of technology used to capture the 3-D surface shape of the body include lasers; projected light; and, latterly, millimetric radio waves and smart phone capture (Ballester et al., 2017) (although the latter maybe useful for size prediction, its suitability for accurate anthropometric clothing-specific studies is not yet clear). Each has its advantages (resolution, cost, automatic measurement extraction, etc.), but new, extended, or enhanced systems are being offered (e.g., 3dMD System-4D). Guidance is also available via conference publications (e.g., Hometrica Consulting, 2018) and in some ISO standards (e.g., ISO 20685 Parts 1 and 2). It is, however, advisable to conduct a benchmarking exercise—especially if it is planned to

• scan subjects with differing heights, sizes, skin shades, and ages;

- use a range of underwear/scan wear—this will depend on the objectives of the study, the requirements of commercial partners, cultural variations in a population, and the selected scanner;
- scan subjects who may not be able to remain still for longer than 8–10 s;
- use survey-specific measurement sets that do not include dimensions that are listed in ISO standards (e.g., ISO 8559-1).

(See Allen et al., 2003, for an example of benchmarking.)

A key issue raised by the need for benchmarking is an assessment of the accuracy of 3-D scan capture, landmarking, and measurement extraction. ISO 20685 Parts 1 and 2 give guidance for reducing error in 3-D scanning and for establishing accuracy of body dimensions by comparing traditional manual measurements with those extracted from scanners. In addition, although such comparisons are still the subject of debate, it is acknowledged that some scanners have now reached a stage of development where they can automatically extract measurements from subjects with a higher accuracy and consistency than those taken by trained anthropometrists. Notwithstanding these advances and debates, what is perhaps as important is the need to agree definitions and measurement positions. As can be seen in earlier work, where waist definitions of ISO 8559:1989 (now reissued as ISO 8559-1) are compared with those of a preferred waist taken during the CAESAR study (Veitch, 2012) and where acceptable upper and lower limits of the waist are compared to define the true height of the waist (Gill et al., 2014).

3.4.2.6   Statistics

The determination of a statistical sample of earlier clothing-specific surveys (e.g., SizeUK) was based on the pioneering work of US military studies (Gordon et al., 1989), and as can be seen in Table 3.3. I., the SizeUK study took place in three geographical regions (the total population within each region was approximately equal to obtain statistical relevant results), and each region was divided into seven age bands.

However, an ISO standard (ISO 15535 General requirements for establishing anthropometric databases) introduced in 2006 and revised in 2012 sets out options for desired levels of relative accuracy and confidence—for stature, chest circumference, and shoulder breadth—with proposed numbers of subjects for the achievement of each level. How those numbers are distributed will depend on the number of regions selected and their relative populations. Account needs to be taken of the homogeneity of the population (e.g., age distribution, ethnicity, and socioeconomic status); a country, for example, that has a homogeneous population might be treated as a single region. A further difference in the ISO standard is the age bands. For example, in the case of the SizeUK survey, the lower adult age band was determined by the (then) UK school leaving age of 16 years, as those subjects could be included in a socioeconomic category. It is suggested in ISO 15535:2012 that adult population data should be collected from those aged over 20 years of age and that young adults should be considered as single-year bands. No upper age limit is proposed, and as can be seen in recent studies (Tables 3.1 and 3.2), there is a wide range of upper ages, but with an ageing global population and the fastest growing group being those 80 years and above (UN 2017 Revision), future anthropometric studies for clothing will need to address these intergenerational clothing requirements. For example, SizeUK had an upper age group of 75 plus (eldest person was 93 years), but future studies will need to include data for centenarians if they are to meet an inclusive design and sustainable strategy for RTW clothing, that is, to provide a choice of garment shape, size, and fit for all members of a population that meets the required sustainable process.

There are several options for subject recruitment, but if software is designed to constantly monitor registration, subjects can be selected in two stages: first, to meet the recruitment strategy (gender, age, and geodemographics), and, second, to ensure other national statistics (ethnicity and socioeconomic grouping) before being invited to a center to be scanned.

3.4.2.7 Model data collection center

A model venue for the collection of data, setting out the orientation of equipment for easy team measurement operations and subject processing, needs to be arranged and tested before any necessary training. A data collection center, whether a static venue or a mobile unit, is likely to comprise

- a reception desk with a PC;
- additional registration facilities;
- an auto height gauge;
- an auto weight scale (e.g., a body composition monitor);
- a set of traditional anthropometric tools;
- a foot, hand, and/or head scanner;
- a whole-body scanner, with dedicated PC and integral changing space;
- storage facilities (e.g., underwear/scan wear, cleaning equipment, and materials).

3.4.2.8 Proofing survey

Crucial to the efficient operation of a survey and to its overall cost is the number of subjects that can be processed per day in a given data collection center or unit. Issues that could arise to affect the operation of a collection center might include the selected equipment, the data collection team, subject ages, the location of the data collection center, and the recruitment of a steady stream of subjects for measurement. To test the process of the layout and efficiency of the equipment and to provide an estimate of the daily subject throughput, it is important to conduct a proofing survey of at least 40 subjects (ISO 20685). Subject recruitment needs to be in accordance with the objectives of the study and in the proportions indicated in current national census for gender, age, and ethnicity.

If a survey is *not* to be conducted using a mobile unit (Fig. 3.4), then the identification and inspection of all venues in selected cities would ideally be completed before the selection and/or training of a team or teams.

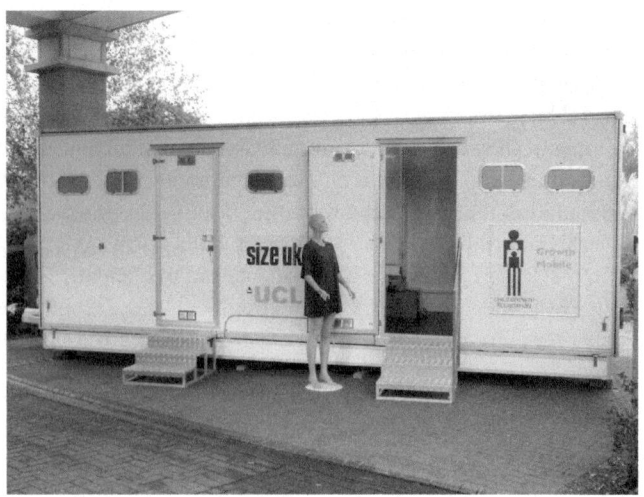

Fig. 3.4 Mobile scanning unit: University College of London and Sizemic Ltd.

3.4.2.9   Selection of data collection team and training

Collection of data may be either one team of experts that travel to each venue or several individual teams recruited and trained for each venue (e.g., university/industry partner venue and team). If the latter course is taken, then the selection of a team and the method of training could be influenced by the selected scanner. For example, in addition to a scanner operative, measurement extraction systems

- may require prior manual landmarking;
- have automatic measurement extraction but with the need for some supplementary manual measurements;
- offer fully automatic landmarking and measurement extraction.

Most scanner systems have training manuals, but if training is required for the collection of manual anthropometric data, courses are available (e.g., ISAK) as well as publications, for example, Marfell-Jones, 2006; Stewart et al., 2011; and ISO standards ISO 8559-1:2017 and ISO 7250-1:2017. If a survey is to be designed for its data to be internationally compatible, guidance can be found in ISO/DIS 20685-1:2018, ISO/20685-2:2015. Whichever scanner system, if training were required, it would be necessary to assemble an appropriate team, prepare a training manual for the selected scanner system, and assemble a set of training materials.

A team of trainers could comprise a computer scientist responsible for logistic software, scanner operatives, and expert anthropometrists. Male and female models that represent the proposed population could be recruited to act as subjects, though it is advisable for all team members to experience the subject measuring sequence.

Proposed training materials could include information related to:

- *data collection center*—health and safety issues for subjects and team;
- *subject registration*—questionnaire and measurement briefing papers, preparation for scanning procedures, and processing sequence;
- *physical training*—team presentations, instructions for logistic software and manual measurement, equipment and scanning system procedures, videos of subject processing, and care and maintenance.

The number of teams and selected scanner would determine the number of days required for training, although the aim would be to ensure that all team members would be multiskilled and hence able to:

- prepare, maintain, and derig a venue;
- welcome, brief, and guide subjects through each stage of a measuring process;
- ensure subjects prepared for measuring process;
- use and maintain all measuring equipment;
- evaluate the capture and accuracy of all manual measurements (i.e., intra- and inter-validation, where necessary);
- assess accuracy of 3-D scan images and the extracted measurements;
- record and store data in accordance with chosen system;
- prepare for the next session and/or derig the venue.

3.4.3 Implementation

Effective recruitment, scheduling, processing, and tracking of subjects are vital to the success of the data collection process, as is the need to ensure the quality of the shape and size data before its automatic uploading to a logistics software system.

3.4.3.1 Recruitment

Having determined some of the key issues in the preparation made for the conduct of a size and shape survey, the success of data collection will depend on having recruited sufficient and appropriate subjects at each venue. Given public interest in clothes sizing, a national survey receives massive amounts of free publicity with a consequent surge in the number of volunteers coming forward to be measured. For example, with SizeUK, 17,000 individuals registered interest through a dedicated website; 11,000 returned registration information and a questionnaire through retail outlets; and, in response to mail shots, 20,000 people telephoned the helpline established for the purpose. However, with subject registration being processed through a central system, daily lists of subjects for each venue/center/unit can be generated well in advance of data collection scheduling.

3.4.3.2 Scheduling

The most straightforward but time-consuming method of collecting data is to have one team to visit all centers, as this only involves recruiting one set of subjects at a time and the same team to set up and take down the set of equipment. If, however, there are to be multiple data collection teams, then a more complex scheduling is required to ensure the transport and installation of equipment and the availability of technical support.

For example, SizeUK used eight centers, three scanners, and three sets of equipment plus one expert support team to complete data collection within a 6-month time and funding frame.

3.4.3.3 Processing subjects

Prior to the arrival of the subjects to be measured, the data collection team (with assistance from the support team) needs to

- set up and check the data collection center,
- calibrate and test all equipment,
- organize ancillary materials (such as scan wear/underwear),
- set up appropriate internal and external signage (e.g., guidance for those arriving, visuals of the scanning process, postures, health, and safety guidance and help in navigating within the center).

On arrival at the center, subjects are welcomed, and the team confirms registration and ensures completion of a data protection document (e.g., confidentiality agreement) before explaining the measuring sequence and the required process needed for each individual item of equipment.

All procedures for the maintenance of hygiene and safety need to be observed throughout preparation, before and after scanning and data collection.

3.4.3.4 Data quality

If the planning and preparation for a survey is carefully considered (e.g., a benchmarking exercise has been undertaken), then in-process validation of scans, landmarking, and measurements, whether manual or automatically extracted, should be easily assimilated into the subject processing sequence. However, the data quality maintenance procedures followed would depend on the selected measurement process and the type of scanners selected for use. For example, validation could include

- intra- and interevaluation of manually placed landmarking and measuring,
- checks to define or modify preset landmarks and measurements (Preiss and Botzenhardt, 2012),
- the visual appraisal of all scans and automatically extracted measurements so that (where, e.g., a subject has moved or has assumed an incorrect posture) a repeat scan can be made.

3.4.3.5 Data tracking

At the end of the measuring process, a team member would need to verify completion using a subject barcode and, if necessary, complete the process with the issue of a formal thank you card, shopping voucher, or agreed industry partner gift (e.g., dressing gown). At the end of each session, the team would

- ensure that data have been successfully transferred to the logistics system,
- clean and close down all equipment in preparation for the next session,
- help the support team at the end of data collection exercise to derig the center and secure equipment and materials for transportation.

3.4.4 Data storage, access and sales

A database can be organized for national, regional, and/or international use, comprise a wealth of data, and be accessed for analysis through a variety of routes. A format and set of contents for organizing a database are proposed in ISO 15535:2012, and although the recommended measurement set would need to be reviewed (see Section 3.4) and legal agreements may prevent raw and/or processed data being available in the public domain, the standard can serve as a useful guide.

3.4.4.1 National database

Storage—Many data are likely to be collected during a clothing-specific study using 3-D technologies. For example, each subject's set of anonymized data could comprise personal data (excluding name and home address), one or more 3-D scans, and as many as 250 extracted and derived measurements. There may also be manual measurements and, if used, information from a body composition monitor and data from a clothing market research life style questionnaire. Each subject's data could occupy 15–20 MB of storage. To protect raw and analyzed data, results need to be held in a secure server, if appropriate, for the study to be accessed using appropriate software tools.

Access—Raw data are difficult to protect and, if commercially sensitive, would not normally be released. SizeUK chose only to release processed data and/or analyzed data. Options for commercial partners to access data could include the use of

- real-time, online software tools
- online national analyzed data
- a bespoke service
- complete population data set

A suite of real-time online software tools, such as those developed by SizeUK, can run on a database server and be used to check data integrity, extract and analyze 3-D body size and shape and associated marketing data, and compile body size measurement charts for selected markets (see Figs. 3.5, 3.6, and 3.7).

Further analysis of the shape of the whole body and other body parts continue to be developed for physical and digital applications (SizeUK; Ball 2011; ISO 18825-2:2016).

In addition to an online suite of software tools, industrial partners may access online analyzed data, request bespoke analysis, and (for companies with the requisite statistical skills) use of the whole data set. There is also a new ISO standard designed to aid individual company data analysis (e.g., ISO 8559-3:2018). However, despite these opportunities, experience has shown that most clothing companies prefer ready-to-use data, such as body size tables, and items developed using 3-D shape analysis—block and graded patterns and physical and digital fit mannequins and avatars (see Fig. 3.1). Such provision allows for consistency of fit throughout the product development and quality assurance process, irrespective of the countries in which clothing is designed and manufactured.

Sale of data—Central to the organization of clothing-specific surveys (such as SizeUK and SizeUSA) is the sale of the data to clothing companies that did not

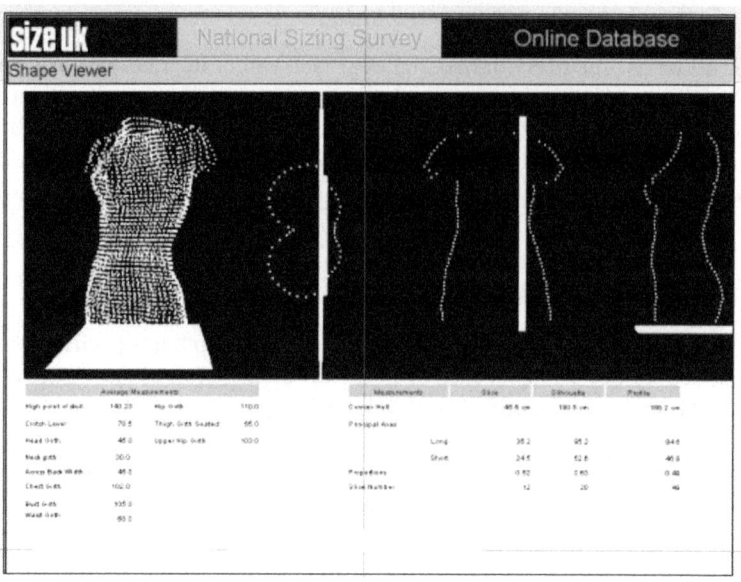

Fig. 3.5 Screenshot captured in 2002 by the SizeUK team, showing web-based software tools for 3D data analysis. The online software allows the user to view (images, *left* to *right*) 3D shape analysis, seated 2D cross section, the front silhouette and side profile; and (in the panel of figures below) a table of average measurements (Ruiz et al., 2002).

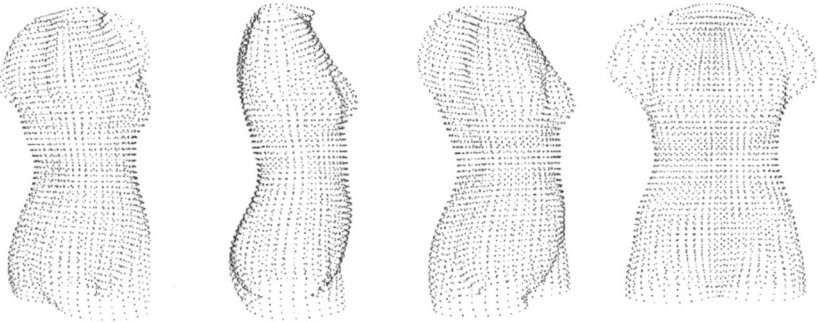

Fig. 3.6 Four views of an average female torso computed from subjects in the preliminary SizeUK survey. The torso is rotated about a vertical axis between views (Tahan et al., 2003).

participate as partners in national studies. This can be done to offset the cost of a survey or to meet cost overruns. It is therefore particularly important to protect data from illicit copying.

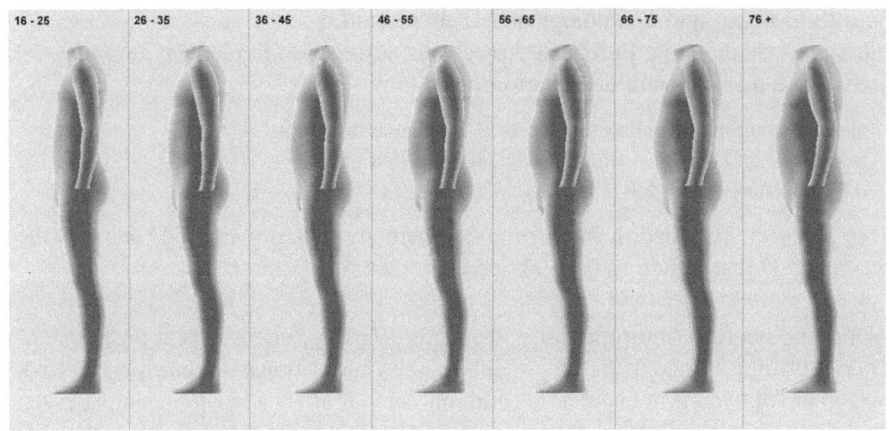

Fig. 3.7 Profiles of the seven age groups of SizeUK male subjects.
Source: SizeUK database.

3.4.4.2 Regional and international databases

Methodologies developed and used for earlier survey data analysis of dimensional size, 3-D body shape, and life style data were limited to propriety systems (e.g., SizeUK). Subsequent publications, referencing data collected from individual national or regional surveys and stored in 1-D, 2-D, or 3-D anthropometric portals, suggest that methodologies for clothing and technological design applications are now included for processing and analyzing anthropometric data, for example, Veitch and Robinette (2006); Trieb et al. (2013).

Portals and standards that have been set up for assembling and/or mining these multinational size and shape survey data include iSize, EUROFIT (2012), WEAR (2012), and ISO 7250-2.

iSize: Clothing and technological dimensional portal

Launched in 2010, this portal is an international body size dimension developed between Human Solutions, the Institut Français du Textile et de l'Habillement (IFTH), and the Hohenstein Institute. The portal stores both manual and digital data collected from more than 100,000 men, women, and children aged from 6 to 75 years, with 44 body dimensions taken from ISO 8559 and ISO 7250. (See above for revisions of these standards.) This portal enables country, age, gender selections, and analysis of body dimensions according to target groups. Size tables and grading intervals can be examined, and some specific 3-D country data are available, for example, SizeGermany Scantars (3-D avatars). These can be downloaded and transferred to CAD software systems such as Assyst. Over 120 companies are now accessing measurement data for global applications (iSize, 2009).

Eurofit: Clothing and technological 3-D shape portal

Initially a collaborative European project was designed to implement an online 3-D body shape measurement data platform enabling

*   designers and industrialists to draw on 3-D shape information,
*   owners of 3-D data to pool their data and receive revenue,
*   IT companies to develop new services.

This platform, launched in 2014, offers opportunity to harmonize 1-D and 3-D data sets and 3-D data handling tools (Trieb et al., 2013).

World Engineering Anthropometry Resource (WEAR): Technological design

This platform was inaugurated for members of an international organization (WEAR) by a group of experts in engineering anthropology. It offers a site that comprises data from over 100 technological design surveys. Access is limited to members who share their data or tools. The group maintains quality control of the WEAR anthropometric databases accessed through the site. WEAR proposes checklists for validity (sampling, subject population, and secular change), comparability (definition of measurements), and accuracy (before, during, and after measurement capture) and suggests ways in which these may be affected (Kouchi and Mochimaru, 2009). There is an Anthropometric Measurement Interface (AMI) with a web-based software tool to facilitate collaboration and data sharing between anthropometric researchers across the globe. It enables users to plan, compare, and search for measurements taken by others. Further additions include tools and methodologies not only for analyzing data collected from 3-D static scan data but also for dynamic modeling using motion capture (Veitch and Robinette, 2006).

ISO/TC 7250-2 technology design

This technical report is designed to be a continually updated repository for the most recent anthropometric data for technological design applications. It contains statistical summaries of body measurements together with database background information for working age people in the national populations of individual ISO member bodies. The data are intended for use in conjunction with ISO standards for equipment design and safety, which require body measurement input, wherever national specificity of design parameters is required (ISO 7250-2: 2010).

3.5   Reflection

Many anticipated problems fail to appear, while others, not envisaged, emerge. Of the surveys completed during the previous two decades, few organizations have published reflections on their studies. This is regrettable as prior experience of designing and implementing surveys could serve to help groups planning to undertake future anthropometric studies. Some comments set out in the succeeding text cover information drawn from reports of two studies: CAESAR undertaken for technological design and SizeUK a clothing-specific study.

3.5.1 The CAESAR study

Concerning CAESAR (summary details in Table 3.2), active participants are reflected on the study and flagged four recommendations.

3.5.1.1 User support

It is vital to get plenty of feedback from prospective users of the data, from the conceptual stage of a study right through to the delivery of results.

3.5.1.2 Equipment

The team should test all equipment before initiating data collection, even where a new item appears to be very similar to one used previously.

3.5.1.3 Recruitment

It is a good idea to use more than one means to recruit subjects. For example, invitations and advertisements may need to be translated into many languages. Allied to this is the need to consult on the recruitment strategy with representatives of the different segments of a sampling strategy.

3.5.1.4 Planning

Be flexible and have a backup plan for every stage and aspect of the study that might go wrong (Robinette and Daanen, 2003).

3.5.2 SizeUK

The experience of the authors outlined earlier leads them to concur with some of the conclusions drawn from the CAESAR study. In addition the following observations are considered:

3.5.2.1 Press and publicity

It is important in creating momentum for a survey that major public interest is generated. All journalists need to be given equal opportunity and access as if some parts of the press are favored; then, others may decline to give coverage or, in extreme cases, to disparage a project.

3.5.2.2 Funding

Most surveys use a mix of government cash and, from industry, a mix of cash and "in-kind" support. It is important to underline the need for prior agreement among partners on what is meant by "in-kind" contributions (e.g., personnel costs) and (if to be offered) to allow for the different levels of popularity between shopping vouchers.

3.5.2.3 Equipment

All potential hardware and software needs to be thoroughly benchmarked particularly as there are updated ISO anthropometric standards for clothing. Unsurprisingly, given the prominence of a national sizing survey, an organizer is likely to come under considerable pressure to use equipment that may not be the best for conducting a survey, or may be persuaded to purchase equipment before the requirements are fully understood. This could distort the survey process and add to overall costs.

3.5.2.4 Data collection

As suggested earlier, it is extremely useful to allocate a unique barcode to each subject on registration for a survey. This will help to ensure that discrete sets of subject data can be tracked and assembled in the database. Further, although a seated scan was captured, there were no resources available to develop the software for automatically extracting measurements. (That capability is now available for clothing-specific applications.)

3.5.2.5 Data analysis

SizeUK undertook extensive basic data analysis on behalf of the industrial partners, and while this produced a mass of statistical data in formats requested, partners really only required RTW body size charts. However, as interest in using online tools has since been received, further tools have been introduced to enable SizeUK license holders to access the SizeUK database from their desktop and carry out customized measurement analysis on a subset of the data based on their customer profile. These tools are user-friendly and have been designed primarily for clothing retailers and brands so that they can understand the overall size distribution of their customers, analyze the relationship between key measurements, and create size tables. In summary, users can

- create and save subsets of the database based on specific demographic and socioeconomic criteria,
- view statistical tables (min/max/mean/standard deviation/whole populations or 5th to 95th percentile),
- generate frequency distribution charts for any measurements,
- view measurement definitions and images,
- produce measurement correlation charts for any two measurements and create size tables,
- analyze the frequency distribution both across the size range and within an individual size.

Sizemic Ltd.

3.5.2.6 Data sales

Most national surveys have typically sold their data, but each has addressed data exploitation differently. For example, SizeUK estimated the value of the data and, depending on the turnover of a client, contributed £40,000, £60,000, or £80,000. This rather complicated the sale of data at a price smaller clients could afford, particularly in the case of negotiations with very small clothing retailers and with individual fashion designers (Bougourd and Treleaven, 2010).

3.6   Future trends

During the last 20 years, there has been a particular focus on the generation of one-dimensional data from 3-D anthropometric surveys. Three-dimensional static data can do much more: it can, as seen in Section 3.3.3, earlier, dramatically improve the process of design and development across the supply chain, enhancing the shape, size, and fit of clothing. However, despite these growing applications, it has been suggested that "a perfect suit is more than static data" (Meixner and Krzywinski, 2011).

There is growing interest in creating and capturing dynamic scan data, that is, data from a subject whose movements range between mild and extreme activities, such as walking and skiing. Initial developments—using captured scan variations and shape analysis tools—led to animated body scans (e.g., Ruto, 2009). Further research, using an anatomically correct 3-D scan, has been created with movements relevant to high-performance sport (e.g., cycling), where different poses can be used to construct 2-D patterns from 3-D scans for close-fitting garments (Meixner and Krzywinski, 2011), while other work explored the 3-D scanning process as a tool to help predict cycling performance (Luke, 2016).

Notwithstanding these 3-D scan developments, interest in 4-D temporal capture is growing (e.g., Cloth Cap). This research describes new techniques to greatly simplify the process of virtual try-on. The researchers used 4-D movies of people (recorded with a 4-D high-resolution scanner from 3dMD), which enabled automatic transfer of 3-D clothing to new body shapes (Black et al., 2017; Pons-Moll et al., 2017).

The approach is to "scan a [moving] person wearing a garment, separate the clothing from that person, and then render it on top of a new person" (Black et al., 2017). The figure in the succeeding text illustrates how "cloth cap" supports a range of applications related to clothing capture, modeling, retargeting, reposing, and try-on (Fig. 3.8).

The full impact of these developments on future size and shape surveys is not yet known, but with four-dimensional, real-time scanning technologies becoming available, it may be that survey databases will be compiled, not only with a range of

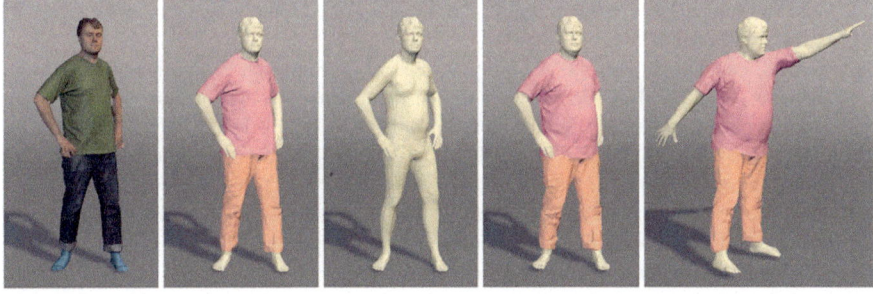

Fig. 3.8 Cloth cap. From *left* to *right*: (1) An example of 3-D textured scan that is part of a 4-D sequence. (2) The multipart aligned model, layered over the body. (3) The estimated minimally clothed shape (MCS), under the clothing. (4) The body made fatter and dressed in the same clothing. (5) This new body shape posed in a new, never seen pose (Pons-Moll et al., 2017).

dynamic postures but also with dynamic sequences. Such data could lead the way to digital design and fit for inclusive clothing design strategies within a more sustainable circular economy—which, with the EU-China agreement indicating a way, could be for a global circular economy.

Acknowledgments

The authors would like to acknowledge the kind assistance of the following:
Sandra Alemany Mut, Universidad Politécnica de Valencia, Spain.
Supiya Charoensiriwath, Thai National Electronics and Computer Technology Centre, Thailand.
Stroud Cornock, Higher Education Consultant, United Kingdom.
Andrew Crawford, Sizemic Ltd., United Kingdom.
Monika Gupta, National Institute of Fashion Technology, India.
Kerry King, TC2, United States of America.
Alexander Kung, TPC (HK) Ltd., Hong Kong, China.
Audrey Lauxerrois, Institut Français du Textile et de l'Habillement, (IFTH) France.
Young-Suk Lee, Chonnam National University, South Korea.
Simone Morlock, Hohenstein Institute, Germany.
Patrick Robinet, Institut Français Textile et Habillement, (IFTH) France.
Darko Ujević, University of Zagreb, Croatia.
Norsaadah Zakaria, Telestia Malaysia, Centre of Clothing Technology, IBE, UiTM, Malaysia.

References

3dMD, 2016. http://www.3dmd.com/. Accessed January 2018.

Allen, R.M., Bougourd, J., Staples, R.A.J., Orwin, C., Bradshaw, M., 2003. Scanner Benchmarking for SizeUK. In: Presented at the 15th Annual Conference of the International Ergonomics Association, Seoul, Korea CD ISBN 89-90838-10-X 98530.

Ashdown, S., Na, H., 2008. Comparison of 3D body scan data to quantify upper body variation in older and younger women. Cloth. Text. Res. J. 26 (4), 292–307.

Ball, R., Luximon, Y., Chow, H.C.E., 2012. Anthropometric study on Chinese head. In: Asian Workshop on 3D Body Scanning Technologies, Tokyo Conference Proceedings. Hometrica, Switzerland.

Ballester, A., Parrilla, E., Piérola, A., Uriel, J., Perez, C., Piqueras, P., Nácher, B., Vivas, J., Alemany, S., 2017. Data-driven three-dimensional reconstruction of human bodies using a mobile phone app. Int. J. Digit. Hum. 1, 361. https://doi.org/10.1504/IJDH.2016.10005376.

Black, M., Pons-Moll, G., Pujades, S., 2017. 4D movies captured people in clothing, creating realistic virtual try-on. In: August 2017 SIGGRAPH 2017, Los Angeles, CA.

BOF/McKinsey, 2018. Fashion in 2018 I 08 Sustainability Credibility. https://www.businessoffashion.com/articles/intelligence/top-industry-trends-2018-8-sustainability-credibility. Accessed June 2018.

Bougourd, J.P., Treleaven, P., 2010. UK national sizing survey—SizeUK. In: d'Appuzo, (Ed.), International Conference on 3D Body Scanning Technologies, Conference Proceedings. Hometrica, Lugano, Switzerland.

Brightpearl Report, 2018. Try Before You Buy: A Returns Tsunami for Retail. https://info.brightpearl.com/returns-tsunami-for-retail. Accessed April 2018.

Bryers, T., 2017. Patagonia Wins Circular Multinational Award at World Economic Forum Annual Meeting in Davos. http://www.patagoniaworks.com/press/2017/1/17/patagonia-wins-circular-economy-multinational-award-at-world-economic-forum-annual-meeting-in-davos. Accessed June 2018.

Burberry, 2018. https://www.bbc.co.uk/news/business-44885983. Accessed July 2018.

Carson, R., 1962. Silent Spring. Houghton Mifflin, Boston.

CEBE, 2002. Building and sustaining a learning environment for inclusive design: a framework for teaching inclusive design within built environment courses in the UK. In: Report of Special Interest Group for Centre for Education in the Built Environment. http://cebe.cf.ac.uk/learning/sig/inclusive/full_report.pdf. Accessed May 2018.

Cilley, J., 2016. How to Use 3D Body Imaging to Innovate in the Apparel or Footwear Industry. Body Labs.

Defra, 2010. Sustainable Clothing Action Plan (Update February 2010). Department for Environment, Food and Rural Affairs, London.

Defra, 2011. Sustainable Clothing Roadmap. Department for Environment, Food and Rural Affairs, London.

Disclosure and Barring Service (DBS), 2018. https://www.gov.uk/government/organisations/disclosure-and-barring-service. Accessed June 2018.

Ellen MacArthur Foundation, 2017a. The Circular Economy 2017. https://www.ellenmacarthurfoundation.org/circular-economy/overview/concept. Accessed June 2018.

Ellen MacArthur Foundation, 2017b. A New Textiles Economy: Redesigning Fashion's Future, (2017). https://www.ellenmacarthurfoundation.org/publications/a-new-textiles-economy-redesigning-fashions-future. Accessed June 2018.

Eurofit, 2012. Integration, Homogenisation and Extension of the Scope of Anthropometric Data Stored in Large E.U. Pools. https://cordis.europa.eu/docs/projects/cnect/6/296116/080/publishing/readmore/Eurofit-factsheet.pdf. Accessed June 2018.

European Commission, 2018a. Circular Economy Implementation of the Circular Economy Action Plan. http://ec.europa.eu/environment/circulareconomy/pdf/circular_economy_MoU_EN.pdf. Accessed June 2018.

European Commission, 2018b. EU-China Memorandum of Understanding on Circular Economy. http://ec.europa.eu/environment/circulareconomy/pdf/circular_economy_MoU_EN.pdf. Accessed July 2018.

Gill, S., Brownbridge, K., Wren, P., 2014. The true height of the waist: explorations of automated body-scanner waist definitions of the TC2 scanner. In: Proceedings of the 5th International Conference on 3D Body Scanning Technologies, Lugano, Switzerland, 21–22 October 2014.

Global fashion industry, 2016. Global fashion industry statistics – International apparel. https://fashionunited.com/global-fashion-industry-statistics. Accessed May 2018.

Gordon, C., Churchill, T., Clauser, C., Bradtmiller, B., McConville, J., Tebbits, I., Walker, R., 1989. 1988 Anthropometric Survey of U.S. Army Personnel: Methods and Summary Statistics (Technical Report, Natick/TR-89/044).

Greenpeace, 2017. PFC Revolution in the Outdoor Sector. https://www.greenpeace.org/international/publication/7150/pfc-revolution-in-outdoor-sector/. Accessed June 2018.

Greenpeace International, 2017. New Report Breaks the Myth of Fast Fashion's so-Called 'Circular Economy'—Greenpeace International 18 September 2017. https://www.greenpeace.org/international/press-release/7517/new-report-breaks-the-myth-of-fast-fashions-so-called-circular-economy-greenpeace/. Accessed May 2018.

HM Government, 2005. Securing the Future: The UK Government Sustainable Development Strategy. The Stationary Office, Command 6467, London.

Hometrica Consulting, 2018. Organizer of Annual Conferences on 3D Body Scanning Technologies. www.hometrica.ch. Accessed February 2018.

INDIAsize, 2018. http://pib.nic.in/newsite/PrintRelease.aspx?relid=176888. Accessed March 2018.

iSize, 2009. Body Dimension Portal. http://www.human-solutions.com/fashion/front_content. php?idcat=151&lang=7&changelang=. Accessed October 2017.

ISO 15535:2012. General Requirements for Establishing Anthropometric Databases.

ISO 18825-2:2016. Clothing—digital fittings—Part 2. Vocabulary and terminology used for attributes of the virtual human body.

ISO 7250-1:2017. Basic human body measurements for technological design—Part 1. Body measurement definitions and landmarks.

ISO 8559:1989. Garment construction and anthropometric surveys.

ISO 8559-1:2017. Size designation of clothes—Part 1. Anthropometric definitions for body measurement.

ISO 8559-3:2018. Size designation of clothes—Part 3. Methodology for the creation of the body measurement tables and intervals.

ISO/20685-2:2015. Ergonomics—3-D scanning methodologies for internationally compatible anthropometric databases—Part 2. Evaluation protocol of surface shape and repeatability of relative landmark positions.

ISO/DIS 20685-1:2018. 3-D scanning methodologies for internationally compatible anthropometric databases—Part 1. Evaluation protocol for body dimensions extracted from 3-d body scans.

ISO/TR 7250-2:2010. Basic human body measurements for technological design—Part 2. Statistical summaries of body measurements from national populations.

ISO14000:2009. Environmental management family (life cycle analysis, LCA).

Kirchdoerfer, K., Mahr-Erhardt, A., Morlock, S., 2011. 3D body scanning—utilization of 3D body data for garment and footwear design. In: d'Appuzo, (Ed.), International Conference on 3D Body Scanning Technologies, Conference Proceedings. Hometrica, Lugano, Switzerland, pp. 150–157.

Kouchi, M., Mochimaru, M., 2009. Quality Control of Anthropometric Databases—Presentation for WEAR. http://wear2.io.tudelft.nl/files/Banff08/Makiko.pdf. Accessed January 2018.

Krzywinski, S., Rödel, H., Siegmund, J., 2005. Virtual product development for close-fitting garments of knitwear with Elastan yarns. In: Second International Conference of Textile Research Division, NRC, Cairo: Textile Processing: State of the Art and Future Developments2, pp. 136–140.

Kung, A., 2012. TPC (HK) Ltd. Hong Kong, China. http://www.tpc-intl.com/index.asp. Accessed October 2017.

Lane, C., 2017. The evolution from static-3d scanning 3dMD CEO presents at wear. In: Wearable Technologies (Wear) 2017 Conference in San Francisco.

Luke, J., 2016. How Body Scanning Can Predict your Cycling Performance. https://www. bikeradar.com/news/article/body-scanning-predicts-performance-of-athletes-48337/. Accessed June 2018.

Marfell-Jones, M., 2006. International Standards for Anthropometric Assessment. International Society for the Advancement of Kinanthropometry, Potchefstroom, South Africa.

Marks and Spencer, 2018. https://corporate.marksandspencer.com/plan-a/business-wide/product-sustainability#248c756273124f1ebd72640eeb4ea1dd. Accessed July 2018.

McCann, J., Bryson, D., 2015. Textile Led Design for the Active Ageing Population. Woodhead Publishing Ltd., Cambridge. UK.

Meadows, H.D., Meadows, D.L., Randers, J., Behrens, W.W., 1972. The Limits to Growth. Potomac Associates, Virginia, USA.

Meixner, C., Krzywinski, S., 2011. Automated generation of human models from scan data in anatomically correct postures for rapid development of close-fitting, functional garments. In: d'Appuzo, (Ed.), International Conference on 3D Body Scanning Technologies, conference proceedings. Hometrica, Lugano, Switzerland, pp. 164–173.

Morlock, S., Schenk, A., Klepser, A., Schmidt, A., 2016. XL Plus Men—New Data on Garment Sizes. Hohenstein Institut für Textilinnovation gGmbH, Boennigheim, Germany.

NDA, 2010. Newsletters on Design for Ageing Well, Summer 2010. published online as part of The New Dynamics of Ageing project, at. http://www.newdynamics.group.shef.ac.uk/design-for-ageing-well-newsletters.html. Accessed June 2018

Nike, 2010. Considered Design—An Environmental Design Tool. published at https://news.nike.com/news/nike-releases-environmental-design-tool-to-industry. Accessed June 2018.

Op de Beeck, P., 2018. Three Sustainable Trends Shaping the Future of the Fashion Industry Viewpoint. https://www.carbontrust.com/news/.../sustainability-trends-future-of-fashion-industry/. Accessed May 2018.

Patagonia, 1980. http://www.patagonia.com/company-history.html. Accessed January 2018.

Pons-Moll, G., Pujades, S., Hu, S., Black, M.J., 2017. Cloth cap: seamless 4D clothing capture and retargeting. ACM Trans. Graph. 36 (4), 73.

Preiss, A., Botzenhardt, U., 2012. Size survey—process chain and available products. In: Asian Workshop on 3D Body Scanning Technologies, Tokyo conference proceedings. Hometricaa, Switzerland, pp. 106–114.

Regulation (EU), 2016. 2016/679 of the European Parliament. EUR-LEX. https://eur-lex.europa.eu/legal-content/EN/TXT/?uri=CELEX:32016R0679. Accessed January 2018.

Research Excellence Framework (REF) UK, 2018. https://www.ref.ac.uk/about/whatref/. Accessed June 2018.

Robinette, K.M., Daanen, H., 2003. Lessons Learned from Caesar: A 3-D Anthropometric Survey. https://www.researchgate.net/.../228556871_Lessons_learned_from_CAESAR_A_3-D_a. Accessed January 2018.

Ruiz, M.C., Buxton, B.F., Douros, I., Treleaven, P.C., 2002. Web based tools for 3D body database access and shape analysis. In: Int. Conf. of Numerisation 3D - Scanning, 2002 Paris France.

Ruto, A., 2009. Dynamic Whole Body Modeling and Animation, Engineering. Doctorate ThesisUniversity of London.

Sayem, A.S.M., Kennon, R., Clarke, N., 2014a. A 3D grading and pattern unwrapping technique for loose fitting shirt Part 1: Resizable design template. J. text. Apparel Technol. Manag. 8 (4 Spring), 2014.

Sayem, A., S, M., Kennon, R., Clarke, N., 2014b. B 3D grading and pattern unwrapping technique for loose fitting shirt Part 2: Functionality. J. Text. Apparel Technol. Manag. 8 (4 Spring), 2014.

Simmons, K.P., Istook, C.L., 2003. Body measurement technique: comparing 3D body scanning and anthropometric methods for apparel applications. J. Fash. Mark. Manag. 7 (3), 306–332.

Size Thailand, Charoensiriwath, S., 2008. A real-time data monitoring and management system for Thailand's first national sizing survey. In: Proceedings of the Portland International Conference on Management of Engineering & Technology 2008 (PICTMET 2008)pp. 856–863.

SizeGermany, 2018. https://www.human-solutions.com/group/front_content.php?idcat=211&idart=2204&lang=2. Accessed January 2018.

Sizemic Ltd., 2012. A Service Developed to Commercialise SizeUK Data. http://www.sizemic.eu/. Accessed October 2017.

SizeNorth America, 2018. How to Become a Sponsor. http://www.sizenorthamerica.com/cms/front_content.php?idcat=4. Accessed January 2018.

SizeUK, 2018. Results from the UK National Sizing Survey. http://www.arts.ac.uk/research/current-research/ual-research-projects/fashion-design/sizeuk-results-from-the-uk-national-sizing-survey/ and http://www.size.org/SizeUKInformationV8.pdf. Accessed January 2018.

SizeUSA, 2018. https://www.tc2.com/size-usa.html. Accessed January 2018.

Stewart, A., Marfell-Jones, M., Olds, T., de Ridder, H., 2011. International Standards for Anthropometric Assessment ISAK: Lower Hutt, New Zealand. Sustainable Apparel Coalition. https://apparelcoalition.org/the-sac/. Accessed May 2018.

Sustainable Brands, 2018. Houdini Sportswear Launches Industry—First Planetary Boundaries Assessment. http://www.sustainablebrands.com/news_and_views/new_metrics/sustainable_brands/houdini_sportswear_launches_industry-first_planetary_b. Accessed June 2018.

Tahan, A., Buxton, B., Ruiz, M., Bougourd, J., 2003. Point distribution of models of human body shape from a canonical representation of 3D scan data. In: Paper presented at conference 3D Human Digitizing and Modeling, Paris, France.

The Higg Index, 2018. https://apparelcoalition.org/the-higg-index/. Accessed June 2018.

Trieb, R., Ballester, A., Kartsounis, G., Alemany, S., Uriel, J., Hansen, G., Fourli, F., Sanguinetti, M., Vangenabith, M., 2013. EUOFIT—integration, homogenisation and extension of the scope of large 3D anthropometric data pools for product development. In: 4th International Conference and Exhibition on 3D Body Scanning Technologies, Long Beach CA 2013.

UK Government Environmental audit committee, 2018. Sustainability of the Fashion Industry Inquiry Launched. https://www.parliament.uk/business/committees/committees-a-z/commons-select/environmental-audit-committee/news-parliament-2017/sustainability-of-the-fashion-industry-inquiry-launch-17-19/. Accessed June 2018.

UN, 1987. Our Common Future, Known as the Grundtland. Report, is available at http://www.earthsummit2012.org/about-us/historical-documents/92-our-common-future. Accessed October 2017.

UN, 2012a. Resilient People, Resilient Planet A report of the UN Secretary-General's High-Level Panel on Global Sustainability. UN, New York ISBN: 978-92-1-101256-9.

UN, 2012b. Report on the Outcome of the Rio+20 Corporate Sustainability Forum. http://new.artmill.eu/data/329-9696-rio-20-ed-fin-report.pdf. Accessed June 2018.

UN, 2015. Transforming Our World: The 2030 Agenda for Sustainable Development A/RES/70/1.

UN, 2017. World Population Prospects. revision https://esa.un.org/unpd/wpp/Publications/Files/WPP2017_KeyFindings.pdf. Accessed February 2018.

Veitch, D., 2012. Where is the human waist? Definitions, manual compared to scanner measurements. Work 41, 4018–4024. https://doi.org/10.3233/WOR-2012-0065-4018.

Veitch, D., Robinette, K., 2006. World engineering anthropometry resource (WEAR) A review. In: National Conference, Sydney Australia. Human Factors & Ergonomics, Society of Australia Inc Uploaded 19 August 2015 https://www.researchgate.net/publication/256706869_World_Engineering_Anthropometry_Resource_WEAR_A_Review. Accessed January 2018.

WEAR, 2012. A Web-Based Resource at. https://bodysizeshape.com/. Accessed June 2018.

Wells, J.C.K., Cole, T.J., Treleaven, P., 2008. Age variability in body shape associated with excess weight: the UK national sizing survey. Obesity 16 (2), 435–441.

WRAP, 2017. Valuing Our Clothes: the Cost of UK Fashion WRAP UK. http://www.wrap.org. uk/sustainable-textiles/valuing-our-clothes%20. Accessed June 2017.

Sources of further information

3D Scanning Conference Series, and Associated Publications of Hometrica. n.d. Contact Dr. Nicola D'Apuzzo, Hometrica Consulting, Lugano, Switzerland.
The International Standards Organization (ISO), 2018. https://www.iso.org/standards.html. Accessed January 2018.

Further reading/viewing

Business Insider, 2016. A First Look at the New Facility Where Under Armour Creates Athletic Apparel of the Future (video). http://www.3dmd.com/a-first-look-inside-the-new-facility-where-under-armour-creates-athletic-apparel-of-the-future-business-insider/. Accessed May 2018.
Chapman, J., 2009. Emotionally Durable Design. Earthscan, London.
Council of Europe, 1981. Convention for the protection of individuals, with regard to automatic processing of personal data. In: Council of Europe, Strasbourg, New EU/UK law introduced 25 May 2018. Available from https://www.which.co.uk/data-protection/uk. Accessed 31 May 2018.
Eileen Fisher, 2018. Circular by Design. https://www.eileenfisher.com/circular-by-designgfa/?___store=en&___from_store=default. Accessed June 2018.
Green peace, 2017. Fashion at the Crossroads. https://www.greenpeace.org/international/publi cation/6969/fashion-at-the-crossroads/. Accessed January 2018.
Green Peace, 2018. Destination Zero: Seven Years of Detoxing the Clothing Industry. https:// www.greenpeace.org/international/publication/17612/destination-zero/. Accessed July 2018.
Kim, J.Y., You, J.W., Kim, M.S., 2017. South Korean anthropometric data and survey methodology: 'Size Korea' project. Ergonomics 60 (11), 1586–1596. https://doi.org/10. 1080/00140139.2017.1329940.
Robinette, K., Veitch, D., 2016. Sustainable Sizing. Hum. Factors J. Hum. Factors Ergon. Soc. 58 https://doi.org/10.1177/0018720816649091.
Shankleman, J., 2013. Updated: Puma closes the loop with first green clothing range. Available from: http://www.businessgreen.com/bg/news/2243190/puma-closes-the-loop-with-first-green-clothing-range. Accessed June 2018.
Smithers, R., 2017. UK households binned 300.000 t of clothing in 2016. The Guardian. Vaude, 2017. Sustainability report. Available from: http://csr-report.vaude.com (Accessed January 2018).
Veitch, D., Fitzgerald, C., et al., 2013. Sizing up Australia—The Next Step Chapters I, 2 and 3. Safe work Australia, Canberra, Australia.

Developing apparel sizing system using anthropometric data: Body size and shape analysis, key dimensions, and data segmentation

4

Norsaadah Zakaria[a,b], *Wan Syazehan Ruznan*[c]
[a]Centre of Clothing Technology and Fashion, IBE, Universiti Teknologi MARA (UiTM), Shah Alam, Malaysia, [b]Malaysian Textile Manufacturing Association (MTMA) and Malaysian Textile and Apparel Centre (MATAC), Kuala Lumpur, Malaysia, [c]Textile and Clothing Technology Department, Universiti Teknologi MARA (UiTM), Shah Alam, Malaysia

4.1 Introduction: The body sizing system

Anthropometric data are obtained from a comprehensive anthropometric survey to understand the body shapes and sizes of certain population. When the body shapes and sizes are analyzed and understood, then the development of the right size clothing for good fitting garments can be achieved. It becomes especially significant when there is a need for mass-produced garments. These types of garments are mainly known as ready-to-wear where the sizes are set for certain body shapes and sizes.

The body sizing system is the method or system used to create a set of clothing for a variety of people in the target market (Xia and Istook, 2017). The most common sizing system in the apparel industry today uses a base size designed for a fit model and graded set of proportionally similar sizes derived from this base size. According to Xu et al. (2002) a size designation is "a set of garment sizes in a sizing system designed to reflect the body sizes of most individuals in a population" (Xu et al., 2002). On the other hand, RTW is the clothing that is manufactured in standard measurements and in size sets. It is designed to be purchased in an appropriate size and worn without further alterations. A sizing system can be as simple as "one size fits all" or S, M, L, or XL or as complex as a system that provides a custom-fitted garment for each individual. However, it has been noted that it is critical for any designer and manufacturer to provide clothing that fits all population (Goldsberry et al., 1996).

The purpose of a sizing system for apparel should be to make clothing available in a range of sizes that fit as many people as possible (McCulloch et al., 1998; LaBat, 1987). Standard sizing system should be able to help the apparel firms in categorizing garments of different sizes in a way that customers will recognize it (Hrzenjak et al., 2013). It also helps customers in identifying the garments that will provide them with a

Anthropometry, Apparel Sizing and Design. https://doi.org/10.1016/B978-0-08-102604-5.00004-4

reasonable fit (Beazley, 1998a). It is very challenging to develop a size chart that can fit range of sizes for apparel. This size chart for clothing industry should effectively fit the variation of existing sizes and shapes of many different diverse population (Xia and Istook, 2017; Pei et al., 2017). Moreover, with different market segments, the problem also relies on accommodating the population for different ready-to-wear clothing stores that carry huge garment types and variations in which the right size and fit are usually compromised (Gill, 2015; Makhanya et al., 2014).

4.1.1   Size charts

The correct and reliable data collected from human measurements will provide real information for the industry about the sizes of their customers (Brownbridge et al., 2018). Individual companies may decide to put labels in their clothing with the bust, waist, and hip sizes, which will help consumers make better selections. Particularly in women's clothing the size label does not mean any particular measurement unless the company publishes a size chart. For women the body sizing systems are produced based on different combination of body measurements, proportions, and heights (Otieno and Fairhurst, 2000). For women's wear, sizes are labeled with numbers that correlate with height, bust, waist, hip, and other measurements. Skirts and pants are sold by size or by the waist measurement. Size categories for men are based on measurements on chest and length. Pants are sized at the waist and inseam. Shirt sizes are based on the measurement around the collar and the sleeve length. Most systems for sizing RTW garments have been based on very limited information. Thus a good sizing system should have detailed information of the key dimensions that are used to derive the size system, and this is called the size designation (Ashdown, 2014).

4.1.2   Size and fit

Size and fit are synonyms. It is reflected in the ready-to-wear (RTW) clothing as the product of a standard sizing system. To get a good fit, RTW clothing should be evaluated from the angle of individual *self and body*. The *self-evaluation* is called *body cathexis* (Manuel et al., 2010). Body cathexis is the evaluative dimension of body image and is defined as positive and negative feelings towards one's body. The dissatisfaction with body cathexis will result to the dissatisfaction with clothing fit (Kaiser, 1997). If one has higher body cathexis, the probability of satisfaction with the fit of clothing is higher (LaBat, 1987). The feeling of dissatisfaction of clothing fit can also be related to dissatisfaction with weight and lowered body cathexis (Song and Ashdown, 2013). A very important first step in determining correct sizing or creating well-fitted garments is obtaining accurate measurements of the specific human body. These measurements are considered the *body evaluation*, which is also known as the *anthropometry* (Gill, 2015). It is the science of human body dimensions. It deals with the methodical and precise measurement of the human body. The sizing systems that have been developed throughout the world are very small in numbers. Most of the world's clothing sizing systems are neither standardized nor related to the average human's body measurements. Therefore the needs for anthropometric studies are significant at this point for each country in the world. At the moment, there are only a handful of countries that have completed their national sizing system such as Korean size, Thailand size, Taiwan size, Japan size, Mexico size, Spanish size, Australia size, United Kingdom size, and North America size. Some of the countries now conducting the anthropometric research are India, the Philippines, and Malaysia.

4.2   Sizing system development and methodology

The process of developing the sizing system for any target market can be observed using the following procedure as shown in Fig. 4.1. The targeted sample for this methodology is demonstrated for the sample of female (7–12 and 13–17 years old). The illustrated procedure and process are conducted for children, and the original data were collected within 6 months. Anyhow the same process can be completed for the adult data too.

In this chapter, Fig. 4.1 illustrated the whole process to develop sizing system as shown later.

There are basically three important stages in sizing system development: Stages 1, 2, and 3. Stage 1 consists of the first process of anthropometric data collection, Stage 2 is the sizing data analysis, and Stage 3 consist of sizing system development. In general, it is imperative to identify important body dimensions for the development of sizing system. This is the first process after obtaining the data collection in which during the first stage of anthropometric survey, different body dimensions are measured on each person to develop sizing system. Then, from these body dimensions, only important body dimensions are selected to cluster sample population into similar body dimensions.

These important body dimensions are known as key dimensions or control dimensions. Key dimensions are the best representative of body shapes and sizes in which the sizing system is created (Adu-Boakye et al., 2012). Key dimensions consist of primary dimensions and secondary dimensions. Two distinct techniques to identify the

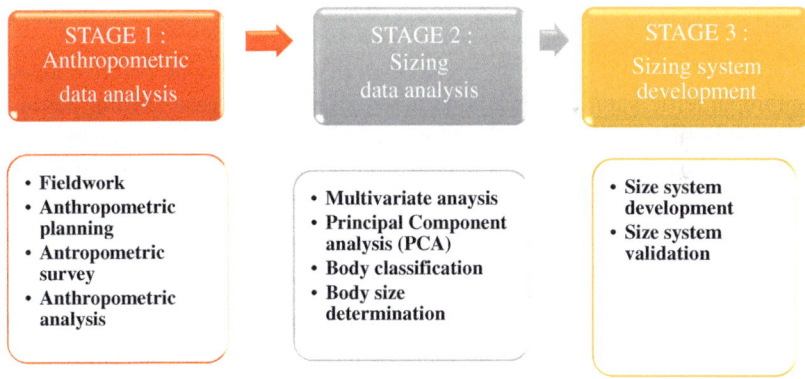

Fig. 4.1 Sizing system development process.

key dimensions are discussed in this chapter, which are the bivariate technique and principal component analysis, which is the multivariate technique. This chapter is designed to show how the anthropometric data are analyzed specifically using both techniques. Using bivariate technique will help to identify visual patterns of the dimensions as it only look at the relationships of two dimensions at one time. It will only give correlational relations between two variables. Using the principal component will give relative importance of the body dimensions. From the total body dimensions, this technique is able to classify them into important groups that suggest the group characteristics (Zakaria, 2011).

All steps and processes of obtaining the key dimensions are demonstrated in the bivariate and multivariate technique sections. In the subchapters the steps to identify the key dimensions are shown. These chapters practically showed the ways to analyze the anthropometric data and the ways to interpret data for the development of sizing system. Fig. 4.1 clearly shows how complex it is to develop a sizing system, encompassing three stages and ten steps from beginning to end. Stage 1 is anthropometric analysis, Stage 2 is sizing analysis, and Stage 3 is sizing system development. All ten steps are explained in the following sections.

Stage 1 is anthropometric analysis. The goal of Stage 1 is to collect body measurements of the sample population and analyze those using simple statistical methods. The purpose of this analysis is to understand the body ranges and variations present in the sample population. This stage consists of four steps—fieldwork preparation, anthropometric planning, the anthropometric survey itself, and anthropometric analysis—each of which is described in the following sections.

4.2.1 Step 1: Fieldwork preparation

This is the first process under anthropometric data analysis. Before obtaining the data the planning on how to conduct the anthropometric survey must be planned carefully. There should be careful consideration to select the team groups for better cooperation and understanding to work together.

4.2.1.1 Paperwork

Fieldwork preparation refers to the preparation that must be done before conducting the anthropometric survey. Preparation activities could include getting permission from the authorities, developing the anthropometric protocol, and training the measurers. When conducting an anthropometric survey, the process of preparing paperwork and getting the access from the authorities might involve asking formal permission from government bodies or agencies or from various related management of companies that owned spaces for public to go like the malls and town hall. The paperwork granting permission to work with different target populations can be challenging, which needs persistence and patience with the process of waiting for authorities to organize them. However, once the permission is granted, it will be a really good experience to conduct the anthropometric survey because you will have a chance to mingle and meet with many people of different walks of life.

4.2.1.2 Training the measurers

The next task is to prepare the measurers, if the anthropometry survey is conducted manually. If the survey is to be done using digital methods, such as a 3-D body scanner, then training will be focusing on using the sophisticated machine. The training for an anthropometric survey for clothing purposes is based on the standard ISO 8559/1989, which defines the terms used for each different body dimension. Under this ISO standard, there are 51 body dimensions to be measured for a clothing system. These body dimensions are divided into three groups: vertical length, width, and girth, as shown in Table 4.1. These dimensions are also divided into the upper and lower body. The 29 dimensions suitable for the upper body or whole body are marked with superscript "a", while the other 20 dimensions (not marked) are categorized as lower body dimensions.

First the measurers have to be briefed about the objectives of the anthropometric survey. They can be introduced to the topic using a PowerPoint presentation, and the objectives were clearly explained laying emphasis on the consistency and precision of measurement process. Each of the trainees should be provided with a copy of the anthropometric manual giving pictures of all the body dimensions. This is followed by a detail explanation of each body dimension followed by a demonstration on a real body.

The measurers should ideally work in pairs and perform hands-on practice on their partners for some days till they are comfortable and familiar with each body dimension. Each measurer uses a form, which lists all the body dimensions to record the measurements. The measurement practice should be continued until the measurers gained confidence and started getting consistent readings.

4.2.1.3 Anthropometric measurement protocols

Anthropometric protocols demonstrate how a manual anthropometric survey can be conducted. The measurement process starts with the subject changing into a tight-fitting garment for better and more accurate body measurements. In every anthropometric survey activity, a team of workers will attempt to finish the targeted number of measurements per day to achieve their daily goal. Manual measurement takes an average of 40 min per subject, and the goal is to measure at least seven people daily. Using the 3-D body scanner, it is much faster: the time from changing clothes to completion of measurement is about 5–10 min per person.

A consistent set of procedures should be employed for the manual measurement process, such as

- fill out demographic data (name, age, gender, and ethnic group),
- measure height,
- measure weight,
- measure upper body dimensions,
- measure lower body dimensions.

One member of the team can measure, while the other records the measurements. All measurements should be taken from one side of the subject's body consistently. After

Table 4.1 List of body dimensions according to ISO8559/1989

Length (vertical)	Width (vertical)	Girth (horizontal)
Height 1. [a]Under arm length	1. [a]Shoulder length 2. [a]Shoulder width	Weight 1. [a]Head girth
2. [a]Scye depth	3. [a]Back width	2. [a]Neck girth
3. [a]Neck shoulder point to breast point	4. [a]Upper arm length	3. [a]Neck-base girth
4. [a]Cervical to breast point	5. [a]Arm length	4. [a]Chest girth
5. [a]Neck shoulder to waist	6. [a]Seventh cervical to wrist length	5. [a]Bust girth
6. [a]Cervical to waist(front)	7. [a]Hand length	6. [a]Upper arm girth
7. [a]Cervical to waist(back)	8. Foot length	7. [a]Armscye girth
8. [a]Cervical height(sitting)		8. [a]Elbow girth
9. [a]Trunk length		9. [a]Wrist girth
10. [a]Body rise		10. [a]Hand girth
11. [a]Cervical to knee hollow		11. Waist girth
12. [a]Cervical height		12. Hip girth
13. Waist height		13. Thigh girth
14. Outside leg length		14. Mid-thigh girth
15. Waist to hips		15. Knee girth
16. Hip height		16. Lower knee girth
17. Crotch		17. Calf girth
18. Trunk circumference		18. Minimum leg girth
19. Thigh length		19. Ankle girth
20. Inside leg length/crotch		
Height 21. Knee height 22. Ankle height		

Lower body dimensions.
[a]Upper body dimensions.

completing the measurements of the day, the forms should be counted and the overall quality of anthropometric measurements checked. Forms with missing data are considered not valid and discarded. The number of subjects measured each day is recorded to ensure that the target number of samples has been achieved and that results are accurate.

4.2.2 Step 2: Anthropometric planning

Anthropometric planning comprises the preliminary study, sample size calculation, and fieldwork coordination. The first purpose of the preliminary study is to test the whole process of measuring, to understand the nature of the survey, and to solve any potential problems before undertaking the real anthropometric survey. The second purpose is to take the measurements needed to calculate the sample size for the anthropometric survey.

4.2.2.1 Preliminary survey

The preliminary survey can be conducted on a small scale and is usually called the pilot study. The sample size can range from 30 to 100 people. The main objective is to collect sufficient body measurements to calculate the sample size needed for the real anthropometric survey.

One common technique that can be used to calculate the sample size for a study is the proportionate stratified random sampling technique (Hair et al., 1998; Bartlett et al., 2001). Proportionate stratified sampling refers to taking the same proportion (sample fraction) from each stratum (Tabachnick and Fidell, 2007). For example, say there are three groups of students: group A with 100 people, group B with 50 people, and group C with 30 people. These groups are referred to as strata. The sample units are randomly selected from each stratum based on proportion. For example, a proportion of 10% from each group (strata) would mean that 10 people were taken from group A, 5 people from group B, and 3 people from group C. The stratum group for this study was based on two groups, age (7–17 years old) and gender (female and male).

A study can have, for example, four demographic variables: age, gender, ethnic group, and geographical area (rural and urban). If the study is focused on two factors such as age and gender, then the proportionate sample size will reflect the distribution of age and gender groups in the real population. The other two parameters, ethnic group and geographical area, can be selected according to simple random technique with the targeted number of subjects calculated from the proportionate sample size (Bartlett et al., 2001).

Data obtained from the preliminary study can be analyzed to calculate the total sample size using the stratified random sampling formula (Eq. 1.1). Then the number of subjects to be sampled from six gender and age groups can be calculated using proportionate sampling based on the actual number of subjects present in the geographical area of interest. The steps are given in detail later.

4.2.2.2 Sample size determination

The sample size for a survey can be calculated using the stratified random sampling formula as shown in Eq. (1.1) (Scheaffer et al., 2005):

$$n = \frac{\displaystyle\sum_{i=1}^{l} N_i^2 \sigma_i^2 / a_i}{N^2 D + \displaystyle\sum_{i=1}^{l} N_i^2 \sigma_i^2} \tag{1.1}$$

where

N = total number of target population.
X_i = input for sample i.
Y_i = output for sample i
x = mean value of input data
y = mean value of output data

The body dimensions used to calculate the sample size are the common key dimensions for a sizing system: height, chest girth, bust girth, waist girth, and hip girth. After figuring the total sample size, the sample size for each of the strata can be calculated using Eq. (1.2), and then the total number of subjects for each age range and gender can be calculated based on the proportionate method formula (Eq. 1.3).

First step:

$$\bar{y}_{st} = \frac{1}{N}[N_1\bar{y}_1 + N_2\bar{y}_2 + \ldots + N_l\bar{y}_l] \tag{1.2}$$

where

N = total population age 7–17
N_1 = total population age 7–12
N_2 = total population age 13–17
$\bar{y}$ = mean of variables for each age group

Second step:

$$D = \left(\frac{0.01 \times \bar{y}_{st}}{2.326}\right)^2 \tag{1.3}$$

Third step:

Calculating the sample size using stratified random sampling (Eq. 1.1)

To calculate the sample size, the age range for the sample population is calculated, and then the total population in the geographical area is calculated. For example, the total population in one state is 823,071[N]. The number of subjects in each age group is then tabulated. Each age group forms a stratum, and the sample for each stratum is calculated using the proportionate method according to the ratio of the real population. Each stratum age [h] is given by

$$n_{h=(N_h/N)*n} \tag{1.4}$$

where n_h is the sample size for stratum h, N_h is the population size for stratum h, N is total population size, and n is the sample size.

For example, the male-to-female ratio in the real population in one state is 51% male and 49% female. For each age range the sample is divided into the corresponding ratio (n) of male and female:

$$n_m = N_h*n \tag{1.5}$$

$$n_f = N_h * n \tag{1.6}$$

where n_m is the sample size for the male stratum, n_f is the sample size for the female stratum, N_h is the population size for stratum h *(age)*, and n is the total sample size.

4.2.3 Step 3: Anthropometric survey—Manual method

A preliminary study is conducted before the main anthropometric survey to check the feasibility of the research approach and to improve the design of the research. ISO standard 8559:1989 (garment construction and anthropometric surveys—body dimension) can be used as a guideline for taking body measurements. In the traditional manual technique, measurement tools to be used include calibrated nonstretchable plastic measuring tapes, height scale with movable head piece, long ruler, elastic 5-meter tapes, and digital weight scale. Since measuring a single subject can take from 20 to 40 min, provision for refreshments for measurers and the subjects should be made to incentivize them. The survey data collected in the form of categorical (demographic data) and continuous data were screened and stored in a standard format.

4.2.3.1 Data entry

All the collected data are keyed into a software such as SPSS or MS Excel. The usual format is to key in the subject's name and data into a row, which is known as a case. The body variables are keyed into the columns. The demographic information (categorical data), gender, ethnic group, age, and geographical area (urban or rural), comes first followed by columns containing numeric body measurements (continuous data).

4.2.3.2 Data screening

Data screening consists of examination for data entry errors, missing data, or outliers. The entire data set is filtered to ensure that there are no errors or missing data. Errors can creep in due to mistakes in keying in the data; these can be rectified by crosschecking with the raw data. The distribution of data can be tested using graphical and numerical methods. The graphical method makes use of histograms, while the numerical assessment is based on values of mean, median, skewness, and kurtosis. Histograms provide a useful graphical representation of the data. Data are normally distributed if the histogram shows a Gaussian distribution. This involves evaluating the bell shape of the data distribution. When tabulating common key dimensions like height, chest girth, bust girth, waist girth, and hip girth, the mean and median values should be the same, while the skewness and kurtosis should show values of 0 and 3, respectively; this indicates that the data are normal (Tabachnick and Fidell, 2007). Skewness refers to the asymmetry of the distribution. If the skew has a negative value, this means the data are skewed to the left; if positive the skew is to the right. Kurtosis refers to the peakness or the flatness of the graph (Hutcheson and Sofroniou, 1999).

4.2.4   Step 4: Anthropometric analysis

The final step of Stage 1 is to analyze the data. The statistical method generally applied at this stage is the descriptive analysis also known as univariate analysis based on simple statistics. Categorical and continuous data can be analyzed as follows:

4.2.4.1   Categorical data

The categorical data are analyzed to understand the demographic profile of the sample population. The first classification to be made often is to divide the population into gender-based subsets, namely, male and female. Frequency distribution curves are plotted by quantity and percentage, and results can be illustrated using tables and bar graphs.

4.2.4.2   Continuous data

Continuous data analysis based on descriptive statistics includes calculation of frequency distributions, range, mean, median, mode, standard deviation, coefficient of variation, and Pearson correlation coefficients to determine the interrelationships between the various body dimensions.

The objective of anthropometric analysis is to profile the demographic data and the continuous data in such a way that the overall patterns of body dimensions are described and one can distinguish between genders and different age groups for selection of key dimensions.

The next section deals with Stage 2—the sizing data analysis.

In this stage the objective is to divide the sample population into smaller groups composed of individuals who have similar key body dimensions. The center panel of Fig. 4.1 shows the phases of Stage 2, which consists of four steps (Steps 5–8). The analysis shown in Stage 2 is only one possible method of determining key dimensions and clustering the sample population. Besides the three methods shown here (PCA, cluster analysis, and decision tree analysis), other methods like bivariate analysis, neural networks, and artificial intelligence can also be used (Kim et al., 2018; Doustaneh et al., 2010).

Step 5 is multivariate analysis, the purpose of which is to test the sampling adequacy of the collected data. In Step 6, principal component analysis (PCA) is employed to reduce all the variables into significant components. In Step 7, cluster analysis is used to segment the sample subjects into homogenous groups with similar body shapes and sizes. In Step 8 the decision tree technique can be applied to classify sample subjects into groups based on profiles and to validate the cluster groups.

4.2.5   Step 5: Multivariate analysis

Prior to applying a PCA, a sampling adequacy test needs to be performed on the data to confirm the appropriateness of conducting PCA to ensure that the data can be factored well (Tabachnick and Fidell, 2007). In addition, Bartlett's test of sphericity can also be used to add a significant value to support the factorability of the correlation matrix obtained from the items.

4.2.6   Step 6: Principal component analysis (PCA)

The objective of using PCA is to reduce the number of variables and to cluster these variables into a more parsimonious and manageable number of groups. Parsimonious means to summarize most of the original information (variance) in a minimum number of components for prediction purposes (Pallant, 2001).

4.2.7   Step 7: Cluster analysis

Cluster analysis is an exploratory data analysis tool used to segment a population into homogenous subgroups. This means that each person in a group shares similar physical traits with others in the group and that people in one group differ from those in other groups.

4.2.8   Step 8: Classification analysis (decision tree)

Decision tree analysis is a data mining technique that is effective for classification (Lin et al., 2008). The classification and regression tree (CRT) technique can be used to verify and classify the sample population according to cluster groups; CRT is used where the data are continuous. The profile of the tree is useful when interpretation of the data set is required. By doing the classification analysis, important variables can be obtained, and a simple profile can easily be extracted from the tree diagram (Viktor et al., 2006).

The last stage described in Fig. 4.1 is Stage 3—the sizing system development.

4.2.9   Step 9: Size system development

The purpose of developing the sizing system is to create sizes for each cluster group that are appropriate to the individual group's range. Two important decisions must be made. The first is to estimate the size roll, which will accommodate most of the target population, and the second is to determine which samples go into the cluster groups obtained from the cluster analysis technique (Bairi et al., 2017). The goal is to accommodate as many people from the target population as possible using one intersize interval.

For the development of the sizing system, the following elements have to be calculated: size range, size interval, size scale, and size roll.

After the selection of the interval range, the classification profile obtained from the decision tree analysis is used as a guide to select samples matching the right body size and shapes. Using this profile the samples are classified according to the body sizes and shapes. The last step is to validate the efficiency and accuracy of the sizing system thus developed.

4.3   Body dimension profiling

Body dimension profiling is the first process of anthropometric data analysis. From the data, simple statistical analysis was conducted to analyze the data, and the data illustrated the mean of height, weight, and important girth measurements such as bust, waist, and hip. This table is important to give a general outlook of the data of certain population as it represents the real sizes and shapes of the population of each country. For the purpose of analysis and details of the anthropometric procedure to develop sizing system, a sample population of female aged 13–17 years old is used for demonstration.

4.3.1   Size and shapes of female (13–17 years old)

Table 4.2 summarizes the data of the critical body dimensions for upper and lower body garments. These tables represent statistics for the measured sample population of female (13–17 years old). From Table 4.2 the tallest teenage girls (13–17 years old) from the sample are 170.6 cm, and the shortest of height is approximately at 131.1 cm. The heaviest girl in is about 96.6 kg, and the thinnest weighs only at 20 kg. The SD of the height and other variables are lower as compared with the girls (7–12 years old), which means that the variation of height and other body dimensions in this age group is lower. Moreover, hip girth and bust girth showed higher variations as compared with height. The 5th to 95th percentiles of the measurement for the sample size are also shown in the table.

4.3.2   Growth distribution of female

In Fig. 4.2 the description of growth was illustrated. Growth trend for teenage girls showed that the growth in vertical body dimensions started from age 13 and becomes steady at the age of 15. From age 15 onwards the teenage girls have stopped growing. Nonetheless, for horizontal dimensions, there is slight growth from age 13 to 15 years old but becomes steady from age 15 and above.

Table 4.2 Extreme values, SD, and percentile values of anthropometric measurements for 13–17-year-old Malaysian female children

Body dimensions		Height (cm)	Weight (kg)	Bust girth (cm)	Hip girth (cm)	Waist girth (cm)
Minimum		131.1	20	58.5	65	49
Maximum		170.6	96.2	121.5	121.5	117
Percentiles	5th	142.61	33.51	69	76.10	55.74
	50th	154.25	47.95	79.5	88	64
	95th	165.01	76.54	98.5	108.44	86
Std. Deviation		6.80	13.07	9.60	9.69	9.80

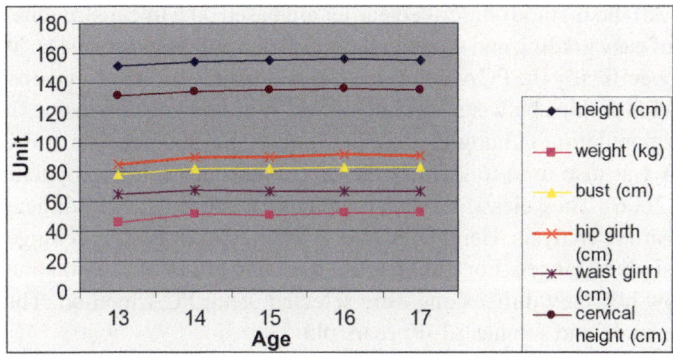

Fig. 4.2 Growth trend of female (13–17 years old).

4.4 Key dimensions and control dimensions' identification using multivariate data techniques

In this chapter, principal component analysis (PCA) technique is used to select key dimensions. The main objective of using PCA technique is to reduce all the variables into small sets of components, which then only significant components are analyzed and named. From the components, key body dimensions will be selected based on the highest relationship with the components. Before running the anthropometric data using the PCA technique, there is a need to do multivariate data examination shown later.

PCA was used to reduce the variables into new significant variables called principal components (PC). In 1985 Salusso-Deonier et al. (1985) developed a sizing system known as principal component sizing system (PCSS) using the PCA technique. However, their application of PCA differed from that of O'Brien and Shelton (1941). In previous research, PCA was applied to reduce the data, and then the components were analyzed for the selection of only one key dimension from each component. But in Salusso-Deonier et al.'s (1985) research, PCA components were applied to the classification of the population. Here the relationship of variables is looked upon in terms of the loading of factors of those variables on each component (correlation between a variable and a component).

If the loading is high, it means that the variable is highly associated with the component. This sizing analysis showed that two components were most important, namely, PC1 as laterality, associated mainly with body girth, arcs, and widths, and PC 2 as linearity, associated with heights and lengths. PCSS also known as principal component sizing system is based on partitioning the PC1 and PC2 geometrically (Salusso-Deonier et al., 1985). PC1 and PC2 behave like the control dimensions in conventional sizing system construction. The height and weight distribution is used to identify the PCSS sizes.

Salusso-Deonier et al. (1985) concluded that PCSS represents better relationship for the sample studied, which classified correctly 95% of subjects within <30 size categories. All the methods described earlier are based on a linear structure that has an advantage of easy grading and size labeling (O'Brien and Shelton, 1941). Multivariate analysis—specifically the PCA technique—is still widely used by many researchers to detect the relationships between variables and in turn find key dimensions by which to classify the population (Chung et al., 2007; Hsu et al., 2006; Gupta and Gangadhar, 2004). PCA was also used to identify the key dimensions for the population (Chung and Wang, 2006). They classified each population according to key dimensions using simple univariate analysis. Height dimension resulted in three height ranges, and bust girth gave six bust ranges. For this chapter, only one group of anthropometric data is used to show how key dimensions were selected using PCA method. The data used here are the male and female 13–17 years old

4.4.1 Multivariate data examination

There are two main measurements conducted in this section prior to PCA, namely, the validity and percentile analysis. All the 50 body dimensions taken from the samples (female age 13–17) are tested for validity and reliability. These measurements are examined prior to PCA. Bartlett's test of sphericity and Kaiser-Meyer-Olkin (KMO) test are performed to ensure the adequacy of sampling for PCA analysis. In addition, Cronbach's alpha is used to determine the consistency and unidimensionality of the data.

The KMO result shows all values are >0.9, and Bartlett's test of sphericity is highly significant since the observed significance level is 0.0000 ($P < .005$). For factor analysis, KMO is supposed to be >.5 and Bartlett test <.05. As shown in Table 4.3, KMO values are 0.973 for male (LaBat, 1987; Otieno and Fairhurst, 2000; Ashdown, 2014; Manuel et al., 2010; Kaiser, 1997) and 0.965 for female (LaBat, 1987; Otieno and Fairhurst, 2000; Ashdown, 2014; Manuel et al., 2010; Kaiser, 1997). According to Raykov (Raykov and Marcoulides, 2008), KMO >0.9 falls into the range of superb or marvelous. Thus it can be concluded that the relationship between the variables is very strong and factor analysis can be carried out. In addition, the results also confirm that the data show sampling adequacy, which means it is likely to factor well for PCA analysis.

4.4.2 Principal component analysis (PCA)

The body dimensions of each sample group are extracted using PCA and varimax rotation. This is a common technique applied on anthropometric data to describe variations in human body in a parsimonious manner by many previous sizing studies

Table 4.3 Sampling adequacy and reliability tests for all sample populations

Tests	Female
KMO of sampling adequacy	0.97
Bartlett's test of sphericity approx. Chi-squared df	31,437.00
sig. ($P < .005$)	1378
	0.000

(Field, 2005; Hsu et al., 2007). Parsimonious means the variation of body dimensions are described using the fewest principal components (PCs) possible (Salusso-Deonier et al., 1985).

4.4.3 The results of components analysis

The results of the extracted components for each sample group are recorded. In general, 50 components are extracted from each sample group to explain 100% of the variance in the data. The summarization of variables proves to be very good as the number of principal components is the same with the number of original variables. This indicates that the original information is not neglected as a variation factor (Salusso-Deonier et al., 1985).

According to Hair et al. (1998) the first few components should extract at least 50% of the variance to prove the usefulness of PCA technique. The study shows that female samples between 13 and 17, with less percentage value, are observed, at 54%. Thus this result indicates variance of >50% in the first component. This indicates that PCA technique is an effective method for this study to obtain a parsimonious solution in describing the variations of body shapes in this sample group.

Furthermore, the extracted factor should at least be 90% of the explained variance to show efficiency (Hair et al, 1998; Salusso-Deonier and DeLong, 1982). From this finding, 90% variance was explained by fourteen principal components for female samples age between 13 and 17. In contrast, study conducted by Hsu et al. (2006) shows that only 60% of total variance for 15 components are found in the study, which is considered low. However, in 2006, the same researcher improved her studies by obtaining a 60% total variance accounted for two components (Gupta and Gangadhar, 2004).

To reduce the numbers of components for a more parsimonious solution, the criterion of retaining components is applied, which are latent root, scree plot, and percentage of accumulated variance. Table 4.4 showed all the components that have been extracted with an eigenvalue greater than one (with bold values), which implies that these components are suitable to be retained.

For the female population aged between 13 and 17 years old, the result is shown in Table 4.4. In this table, six components (Components 1, 2, 3, 4, 5, and 6) show an eigenvalue >1. The first component shows eigenvalue of 27.0% followed by 7.5%, 2.0%, 1.9%,1.1%, and 1.0% in which the total rotation sums of squared loadings account for 81.1%.

For criteria 2, which is the scree plot, the results are shown in Fig. 4.3. An elbow where the curve merges into a straight line can be seen at Components 2 and 3 (which are marked by red circles). The component at that merge location is the point to stop. From Fig. 4.3, the elbow is obvious from PC2 and PC3, and thus PC1 to PC3 are retained.

The third criteria for retaining values depend on the percentage of accumulated variance. Based on the examination of eigenvalue and scree plot, three components are retained for sample's group with rotation sums of value of 76.7% for female (LaBat, 1987; Otieno and Fairhurst, 2000; Ashdown, 2014; Manuel et al., 2010; Kaiser, 1997).

Table 4.4 Principal component analysis extraction for female (LaBat, 1987; Otieno and Fairhurst, 2000; Ashdown, 2014; Manuel et al., 2010; Kaiser, 1997)

| | Initial eigenvalues | | | Rotation sums of squared loadings | |
Component	Total	% of variance	Cumulative %	% of variance	Cumulative %
1	27.0	54.0	54.0	39.3	39.3
2	7.5	15.0	69.0	20.8	60.0
3	2.0	4.1	73.1	7.5	67.6
4	1.9	3.7	76.9	6.6	74.1
5	1.1	2.3	79.1	3.6	77.8
6	1.0	2.0	81.1	3.4	81.1

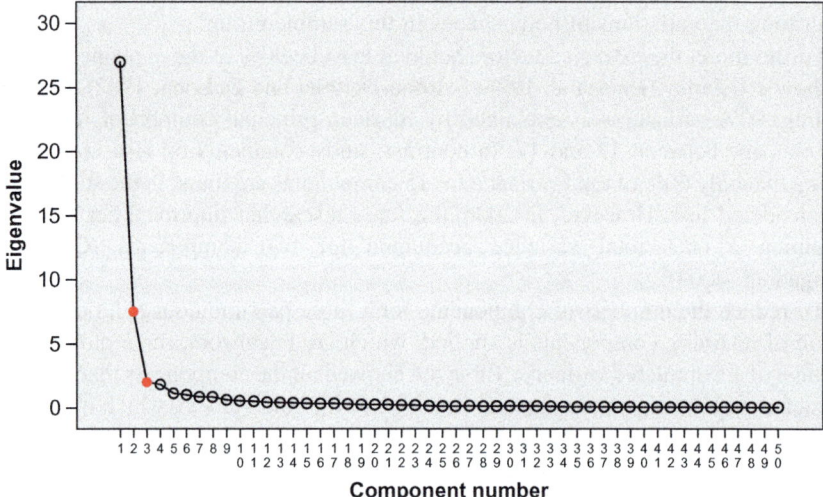

Fig. 4.3 Scree plot of female (13–17 years old).

It can be observed that the cumulative percentage is reduced when fewer components are retained. However, at this point, the numbers of components are not finalized yet until all the factor loadings of each component are examined, which clearly distinguish those variables that correlate highly with each component.

In general, previous studies have shown that body dimensions are accumulated into its own component, which can be interpreted as one type of measurements such as the length, girth, or width. These tables show the results of extracted variable components that show the factor loadings for all the 50 anthropometric dimensions.

After analysis of factor loading, the current findings yield two significant components, which consist of variables of the highest factor loading. Thus two components are retained, namely, the principal component 1. (PC1) and principal component 2 (PC2) for each sample groups. For sample group male and female (LaBat, 1987; Otieno and Fairhurst, 2000; Ashdown, 2014; Manuel et al., 2010; Kaiser, 1997), 44 variables and 42 variables, respectively, are grouped into two components. PCA technique has successfully achieved the goal of getting a parsimonious group of variables, which extracted almost all of the variables in two components.

The two retained components gave cumulative percentage as follows: female (LaBat, 1987; Otieno and Fairhurst, 2000; Ashdown, 2014; Manuel et al., 2010; Kaiser, 1997) with 60.0%. The present findings seem to be consistent with other research, which commonly retained two final components. However, the cumulative percentage for these two components was found differently in different studies ranging from 65% to 81% (Xia and Istook, 2017; Xinzhou et al., 2018).

Consequently, it was found that the first component (PC1) consists of all girth dimensions including bust girth, chest girth, upper arm girth, hip girth, and waist girth and few width dimensions like back width and shoulder width. The second component (PC2) consists of all length dimensions such as height, cervical height, upper arm length, and arm length. This finding is found similar to many previous sizing studies where two components represent the girth and length factors (Hsu, 2008, 2009; Beazley, 1998b).

4.4.4 The results of factor loadings analysis

In this section, all the body dimensions with high factor loading (≥ 0.75) are listed in Table 4.5 for one sample female group. The aim of this section is to select the key dimensions. It is apparent from this table that most of the variables are highly correlated to the individual components, which can be seen from the high factor loadings. Overall, almost all variables are reduced into two factors.

As can be seen from Table 4.5, upper arm girth and waist girth are distinguished as the strongest variables correlated to girth factor for male samples (age 13–17). In contrast, bust girth has the highest factor loading for female samples age 13–17. Under arm length and inside leg length are noted to have the highest factor loading correlated to length in males aged 13–17 as compared with arm length and hip height in females aged 13–17. Hence PCA analysis for each sample group is shown later.

4.4.4.1 Female samples (age 13–17)

For female samples, as can be seen from Table 4.5, 28 variables were found correlated to the girth and length component as compared with 26 variables in female samples. On the other hand, 16 and 15 variables are loaded on length girths for males and females, respectively.

Table 4.5 Principal component analysis with varimax rotation for females (13–17 years old)

Body types	Variables	PC 1 (girth)	PC 2 (length)
Upper body (girth)	Height		0.77
	Weight	0.93	
	1. Neck girth	0.81	
	2. Back width	0.81	
	3. Bust girth	0.93	
	4. Upper arm girth	0.93	
	5. Armscye girth	0.90	
	6. Elbow girth	0.85	
	7. Wrist girth	0.83	
Lower body (girth)	8. Waist girth	0.94	
	9. Hip girth	0.93	
	10. Thigh girth	0.92	
	11. Mid-thigh girth	0.91	
	12. Lower knee girth	0.91	
	13. Calf girth	0.88	
	14. Knee girth	0.84	
	15. Crotch	0.78	
Upper body (length)	16. Arm length		0.84
	17. Under arm length		0.82
	18. Cervical to breast point		0.8
	19. Neck shoulder to breast point		0.76
	20. Trunk circumference		0.77
Lower body (length)	21. Hip height		0.91
	22. Waist height		0.87
	23. Knee height		0.84
	24. Inside leg length		0.84
	25. Out leg length		0.83
	26. Thigh length		0.82
	Total proportion (%)	39.3	60.0

4.4.5 Identifying key dimensions

From the table earlier, PCA method confirms that those variables shown in Table 4.5 prove to be the significant dimensions often taken in any anthropometric survey. The key dimensions commonly acknowledged in other literatures are also found significantly in this study, namely, waist girth, bust girth, chest girth, and height. For example, bust girth has the highest factor loading for female sample age 13–17. However, in other studies, upper arm girth is noted as the strongest variable correlated to girth factor for male samples (age 13–17) that has not previously been described in any sizing studies. Moreover, under arm length and inside leg length are noted to have highest factor loading correlated to length in males aged 13–17 and as compared with arm length and hip height in females aged 13–17. Under arm length is also another variable found to have high relationship with length, which is not common to other sizing studies.

Overall the girth variables chosen as the key dimensions are chest and bust girths for the upper body. In contrast, hip girth is chosen for the lower body key dimensions. The selection of these body dimensions as the key dimensions confirms the findings from other research conducted in which the same common dimensions' girth is found to be significant as the key dimensions for the upper body while for the lower body the hip girth is chosen instead of waist girth. The hip was a better choice as compared with waist girth for some of these reasons; it was mentioned as a variable that has a variety of relationship with the upper and lower torso (Gordon, 1986). It was found out that the hip was a better selection for lower garment simply because this dimension cannot be easily adjusted after it has been made (Gordon, 1986; Otieno and Fairhurst, 2000). From previous studies it has been noted that the hip was a more stable measurement and found to have high correlation with girth component and therefore determined to be the key dimensions for the lower body (Ashdown, 1998).

In general, for example, Table 4.5, height was selected as the key dimensions for both the upper and lower body. Height is selected based on the high factor loading. The current study found that height represents almost 77%, 0.77 for length component for female samples between age 13 and 17. This is consistent with the result from correlation analysis in the other studies that shows height is very strongly correlated with other length variables. From the opinion of other researchers, height is a must to be incorporated in a sizing system (James and Stone, 1984). In addition, according to James and Stone, height was found an advantage especially for teenagers as it could be easily measured in retail shops where a height chart could easily be obtained. As for adults the body dimensions that can be considered the key dimensions are the chest, bust, waist, and legs. Height is also regarded as a better estimator of size rather than age for children as compared to adults (Simmons and Istook, 2003). Hence height is the most suitable dimension to cluster the samples into different groups.

All the key dimensions mentioned earlier are finally selected for the classification of the sample population. These key dimensions are selected based on the high relationship with the main body measurements, girth, and length. Furthermore, it has been mentioned in previous studies that key dimensions must be convenient to take (Winks, 1997). This indicates that the selected key dimensions should be the ones that can be measured easily and practical especially when it comes to children (Xinzhou et al., 2018). In addition, it is also being stated that if these measurements will be used as the size coding, the customers should be very familiar with the key dimensions, for example, chest girth or bust girth for the upper body as compared with upper arm girth.

The next step after determining the key dimensions will be to classify the sample population into the same sizes using the key dimensions. Key dimensions are selected because those body dimensions are significant, very convenient, and familiar for consumers to measure.

Table 4.6 Cluster groups for the upper body: female samples (age 13–17)

	Cluster 2	Cluster 3	Cluster 1
n	120	101	80
Mean height (cm)	150.1	159.0	155.2
Range (cm)	142.7–156.0	153.3–164.5	145.0–164.0
Mean bust girth (cm)	76.6	79.6	91.8
Range (cm)	69.5–86.7	70.6–86.7	85.7–98.5
Body type	Small	Medium	Large

4.5 Body size classification using cluster analysis

The objective of doing cluster analysis is to group the sample population into homogenous groups. Two variables were calculated using the PCA method and then were used as the key dimensions to group the sample subjects. Six separate cluster analyses were run generating participant cluster membership from two to seven grouping categories. Each K-means cluster result was evaluated to determine the ideal number of grouping categories. It is found that three and four are the most ideal because the cluster groups were distinct from each other. However, for this research, three groups were chosen and considered practical for size clustering of the children's wear. K-means cluster technique successfully extracts three distinct cluster groups (Table 4.6). Fig. 4.4 illustrated the distribution of height (y-axis) versus bust girth (x-axis) for all three clusters. The profile of each cluster is defined as this: Cluster 1 with 134 samples, tall stature with medium to big bust; Cluster 2 with 194 samples, average height with small to medium bust; and Cluster 3 with 197 samples, short with small bust.

4.5.1 Female sample size (13–17 years old): Clustering for upper body

Table 4.6 shows the distribution of females (age 13–17) according to height and bust girth, which is meant for the upper body. This table shows that most of the sample population falls under Cluster 2, small size. The second highest distribution of sample size is in Cluster 3, medium size. The last cluster group, Cluster 1, is the large size.

The division of female samples (age 13–17) is depicted in Fig. 4.4. Three clusters are evident. Cluster 1, large size, contains samples that are short to tall with large bust measurements. For Cluster 2, the samples belong to the small-size body type with short to average height and small to average bust. In Cluster 3, medium size, the samples are average to tall with small to average size bust.

Generally, from cluster analysis, three distinct groups were obtained. As can be seen from Table 4.7, the small-size cluster (Cluster 2) has most of the sample size, and most of this cluster is age 13. Ages are evenly distributed in the medium body type, whereas females age 16 are mostly found in the large-size cluster group.

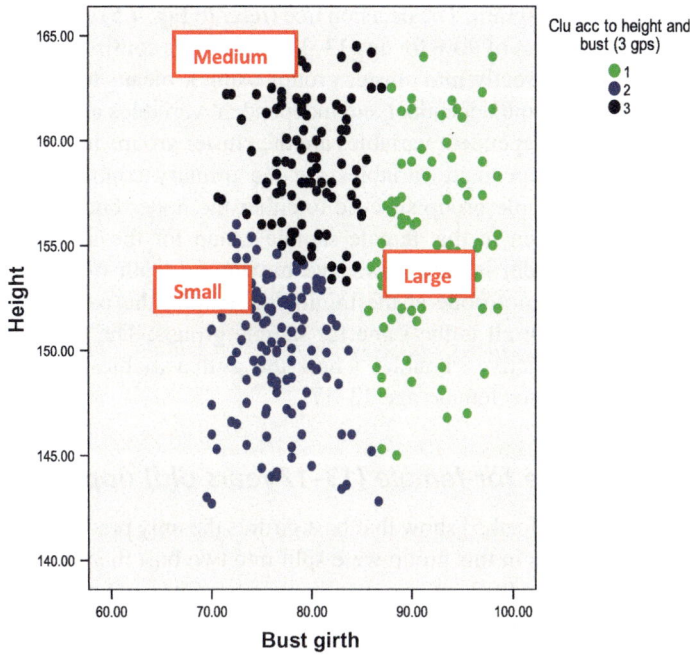

Fig. 4.4 Scatter plot of age groups according to height and bust for females age 13–17.

Table 4.7 Age versus cluster groups for the upper body (females age 13–17)

Age	Cluster according to height and bust (three groups)			Total
	Small	*Medium*	*Large*	
13	36	14	15	65
14	22	24	18	64
15	25	19	13	57
16	17	20	23	60
17	20	24	11	55
Total	120	101	80	301

The next step is to verify that these cluster groups are distinct from each other by using the decision tree technique.

4.6 Decision tree analysis

In this study the decision tree is obtained using the regression tree since all the data are continuous. The objective of doing decision tree is to verify whether the cluster groups are classified properly and the profile of the clusters can be easily and accurately obtained from the tree diagram. The decision tree (refer to Fig. 4.5) classified the cluster groups with correctness of 90% for age 13–17. Hence this confirms that the sample groups are classified correctly into cluster groups using k-means technique.

For this section the same dependent and independent variables are used to classify the cluster groups. The dependent variables are the cluster groups for each individual sample group. The independent variables are the primary control variables (key dimensions) of each sample group specific to either the upper body or lower body. The observations are seen in this female sample group for the upper body where the rule of this tree model is as follows: the maximum depth of the tree extends two levels beneath the root node at the minimum sample; the parent node is 100, and the child node is 50; all is the same for sample groups. The same observation is also seen in the predictive variables where the girth variables are identified as the predictive variables for female age 13–17.

4.6.1 Decision tree for female (13–17 years old) upper body

The results of the tree (Fig. 4.5) show that bust girth is the only predictor for the large size. The female samples in this group were split into two bust measurements. Those with bust measurements >86.3 cm are most likely to be clustered under the large size. Those with bust measurement <86.3 cm and >161.4 cm tall are likely to be grouped as the medium size. Those <154 cm tall with bust <86.2 cm are likely to be grouped under small size. The profile of the clusters is shown in Table 4.8.

This tree verified that the cluster groups were correctly segmented using the cluster analysis method. The next step is to develop sizing system based on these cluster groups.

4.7 Body size determination for sizing system

In this study the sizing system was designed according to the design limit that accommodated 90% of the population. Using design limit means taking into consideration the values that might have an impact on the total reliability and practicality of clothing production for manufacturers (Abtew et al., 2017). Designing according to limit also prevents sizes to be produced unnecessarily and impractical use for only few people. The sizing system is built according to the percentile values of 5th and 95th.

4.7.1 Female sizing system 13–17 years old for upper body garments

Fig. 4.6 showed the distribution of sizes according to the body types for female aged between 13 and 17 years old; the control variables for upper body are height and bust girth. The size interval for height is 8 cm and for bust and girth is 6 cm. The range for height in this population is 142–166 cm, which is the same for the upper and lower body. The range is divided into five subgroups using an interval of 6 cm, which is 142, 148, 154, 160, and 166 cm. For the bust the range is from 70 to 98 cm; this range

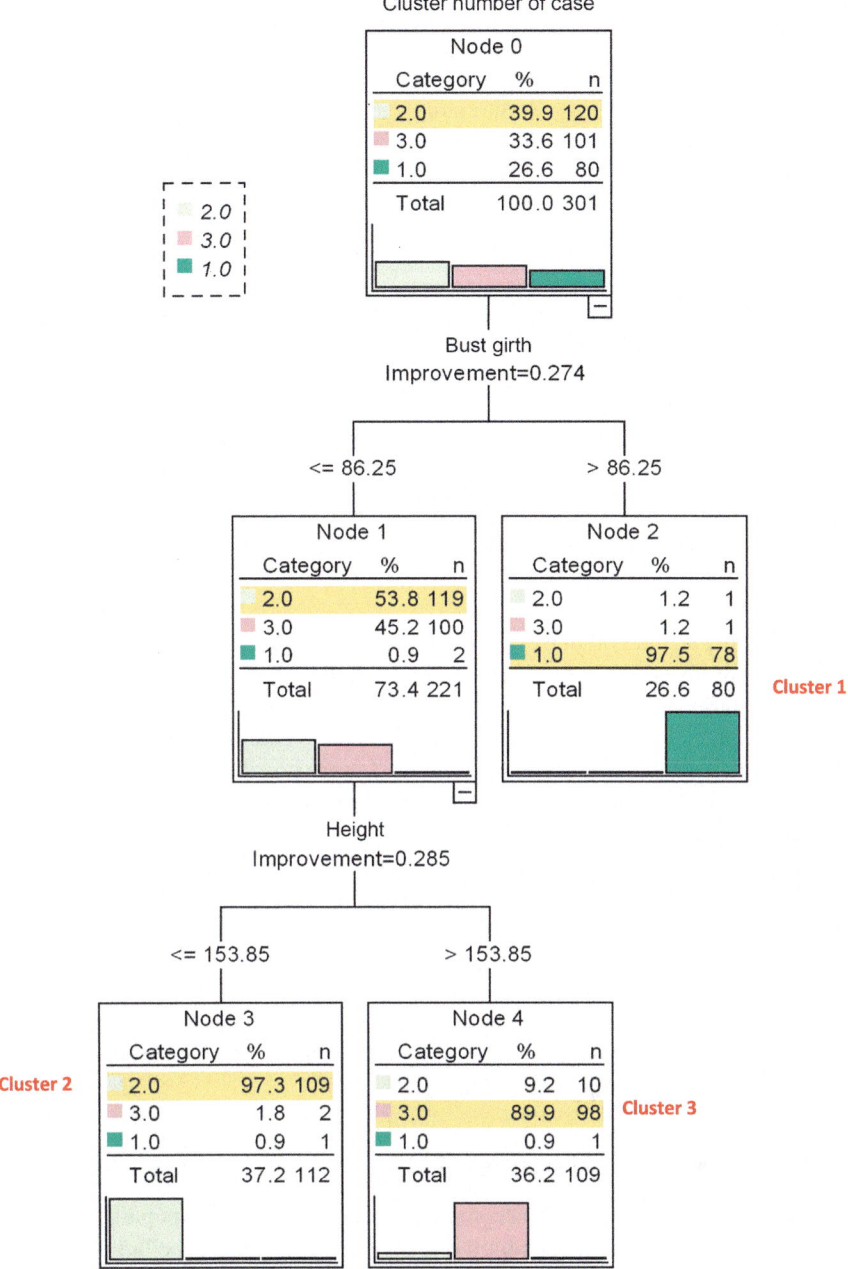

Fig. 4.5 Decision tree for female (13–17 years old) upper body.

Table 4.8 Profile of upper body types for female 13–17

Node	Body type	Classified rule	n =
1 and 3	Small	Bust girth ≤86.3 cm and height ≤153.9 cm	109
1 and 4	Medium	Bust girth ≤86.3 cm and height ≥153.9 cm	98
2	Large	Bust girth ≥86.3 cm	78

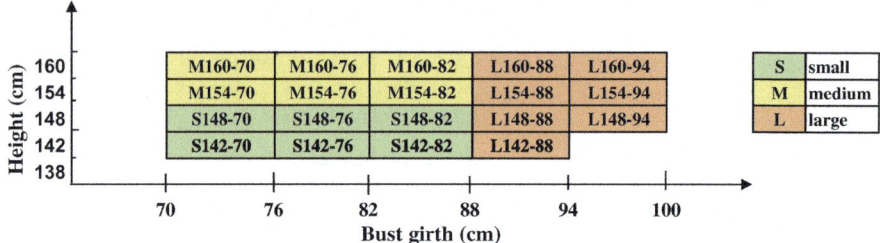

Fig. 4.6 Distribution graph of height versus bust girth for upper body, females (13–17 years old).

is divided using an interval of 6 cm: 70, 76, 82, 88, 94, and 100 cm. With the size interval, the height gives five subgroups, whereas the hip gives six subgroups totaling to 30 sizes designed to accommodate the female samples (age 13–17).

Fig. 4.6 indicates the outcome of the body type selection based on the classified rules. In total, there are 19 sizes suitable for the sample population. Six sizes are developed for small-size samples, six sizes for medium-size samples, and another seven sizes for large-size samples.

There are 30 sizes designed to accommodate this entire sample size, but the outcome indicates that only 19 sizes can cover the whole samples in which it means that the samples are found to be in one of these 19 sizes. This result is proof that the size interval is accurate and so does the rule that dictates which samples get into what sizes based on the IF/THEN statement. However, the coverage of the samples is not known yet as it is calculated in the next section. Lesser size rolls can indicate the size system is good if only the coverage of the samples is high above 80% as compared to lesser size rolls and less percentage of coverage.

4.7.2   Female size validation 13–17 years old for upper body

Table 4.9 proposes 19 sizes to accommodate the entire sample population for this group: six each for small and medium body types and seven for the large body type. Almost 40% of the samples were accommodated in the small body type, 37.5% in the medium, and 21.9% in the large. The result shows 99.3% coverage for this sample group. In addition, the aggregate loss for each body type is excellent as they show much lower values than the ideal, and the average aggregate loss of fit is excellent at 2.5 cm.

Table 4.9 Size validation for upper body, female (age 13–17)

Body type	Size group	n (%)	Mean aggregate loss (cm)
Small	6	120 (39.9)	2.6
Medium	6	113 (37.5)	2.7
Large	7	66 (21.9)	2.6
Total	19	299 (99.3)	2.6

Table 4.10 Size distribution for upper body, female (age 13–17)

Size roll	Body type	Bust girth	Height	n	%	Aggregate loss (cm)
		Key dimensions (cm)		Accommodation rate		
1	Small	70–75.9	142–147.9	12	4.0	2.7
2			148–153.9	32	10.6	2.7
3		76–81.9	142–147.9	10	3.3	2.6
4			148–153.9	45	15.0	2.6
5		82–87.9	142–147.9	7	2.3	2.5
6			148–153.9	14	4.7	2.6
7	Medium	70–75.9	154–159.9	14	4.7	2.9
8			160–165.9	5	1.7[a]	2.8
9		76–81.9	154–159.9	38	12.6	2.7
10			160–165.9	23	7.6	2.6
11		82–87.9	154–159.9	18	6.0	2.7
12			160–165.9	15	5.0	2.6
13	Large	88–93.9	142–147.9	2	0.7[a]	2.5
14			148–153.9	10	3.3	2.5
15			154–159.9	21	7.0	2.7
16			160–165.9	6	2.0	2.6
17		94–99.9	148–153.9	8	2.7	2.6
18			154–159.9	13	4.3	2.8
19			160–165.9	6	2.0	2.7
TOTAL				299	99.3	2.05

[a]Percentage is too low to be considered as a size to manufacture.

Table 4.10 shows an efficient size table as most of the samples were assigned to one of the proposed sizes. The efficiency of the table is confirmed high as only two sizes were <2% coverage. In addition, the aggregate loss for the entire size table shows an excellent value of 2.5 cm, well below the ideal value, meaning that the proposed sizes are accurate for this group.

Table 4.11 Summary table of size measurements for female (7–17 years old)

Gender—female Age range—13–17 years old		
Category		Upper
Key dimensions		Height and bust girth
Cover rate		**99.30%**
Size interval (cm)		8
Number of sizes in each body type	Big	7
	Medium	9
	Small	8
Total of sizes		**24**

4.8 Summary of sizing system

The whole process of developing the sizing system showed the result for the female aged between 13 and 17 years for upper body garment as shown in Table 4.11.

Table 4.11 finally presents the summarized results for the sizing system developed for teenage girls between 13 and 17 years old. From the table the coverage of each sizing system is high with >95%. The total number of sizes developed to cover 99% of the measured sample is 24. This is the final process of a sizing system in which the whole flow was demonstrated using a sample data. The success of a size system can be measured by statistical analysis and wearable analysis. The research conducted from this chapter has shown the statistical analysis only. The future development will be conducting the wearable test on the selected data to prove the sizing system is workable to be implemented.

4.9 Future trends

This chapter has successfully shown how to conduct analysis of the anthropometric data in which to develop an accurate garment sizing system for the RTW garment industry. The purpose of developing a sizing system is to produce garments in sizes that can accommodate a majority of customers within a set of fixed sizes. Without a sizing system that is able to generate an appropriate range of sizes for each size designation, producing good-quality well-fitting garments is impossible, and the overall objective of mass production cannot be met. When a sizing system is introduced, researchers must relate it to the understanding of fit. An accurate sizing system must be built based on actual anthropometric data as the understanding of body sizes and shapes is the only way to cater to the needs of consumers. The method of producing a sizing system impacts the efficiency of that sizing system.

The research that is based on theory and practice of sizing system development has progressively evolved from 1940 to the present. This means that a sizing system needs a lot of improvement to be efficient. For mass production purposes the sizing system must be flexible in nature; if there is a need to reduce the number of sizes to make mass production more efficient, the accommodation rate should not be negotiated. Every manufacturer understands that there is a need to accommodate most of their customers with a high-coverage sizing system.

Sizing system development began decades ago using only simple bivariate methods. As time went on the sizing systems evolved to incorporate many highly intelligent methods such as data mining, neural networks, and SOM (Ng et al., 2007). This was made possible by the experience of >70 years of exploring and understanding how to develop sizing systems that are efficient for manufacturers, customers, and retailers. It is amazing that the study of sizing systems is still ongoing today—it seems that the study of anthropometric data, sizing systems, and size designations never stops (Otieno et al., 2016; Yang et al., 2015). Many discoveries have been made, and the weaknesses of different sizing systems are being discussed to develop new methods as old methods become obsolete. It is anticipated that newer sizing systems using newer advanced analysis techniques will produce better and better sizing systems resulting in a greater goodness of fit for clothing customers (Hsu et al., 2010).

Moreover, researchers are still actively searching for ways in which to improve the efficiency of clothing sizing systems, many of which lie in the improvement of sizing validation. The key efficiencies lie between the accommodation rate and size roll. New advanced intelligent techniques are being applied to produce better sizing systems with higher accommodation rates and lower size rolls. New methodology like artificial neural networks and genetic algorithms is some of the intelligent machine learning techniques that may prove useful in creating a predictive model for finding the right sizes for the right body shapes (Adu-Boakye et al., 2012). This is very important to garment manufacturers, as a better model means that they can produce fewer sizes and still accommodate a majority of the population. This would yield tremendous benefits for both consumers and retailers, since such a model satisfies both parties.

Lastly the global issues of today have taken a toll on the whole concept of manufacturing garments. With the initiative of the sustainable development goals from UN, we are moving towards a better world. Most of the countries in the world are already realizing the concept of "sustainable fashion" or "ethical fashion." In driving towards this initiative, sizing system development is considered one of the most important agenda for every country if we are adopting the concept of ethical fashion. Understanding the body size and shapes of one nation will result to developing the right sizing system to ensure the garment fits the body well and thus will reduce the wastage of fabrics. Wastage of fabric can happen because the sizes that we develop for different master blocks are not based on the anthropometric data. Thus the garments will need alterations, and more fabric wastage will be produced (Adu-Boakye et al., 2012; Naveed et al., 2018).

4.9.1 The importance of fit and sizing system

Commercial market trends indicate that consumers today expect better fitting products and sizing systems that address niche market needs (Yang et al., 2015). A garment that catches your eyes in the store may seem like a good choice to you for that instance, but you cannot be sure until you try it on for the right fit and size. This situation is very common when it comes to buying the RTW clothing. Sometimes it could be frustrating and challenging. Sizes are not always consistent even within the same brand. Why does this happen? Each of RTW manufacturers has its own standard size measurements because of several factors. No matter what size is marked on the label, what really counts is how the garment fits. How do you recognize that your clothes don't fit you well? Check for wrinkles and bulges that indicate a poor fit. A garment that fits well is more likely to be worn and more comfortable (Hsu et al., 2010). Regardless of how one might perceive "fit" to be good or bad, it is impossible to meet the consumer's perception of good fit without a set of accurate measurement (Adu-Boakye et al., 2012). This calls for a good body sizing system that is based on human measurements considering the different body shapes, sizes, and proportions. As a conclusion, in this competitive market, retailers and manufacturers must be responsive to rapidly changing consumer demands. One of the consumers' demand today is the need of garment sizes that fit them well.

4.9.2 Benefits of fit and body sizing system

- Fit is the highest determinant for apparel purchase in today's market; therefore the ability of the manufacturers to produce a well-fitted garment will definitely enhance sales.
- Manufacturers will be more successful in this increasingly competitive global environment by offering quality clothing with good fit and size.
- Customers no longer need to spend long hours trying garments for good fit or finding the right size when shopping for clothes.
- Shopping for clothes will be much easier and fun; shoppers do not have to guess the size and fit because with the right size, it should fit well.
- Improved customer satisfaction will be resulted to less exchange and returns of clothing.
- A good fit will lead to satisfied customers and therefore will ensure customer loyalty in the long run and ultimately will enhance business performance of the retailers.
- Majority of shoppers are willing to pay higher for quality clothing with good fit and size.

4.10 Sources of further information and advice

1. The Limited. The One Fit. Retrieved from: http:/www.thelimited.com/womens-clothing/jeans-denim/jeansdenim-shop-by-legshape. (Accessed 30 July 2016).
2. Nordstrom Incorporated. Women's Dress Fit Guide. Retrieved from: http:/shop.nordstrom.com/c/womensdresses-fit. (Accessed 30 July 2016).
3. Wells, B., 1983. Body and Personality. Longman, London, 1983, pp. 54–58. Sheldon, W.H., Stevens, S.S., Tucker, W.B., The Varieties of Human Physique: An Introduction to Constitutional Psychology. Harper & Brothers, New York, 1940.
4. Friedman, A., 2019. Alvanon, human solutions aim to update sizing standards to fit today's teens. Retrieved from: www.sourcingjournal.com. (Accessed 12 March 2019).
5. Applegate, J., 2019. Virtual fit is dead. www.linkedln.com. (Accessed 12 February 2019).
6. Hayes, E., 2019. Apparel fit and inclusivity. www.linkedln.com. (Accessed 12 February 2019).
7. Scott, E., 2018. Addressing garment fit across a global population. www.fashionshouldempower.com.

References

Abtew, M.A., Yadav, M., Singh, N., 2017. Anthropometric size chart for ethiopian girls for better garment design. J. Fash. Technol. Text. Eng. 5 (2), 1–11.

Adu-Boakye, S., Power, J., Wallace, T., Chen, Z., 2012. Development of a sizing system for Ghanaian women for the production of ready-to-wear clothing. In: Paper Presented at the 88th Textile Institute World Conference 2012, Selangor, Malaysia, 15–17th May 2012.

Ashdown, S.P., 1998. An investigation of the structure of sizing systems. J. Cloth. Sci. Technol. 10 (5), 324–341.

Ashdown, S.P., 2014. Creation of ready-made clothing: the development and future of sizing systems. In: Faust, M., Carrier, S. (Eds.), Designing Apparel for Consumers: The Impact of body Shape and Size. Woodhead Publishing, Philadelphia, pp. 17–34.

Bairi, S.B., Salleh, M., Syuhaily, N., Osman, S., 2017. Development of female adolescents clothing sizing based on cluster analysis classification. In: Paper Presented at the International Business Management Conference (IBMC 2017), Langkawi.

Bartlett, J.E., Kotrlik, J.W., Higgins, C., 2001. Organizational research: determining appropriate sample size in survey research. Inf. Technol. Learn. Perform. J. 19, 43–50.

Beazley, A., 1998a. Size and fit: formulation of body measurement tables and sizing systems—Part 2. J. Fash. Mark. Manag. 3 (1), 260–284.

Beazley, A., 1998b. Size and fit. Formulation of body measurements tables and sizing systems. J. Fash. Mark. Manag. 2, 1998.

Brownbridge, K., Gill, S., Grogan, S., Kilgariff, S., Whalley, A., 2018. Fashion misfit: women's dissatisfaction and its implications. J. Fash. Mark. Manag.: Int. J. 22 (3), 438–452.

Chung, M.J., Lin, H.F., Mao, J.J., Wang, J., 2007. The development of sizing systems for Taiwanese elementary-and high-school students. Int. J. Ind. Ergon. 37, 707–716.

Chung, M.J., Wang, M.J., 2006. The development of sizing systems for school students. In: 36th International Conference on Computers and Industrial Engineering, Taipei, Taiwan.

Doustaneh, A.H., Gorji, M., Varsei, M., 2010. Using self-organization method to establish nonlinear sizing system. World Appl. Sci. J. 9 (12), 1359–1364.

Field, A., 2005. Factor analysis on SPSS. In: C8057 (Research Methods II): Factor Analysis on SPSS, pp. 1–8.

Gill, S., 2015. A review of research and innovation in garment sizing, prototyping and fitting. Text. Prog. 47, 1–85.

Goldsberry, E., Shim, S., Reich, N., 1996. The development of body measurement tables for women 55 years and older (part 1). Cloth. Text. Res. J. 14, 108–120.

Gordon, C.C., 1986. Anthropometric sizing and fit testing of a single battle dress uniform for US army men and women. In: Barker, R., Coletta, G. (Eds.), STP900-EB Performance of Protective Clothing. ASTM International, West Conshohocken, PA, pp. 581–592.

Gupta, D., Gangadhar, B.P., 2004. A statistical model for developing body size charts for garments. Int. J. Cloth. Sci. Technol. 16, 458–469.

Hair, J.F., Anderson, R.E., Tatham, R.L., Black, W.C., 1998. Multivariate Data Analysis, fourth ed. Prentice Hall, Upper Saddle River, NJ.

Hrzenjak, R., Dolezal, K., Ujevic, D., 2013. Development of sizing system for sizing aged 6 to 12 years in Croatia. Coll. Antropol. 37 (4), 1095–1103.

Hsu, C.H., 2008. Applying a bust -to-waist ratio approach to develop body measurement charts for improving female clothing manufacture. J. Chin. Inst. Ind. Eng. 25, 215–222.

Hsu, C., 2009. Data mining to improve industrial standards and enhance production and marketing: an empirical study in apparel industry. Expert Syst. Appl. 36, 4185–4191.

Hsu, H.C., Lin, H.F., Wang, M.J., 2006. Development female size charts for facilitating garment production by using data mining. J. Chin. Inst. Ind. Eng. 24, 245–251.

Hsu, H.C., Lin, H.F., Wang, M.J., 2007. Developing female size charts for facilitating garment production by using data mining. J. Chin. Inst. Ind. Eng. 24, 245–251.

Hsu, C.H., Tsai, C.Y., Lee, T.Y., 2010. Neural network to develop sizing systems for production and logistics via technology innovation in Taiwan. In: New Aspects of Applied Informatics, Biomedical Electronics & Informatics and Communications. ISBN 978-960-474-216-5, pp. 421–424.

Hutcheson, G.D., Sofroniou, N., 1999. The Multivariate Social Scientist: Introductory Statistics Using Generalized Linear Models. Sage, Thousand Oaks, CA.

James, R., Stone, P., 1984. Children's Wear Sizing Survey. Clothing and Allied Products Industrial Training Board, Leeds, UK.

Kaiser, S.K., 1997. The Social Psychology of Clothing: Symbolic Appearance in Context, second ed. Fairchild Publications, New York, pp. 108–109.

Kim, N., Song, H.K., Kim, S., Do, W., 2018. An effective research method to predict human body type using an artificial neural network and a discriminant analysis. Fibers Polym. 19 (8), 1781–1789.

LaBat, K.L., 1987. Consumer Satisfaction/Dissatisfaction with the Fit of Ready-to-Wear Clothing. Unpublished doctoral thesis, University of Minnesota.

Lin, H.F., Hsu, C.H., Wang, M.J., Lin, Y.C., 2008. An application of data mining technique in developing sizing system for army soldiers in Taiwan. WSEAS Trans. Comput. 7, 245–252.

Makhanya, P.B., de Klerk, H.M., Adamski, K., Mastamet, A.M., 2014. Ethnicity, body shape differences and female consumer's apparel fit problems. Int. J. Consum. Stud. 38 (2), 183–191.

Manuel, M.B., Connell, L.J., Presley, A.B., 2010. Body shape and fit preference in body cathexis and clothing benefits sought for professional African-American women. Int. J. Fash. Des. Technol. Educ. 3 (1), 25–32.

McCulloch, C.E., Paal, B., Ashdown, S.P., 1998. An optimization approach to apparel sizing. J. Oper. Res. Soc. 49, 492–499.

Naveed, T., Zhong, Y., Hussain, A., Babar, A.A., Naeem, A., Iqbal, A., Saleemi, S., 2018. Female body shape classifications and their significant impact on fabric utilization. Fibers Polym. 19 (12), 2642–2656.

Ng, R., Ashdown, S.P., Chan, A., 2007. Intelligent size table generation. In: Proceedings of the Asian Textile Conference (ATC), 9th Asian Textile Conference, Taiwan.

O'Brien, R., Shelton, W.C., 1941. Body measurements of American boys and girls for garment and pattern construction. In: US Department of Agriculture Miscellaneous Publication 366, M. P. 454. US Department of Agriculture, Washington, DC, p. 141.

Otieno, R., Fairhurst, C., 2000. The development of new clothing size charts for female Kenyan children. Part I: using anthropometric data to create size charts. J. Text. Inst. 91, 143–152.

Otieno, A.O., Mehtre, A., Mekonnen, H., Lema, O., Fera, O., Gebeyehu, S., 2016. Developing standard size charts for Ethiopian men between the ages of 18-26 through anthropometric survey. J. Text. Apparel Technol. Manag. 10 (1), 1–10.

Pallant, J., 2001. SPSS Survival Manual. Open University, Maidenhead.

Pei, J., Park, H., Ashdown, S.P., Vuruskan, A., 2017. A sizing improvement methodology based on adjustment of interior accommodation rates across measurement categories within a size chart. Int. J. Cloth. Sci. Technol. 29 (5), 716–731.

Raykov, T., Marcoulides, G.A., 2008. An Introduction to Applied Multivariate Analysis. Routledge.

Salusso-Deonier, C.J., DeLong, M.R., 1982. A multivariate method of classify in body form variation for sizing Women's apparel. Cloth. Text. Res. J. 4, 38–45.

Salusso-Deonier, C.J., DeLong, M.R., Martin, F.B., Krohn, K.R., 1985. A multivariate method of classifying body form variation for sizing women's apparel. Cloth. Text. Res. J. 4, 38–45.

Scheaffer, R.L., Mendenhall, W., Ott, L.R., 2005. Elementary Survey Sampling. Cengage Learning.

Simmons, K.P., Istook, C.L., 2003. Body measurement techniques: a comparison of three-dimensional body scanning and physical anthropometric methods for apparel application. J. Fash. mark. manag. 7, 306–332.

Song, H.K., Ashdown, S.P., 2013. Female apparel consumers? Understanding of body size and shape. Cloth. Text. Res. J. 31 (3), 143–156.

Tabachnick, B.G., Fidell, L.S., 2007. Using Multivariate Statistics, fifth ed. Allyn and Bacon, Boston, MA.

Viktor, H.L., Paquet, E., Guo, H., 2006. Measuring to fit: virtual tailoring through cluster analysis and classification. In: Knowledge Discovery in Databases: PKDD 2006. vol. 4213. Springer, Berlin/Heidelberg, pp. 395–406.

Winks, J.M., 1997. Clothing Sizes: International Standardization. The Textile Institute, Manchester.

Xia, S., Istook, C., 2017. A method to create body sizing systems. Cloth. Text. Res. J. 35 (4), 235–248.

Xinzhou, W., Kuzmichev, V., Peng, T., 2018. Development of female torso classification and method of patterns shaping. Autex Res. J. 18 (4), 419–428.

Xu, B., Huang, Y., Yu, W., Chen, T., 2002. Three-dimensional body scanning system for apparel mass customization. Cloth. Text. Res. J. 26 (3), 227–252.

Yang, J., Kincade, D.H., Chen-Yu, J.H., 2015. Type of apparel mass customization and levels of modularity and variety: application of theory of inventive problem solving. Cloth. Text. Res. J. 33 (3), 199–212.

Zakaria, N., 2011. Sizing system for functional clothing- uniforms for school children. Indian J. Fibre Text. Res. 36 (4), 348–357.

Further reading

Pedhazer, E.J., Schmelkin, L.P., 1991. Measurement, Design, and Analysis: An Integrated Approach. Lawrence Erlbaum Assoc, Hillside, NJ.

Apparel size designation and labeling

M.-E. Faust
Philadelphia University, Philadelphia, PA, United States, Université du Québec à Montréal (Uqàm), Montreal, QC, Canada

5.1 Introduction

Ever since people began wearing ready-to-wear clothes, retailers have desired to help customers in finding the garment that would suit them best. One common way of doing this was to attach to the garment what is called today the size designation. As helpful as it could be, it seems that size designation doesn't provide consumers that much help anymore. According to some researchers, women, particularly in Western society, have a hard time finding the garment that suits them best with the sole information of the actual size designation. One reason is because size designation, which provides a single number 8, 10, 12, etc., was initially supported by underlying body measurements and body shapes, but the meaning seems to have been lost over the years. Additionally, at the beginning of the ready-to-wear era, women knew that ready-to-wear garments didn't exactly meant "ready-to-wear" but also meant that it may require alterations, which most women knew how to do at the time.

This being mentioned, it seems obvious that the garment industry needs to adjust for three major reasons. First, it appears that people have lost knowledge of what the measurements behind the size number are. Second, women nowadays take for granted the meaning of ready-to-wear, and only a few are able and willing to alter their garments. Lastly, many companies are now distributing worldwide, which makes it even more challenging for both parties: retailers/brands from different parts of the world and consumers from different ethnicities, with different shapes, or living in a different geographic area. The solution may reside in a more detailed size designation. This chapter will start by covering briefly how size designations were developed particularly in North America and what is the perceived value today. Then, it will present what could be, according to our research, an interesting global size designation and how it could be implemented. This goes without mentioning one of the most important challenges: what it should look like and what should be written on the size designation.

5.2 The importance of size designations

It is not rare nowadays to read that size designations in the United States have not evolved properly to serve their intended purpose, to help consumers find garments that suit them best. The same is true in Canada where those apparel size designations were

Anthropometry, Apparel Sizing and Design. https://doi.org/10.1016/B978-0-08-102604-5.00005-6

defined based on the anthropometric database of the United States of America. One couldn't argue that when garment specification measurements are done using forcing data, it is not surprising that there are unhappy shoppers. But why is it so in America? To get a better understanding, we reviewed the evolution of sizing and the size designation systems from the United States of America and those of its neighboring country, Canada. We provide details as to why Canadians' sizing system was calqued on US anthropometric data. Moreover, we provide information such as the time period when the national survey was conducted and the sample used.

5.2.1   The development and evolution of size designations in North America

Until the 18th century, women's garments were custom-made (Fan et al., 2004). The fit was personalized (Workman, 1991). The first nested patterns for women's wear appeared in 1820–40 (Kidwell, 1979) and the grading systems years later (Bryk, 1988). Ebenezer Butterick and James McCall were the first to market nested patterns in the 1860s, initiating the ready-to-wear industry (Burns and Bryant, 2002). The key measurement points were defined then as the bust, selected because it had proven to be useful for Europeans (Workman, 1991), and the waistline. Paper patterns were graded on the bust circumference for blouses and dresses, and the waistline was used for the skirts (Kidwell, 1979 as cited in Schofield and Labat, 2005). Most bodice measurements were based on the bust. During that era, a proportional grading system was used in patterns to create different sizes. Workman (1991) shows that the basic sample was graded at "36," which meant that it was suitable for a 36-in. (91.44 cm) woman's bust. It was then adjusted with 2-in. (5.08 cm) linear decrements or increments (Ashdown, 1998). The proper fit of the finished garment depended on the dressmaker's skill (Schofield and Labat, 2005).

5.2.2   Size designation according to the identification numbers in catalogs

Among the first to offer and distribute garments, having a big impact on today's size designation, were catalog distributors. Thus at the very beginning of the 20th century, the *Sears Roebuck and Co. Catalog* offered garments with size designations ranging from 32 to 42. These were referring to 32-in. (81.28 cm) to 42-in. (106.68 cm) bust circumference (Workman, 1991). In addition, it stated that a garment with a 32-in. (81.28 cm) bust size designation should be suitable for young ladies of 14 years of age; a 34-in. (86.36 cm) size designation should be suitable for young ladies aged 16 years, and a 36-in. (91.44 cm) size designation should be suitable for young ladies of 18 years (Sears Roebuck and Co. 1902 Catalogue, 2002). It also specified that these garments were suitable for young ladies since the construction was different from those for women (Workman, 1991). They also associated 14-, 15-, and 16-in. (35.5, 38.1, and 40.6 cm, respectively) neck measurements to the bust measurements. A 14-in. neck, for example, should be equivalent to a 32–34-in. bust. One should know

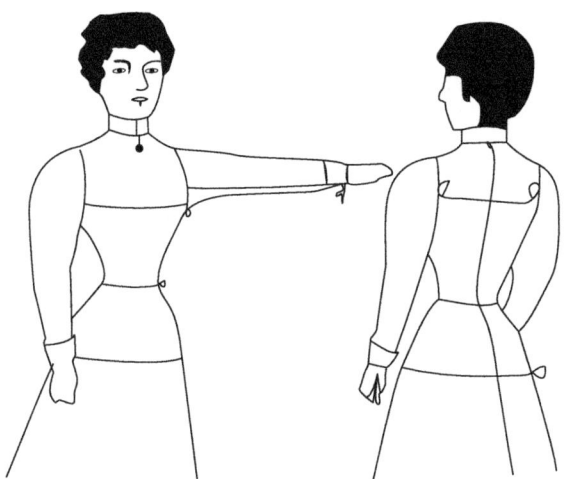

Fig. 5.1 An illustration from the 1902 Sears Roebuck catalog.

that ladies' skirt size designations varied from a 22-in. (55.88 cm) to a 29-in. (73.66 cm) waist. The size designation reflected one or the other: (1) the critical body measurement points and the age of the wearer (Swearingen, 1999 in Fan et al., 2004) as was the case for young children or (2) the associated neck size. The idea was to make sure a woman at home would measure herself and compare her measurements with the ones shown in the catalog index (see Fig. 5.1).

In the United States, besides the *Sears Roebuck and Co. Catalog,* ready-to-wear was also sold through catalogs such as *Montgomery Ward* and in the nascent urban brick-and-mortar department stores. Canadian retailers followed the trend and also distributed millions of their catalogs throughout Canada (Bernier et al., 2003, p. 77).

5.2.3 The need for size designation for ready-to-wear clothing

In the decades following World War I saw the introduction of department stores selling ready-to-wear all through North America. In cities like Philadelphia, New York, or Montreal, department stores such as Macy's (NYC); Wanamaker (Philadelphia); or Dupuis Frères, T. Eaton, or Simpson (Montreal, Canada) started to introduce ready-to-wear, impacting on people's shopping habits. According to Kidwell (2001) the modern age of apparel production and the democratization of clothing had begun. Burns and Bryant (2002) stated that "separates for women" worn by young girls (two-piece outfits combining a blouse and a skirt) created by Gibson became very popular and thus had a large influence on the women's ready-to-wear industry. Although ready-to-wear was a cheap interpretation of the current fashion trend, women followed this trend even if, as mentioned earlier, they knew they would need to alter these garments to obtain an appropriate fit (Cooklin, 1990). Each manufacturer did its best to develop its grading system, but lacking adequate scientific data made it difficult for everyone (O'Brien and Shelton, 1941). Schofield and Labat's (2005) studies reveal that each manufacturer reinvented the process although one point seems to have been consistent: the key position where to measure. Most of them agree on the bust, waist, and hips. Because there were no size designations developed yet, manufacturers used their own, which were based on their predecessors (Gould-Decauville et al., 1998).

5.2.4 Mass production and the establishment of standards

At the end of the 1930s, mass production for armed forces' uniforms triggered the establishment of standards in many areas (O'Brien and Shelton, 1941). According to O'Brien and Shelton (1941, p. 1) "no scientific study of body measurements used in the construction of women's clothing has ever been reported. As a result, there were no standards for garment sizes." Meanwhile, merchandise returns in the stores due to poor fit were high, and it was not unusual for the necessary alterations to increase the cost of a ready-to-wear garment by as much as 25% of its original cost (Winks, 1997). Mass production, variations in measurements, and size designation, in addition to the ones presented in the catalogs, all led to a high percentage of returns and the need for a standardized sizing system (Yu, 2004). Therefore the *Women's Measurements for Garment and Pattern Construction* (WMGPC) was conducted. Indeed, between July 1939 and June 1940, >10,000 women were measured (14,698 women exactly). The survey took place in eight states. And only White Caucasian women were used for this anthropometric survey. The results were then published by the *US Department of Agriculture* in 1941. The WMGPC reported the objective was "to provide measurements which could be used for improving the fit of women's garments and patterns." For the first time, anthropometric measurements were used to help the apparel industry standard sizes and size designation (O'Brien and Shelton, 1941; Goldsberry et al., 1996a, b; Chun-Yoon and Jasper, 1996; Ashdown, 1998; Workman and Lentz, 2000; Burns and Bryant, 2002; Fan et al., 2004; Schofield and Labat, 2005). More importantly the survey report stated that the data would best be used for sizing charts divided, among other things, into "short," "regular," and "tall."

5.2.4.1 Development of size designations

Body measurements of the national survey were compiled in order to develop size designation and to facilitate commercial communication. These resulted from the cooperation of various interest groups, which fulfilled a perceived need within the industry (Kadolph, 1998). Furthermore, in 1945, the *Association for Mail-order Sales* recommended a standard way for labeling size designations for the garment industry. It was in 1958 that the US Department of Commerce published the *Body Measurements for the Sizing of Women's Patterns and Apparel*, the commercial standard *"CS215-58."* Although interesting and useful the statement was as follows: "The adoption and use of a commercial standard is voluntary." Then, its primary and secondary goals were:

1. To provide standard classification, size designations, and body measurements for consistent sizing of women's ready-to-wear apparel (misses', women's, juniors', etc.) for the guidance

of those engaged in producing or preparing specifications for patterns and ready-to-wear garments. The measurements given in this standard are body, not garment, measurements.
2. To provide the consumer with a means of identifying her body type and size from the wide range of body types covered and enable her to be fitted properly by the same size regardless of the price, type of apparel, or manufacturer of the garment (US Dept. of Commerce, 1958, p. 1).

The CS215-58 reports the scopes as follows: four classifications of women: "misses," "women," "half sizes," and "junior." This was followed by groups defined as "short," "regular," and "tall." Then, again with four subgroups within each of these groups: "bust-hip," "slender," "average," and "full." The report itself proposed various possible applications, definitions of the measurement points, measuring methods, sizing charts, and the percentage of women in each of these classes. The size number and symbols were combined to make the complete size designation. For example, "14T −" would refer to a size 14 bust, T for tall in height, and "−" for slender hip type, whereas a size designation of "14R" would mean size 14 bust (with its underlying measurements) and R for regular in height and an average hip type or again "14S +," which would refer to a size 14 bust, S for short in height, and "+" for full hip type. One last detail of the report was that the junior classification was based upon interpolations of portions of the data used in the development of the misses' classification and therefore has traditional odd numbers for size designations. So instead of referring to a size 14 as for misses', it would be a junior size 13. It was then recommended that in order to assure purchasers that garments conform to this system, such garments be identified by a sticker, tag, or a hanger or other label carrying the type of information presented earlier: 14T − or 14R or 14S +. Soon after the CS215-58 was developed, several countries built their sizing standards or published reports on the subjects. The *BS1345* of the *British Standards Institution* was published in 1945; a survey by the *British Board of Trade* stated the need for 126 sizes to cover its female population. In 1950 the *DS923* of *Denmark Standards Association* came out with similar recommendations. In 1954 an anthropometric study was conducted by the *Polish Academy of Science*. Between 1954 and 1959 the United Kingdom provided its report on the anthropometric measurements of military personnel. In 1957 the USSR conducted a survey, etc. Canada was no exception in publishing its report, but it was based on the American population survey. In 1968, 17 countries formed the *International Organization for Standardization (ISO)* and implemented the TC133 "technical committee" entitled "sizing systems and designations." They believed that commercial standards would better serve their purpose if common to all countries. Back in the United States, the *Voluntary Product Standard (PS 42-70)* for pattern development and grading (with increments of 1 in. in circumference and 1½ inches in height measurements for each size) was updated and published in 1971 "as a revision of the CS215-58" (US Dept. of Standards, 1971, p. 1). At the very beginning, it states (par. 1) that:

The objective of a Voluntary Product Standard is to establish requirements that are in accordance with the principal demands of the industry and, at the same time, are not contrary to the public interest.

Therefore the classifications that were defined earlier such as 14S+ or 13T− were now modified/simplified. The new standard included body measurements for four classifications as follows: juniors' misses', women's, and half sizes for shorter women and subclassifications as detailed below:

- juniors' petite sizes ranging from 3P to 15P
- juniors' sizes ranging from 3 to 7
- misses' petite sizes ranging from 8P to 18P
- misses' sizes ranging from 6 to 22
- misses' tall sizes ranging from 10T to 22T
- women's sizes ranging from 34 to 52
- half sizes ranging from 12½ to 26½

In addition, it was written that standards were subject to review at any time and that it remained the responsibility of the users to judge its suitability for their particular purpose. Let's now have an overview of what happened in Canada.

5.2.5   Canadian General Standards Board (CGSB)

The Canadian General Standards Board (CGSB) has, over the decades, also produced size designations linked to data charts. It is stated at the very beginning of the report that:

> The principal objects of the Council are to foster and promote voluntary standardization as a means of advancing the national economy, benefiting the health, safety and welfare of the public, assisting and protecting the consumer, facilitating domestic and international cooperation in the field of standards. (CAN/CGSB-49.203-M87, p. 1).

The CGSB also wrote that:

> This standard describes the abridged Canada Standard sizing system which may be used as a guide in choosing the sizes of women's wearing apparel. The standard contains a selection from the complete system of sizes which is of greatest commercial interest. The sizes are identified by code numbers that correspond as closely as possible to current trade practice. Size identifications based on size indicator body dimensions are also given (CAN/CGSB-49.203-M87, p. 1).

One can also read in the CAN/CGSB-49.201-92 that the source measurements of the database that served to do the Canadian women apparel size designations were taken "a few years ago on a population of about 10,000 American women aged between 18 years old and 80 years old." In the footnote, it is written that the body measurements were taken in 1939 and 1940 in the United States, Miscellaneous Publication No. 454, *US Department of Agriculture* in 1941. Basically the subjects who served for the establishment of the Canadian size designations were the female sample that served to generate the American size designation. The CGSB added that a label with size number could be accompanied with a few key measuring points such as bust and waist and that these girth measurements should be in centimeters. It also promoted the use of a pictogram to support the size designation with specific body measurements. It proposed horizontal measurements for garments such as bust girth, waist girth, and hip girth and in addition to inseam pants.

5.2.6 Pictograms, body chart standards, and other proposed size designation systems

In the 1970s ISO developed and proposed a new way of presenting size designations. Again, it was based on key body measurements, but this time, it included a body pictogram. One of the biggest challenges of the time was which system should be used for size designation: centimeters or inches? This question was never answered, and the adoption of the pictogram never came to life. Years later, other developed countries came up with their own body chart standards: Switzerland in 1972 and *PC3137* and *PC3138* in the USSR (1973). Not long after, similar systems were proposed in Germany (1983); measurements of 9402 subjects were taken, and they concluded that 57 sizes were needed to cover 80% of their population. Such standard body charts would have been too cumbersome to be useful (Yu, 2004). Since then, from time to time, surveys were updated. It was the case in the United States as the relevance of sizing charts for market segments such as women aged 55 and over was questioned. Six thousand American women aged 55 and older were measured, which served to develop the *ASTM D5586* in 1995.

5.2.7 Updated anthropometric data

More recently, many people working in the industry started to feel a need to update national anthropometric data. Several major initiatives begun in the 1990s using the new *3D body scanner* technology to accomplish this task. Thousands of volunteer subjects of all ages were scanned in Asia, Europe, and America. Between 1992 and 1994, many subjects were scanned in Japan. From 1999 to 2002, many were scanned in the United Kingdom. At the beginning of the new century, the same was done in the United States. Some participants/sponsors who funded the project used the database to update their own internal specific size designations. Although numerous studies have been done and many articles have been written over the past century to understand garment size designation, its satisfaction, or dissatisfaction and although many argue that the actual size designation needs to be designed to be suitable for everyone, the consensus is not here yet. While major corporations focus on their target market, many of them do not desire a specific size designation. Why? Some argue it is because manufacturers/brands use their size designation as a marketing tool, well known today as "vanity sizing" to flatter their consumers. Some argue that vanity sizing makes women feel good about themselves, puts them in a better mood for their shopping experience, and increases loyalty.

5.2.8   Retrospective

In retrospect, it appears that few manufacturers used the numerous anthropometric surveys and size charts issued from 3D national campaign. Some set their own size designation to serve their own target customers (Burns and Bryant, 2002); since adhesion to standards is voluntary—recall Workman and Lentz (2000)—some continuously reinvent their own. Finally, notwithstanding the methods used to define size designations, manufacturers use the same key body measurements to size their garments (e.g., waist girth, hip girth, and crotch height) (Beazley, 1997), and most of them use the same numerical size designation system (6, 8, 10, ..., 24 or even 0 and 00) yet making it problematic today since they all refer to their own size charts. It is clear that size designation should help the consumer identify a well-fitting garment, but because of the tremendous variations currently existing in database analysis and actual population measurements, the actual size designation is questionable. Moreover, because of the dissatisfaction it creates, some steps need to be taken to provide a comprehensive size designation that would satisfy manufacturers/brands and be suitable for consumers. On this point of view, we believe that the size designation, as it was initially proposed, combining size number, letter, and symbols such as shown above 14T −, added to the idea presented by CGSB and ISO, for example, showing a pictogram specifying to which body measurements it should suit best, would probably be the best combination. Manufacturers, brands, or retailers could continue to serve one specific target market's shape and size. They would just need to add more details on their size designation label, making it more universal.

5.3   The key elements for an international size designation

As shown before the initial idea of size designation was based on different clusters defined by similar underlying measurements taken at different points on the body: bust, waist, and hip, for example. As also presented earlier, it appears that manufacturers didn't adhere to these standards and did their size designations preferring to define their own for their target market. Yet, Ashdown (1998) argued that manufacturers and retailers try to fit as much as possible into a small number of sizes. Keeping this in mind, in a previous study, Faust and Carrier (2009) validated and defined body key points for size designations. Their results, although providing information mostly for the lower body of the female population, were based on a 3D body scanner and anthropometric data from a national survey conducted in the United States. Their analysis was done clustering a multitude of measurements (over 200 body measurements) extracted from 3D body scans of > 6000 women. Results of their analysis first showed correlations between weight and girth circumferences of the individuals of the sample. Weight was highly correlated with waist girth, hip girth, and thigh girth, each of them with Pearson's correlation coefficient of, respectively, 0.91, 0.95, and 0.87. Total height of women was highly correlated with high hip height (Pearson's correlation coefficient of 0.81) and high hip height highly correlated with waist height (Pearson's correlation coefficient of 0.98). Girth circumferences were also highly correlated. As for example, waist girth and hip girth had 0.96 Pearson's correlation coefficient. Then, waist girth and the high hip girth were correlated with Pearson's correlation coefficient of 0.90. Interestingly, no significant correlation was found between the height and weight. Thus no significant correlations were found between any height and the girth measurements. Results showed Pearson's correlation coefficient of 0.29 between height and weight, whereas the results of the hip height and the hip girth showed 0.17 Pearson's coefficient. In other words, one can be tall and big or tall and slim or again short and big or short and slim. Table 5.1 shows some of these Pearson's correlations. Variables are as follows: h refers to height, w refers to weight, and g refers to girth. Then, B refers to bust; W refers to waist; H refers to hip and HH high hip; HT refers to high tight, whereas T refers to tight; and lastly, C refers to crotch. Then, when letters are put together as it is the case for hHH, it refers to height of the high hip.

As mentioned in the literature, to be effective, size designation needs to provide a small number of sizes, so manufacturers/brands would use it. At the same time, it needs to be distinguishing enough that consumers could find the appropriate garment that suits them best. Taking only those mostly related to the lower part of their body, with the use of software such as SPSS and Statistica, Faust and Carrier (2009) clustered these individuals into 11 groups. The number of groups was based on a reasonable amount of size designations defined by one retailer/brand. These size designations on the market run from 4 to 22 with a size 16 and one 16W for a total of 11 designated sizes. When clustering into only 11 linear sizes, Faust and Carrier (2009) argue that while the amount of numbered sizes may be interesting, it provides no indication of the underlying shape. To be effective, it needs to be split into two or three shapes/groups similar to the original sizing system where + and − were added to the size number. Moreover, since there is no correlation between height and weight or height and any girths and since women today wear pants, they argue that it is important to provide the length onto the size designation. According to Rasband and Lietchy (2006), the best way to provide the length of a pants is by measuring the inseam.

5.3.1 Clustering according to similar height, size, and shape

By observation, one can see that people differ in size and shape (Rasband and Lietchy, 2006). According to Patterson (2012) Afro-Americans differ in size and shape from their White Caucasian American counterparts.

Patterson (2012) argues that even if Afro-Americans would try, there is no way that they could be close in size and shape to their White counterpart. "Of course, the way fat is treated in the black community only reflects how fat is treated in mainstream culture and the fashion community. However, as 'curvy'—not too fat, now—is becoming more acceptable in the fashion world, it's clear the main shade of acceptable curvy is White" (Tasha Fierce in Patterson, 2012). With this in mind the author looked not only at the relation between ethnicity and body measurements but also at differences between age groups.

Table 5.1 Pearson's correlation coefficient: height, weight, and circumference correlation matrix.

	h	*w*	*g*B	*g*W	*g*HH	*g*H	*g*HT	*g*T	*h*W	*h*HH	*h*H	*h*C
h	1.00											
w	0.29	1.00										
*g*B	0.11	0.91	1.00									
*g*W	0.10	0.91	0.94	1.00								
*g*HH	0.09	0.92	0.91	0.96	1.00							
*g*H	0.18	0.95	0.87	0.90	0.94	1.00						
*g*HT	0.24	0.86	0.76	0.75	0.80	0.90	1.00					
*g*T	0.20	0.87	0.73	0.72	0.76	0.85	0.96	1.00				
*h*W	0.79	0.35	0.17	0.12	0.14	0.29	0.34	0.29	1.00			
*h*HH	0.81	0.39	0.22	0.17	0.17	0.30	0.37	0.31	0.98	1.00		
*h*H	0.52	0.59	0.58	0.58	0.58	0.51	0.40	0.34	0.62	0.67	1.00	
*h*C	0.80	0.08	−0.08	−0.12	−0.14	−0.03	0.10	0.06	0.85	0.87	0.49	1.00

5.3.1.1   Difference in height according to ethnicity and age groups

Faust and Carrier's (2009) research found differences between the average size and shapes of White Caucasian American, African American, Hispano-American, and Asian American. While almost 50% of White Caucasian American and African American height measurement is between 5'4″ (164 cm) and 5'7″ (173 cm), this percentage drops to only 25% for Hispano-American and Asian American. Although when looking at what is commonly called petite (<5'4″ or again 164 cm), the percentage of these last two rises to 71% for Hispano-Americans and to close to 70% for Asian Americans compared with only 3% for African Americans. Without any doubts, African Americans are on average the tallest. Another interesting difference clearly appeared between age groups. According to this previous study (Faust and Carrier, 2009), it seems that 65% of the women 66 years old and over are shorter than 5'4″ (164 cm). This number drops to 55% for those 56–65 years old, to 49% for those 46–55 years old, to 47% for those 36–45 years old, and to 45% for women between 18 and 35 years old. On the other hand, tall, which here refers to 5'7″ (173 cm) and more, is in higher percentage in younger women, and it decreases as the age group increases. Table 5.2 summarizes these numbers.

Since there is a high correlation within heights and because there are differences between ethnicities, it appears interesting to show the most common inseam heights per ethnicities and age groups. The most common inseam measurement varies between 27 in. (68 cm) and 31.5 in. (80 cm). More specifically, one could argue that an African American's pants inseam should be longer than those for an Asian.

Referring to Table 5.3 an average inseam length of 30 in. (76 cm) should satisfy 50% of each ethnic group (Caucasian, Hispano, and Asian). Moreover, one could argue that another 20% of them would just need to have the inseam altered so it could suit them. In those cases a small alteration would be needed; on the other hand, this inseam length would serve <40% of the African American population. To satisfy a large group of African Americans, as a target market, one would need to offer longer pants legs. A similar phenomenon appears for different age groups where the most common inseam measurement varies between 27 (68 cm) and 31.5 in. (80 cm). More specifically, one could argue here that youngsters' pants inseam should be longer that those for more mature aged women. An average inseam length of 30 in. (76 cm) should satisfy over 40% of each age group. Again, one could argue that another 15%–20% of them would just need to have the inseam altered so it could suit them. However, this percentage varies when looking at the 18–25 years old, whereas the majority would need pants with longer inseams.

5.3.1.2   Difference in girth measurements according to ethnicity and age groups

The same analysis was done with girth measurements. When comparing the sample used for US size, the tendency of the waist and hip measurements appears to be different according to ethnicity. Fig. 5.2 shows waist and hip measurements for each ethnicity and its average as discussed before.

Table 5.2 Height distributions according to ethnicity.

	Ethnicity (percentage per group)					Age groups (percentage per group)						
	Caucasian American	African American	Hispano-American	Asian American/others		18–25 years	26–35 years	36–45 years	46–55 years	56–65 years	66 years and over
Petite $x < 5'4''$ (164 cm)	39	37	71	68		45	45	47	49	55	65
Regular 5'4'' (164 cm) $\geq$ $x \geq 5'7''$ (173 cm)	47	47	25	27		43	43	41	41	36	27
Tall $x > 5'7''$ (173 cm)	13	15	3	4		11	11	11	9	8	7

Table 5.3 Comparison of Afro-American and Asian American inseam distribution.

	Ethnicity (percentage per group)				Age groups (percentage per group)						
	Caucasian American	African American	Hispano-American	Asian American/ others	18–25 years	26–35 years	36–45 years	46–55 years	56–65 years	66 years and over
27–28.5 (68–72 cm)	17	3.26	18	18	9	11	16	18	27	20
28.5–30 (72–76 cm)	50	34	50	50	42	47	48	50	50	45
30–31.5 (76–80 cm)	30	48	26	25	40	36	31	27	22	31
31.5–33 (80–83 cm)	3	13.25	3	2	8	5	4	4	2	1

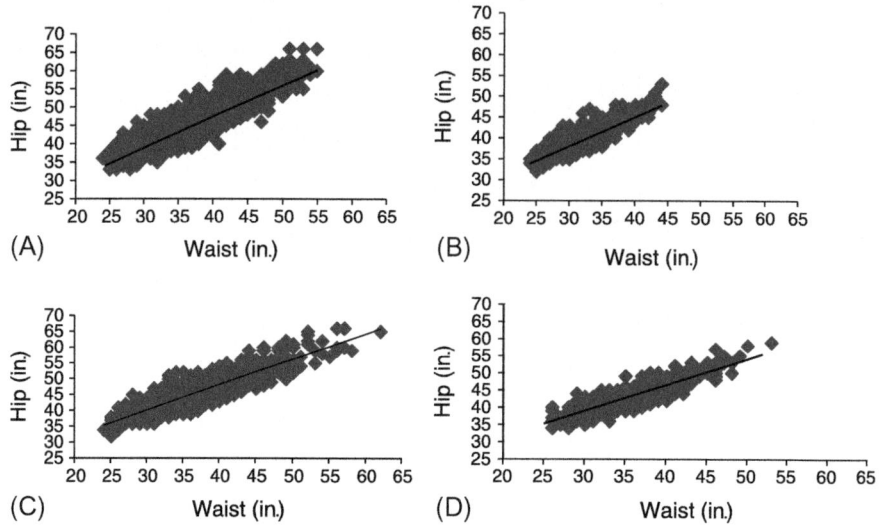

Fig. 5.2 Female waist and hip measurements according to ethnicity. (A) White Caucasian, (B) Asian American, (C) Afro-American, and (D) Hispanic.

The results present a clear distinction between each group. If one compares the African American female population living in the United States versus the Asian female population living in the United States, the range of girth sizes and shapes are obviously different. Fig. 5.3 compares each one with the percentage associated to each waist and hip measurement: the darker the cells, the higher the percentage.

Similar to the different ethnic groups, we found the same types of results for the different age groups. Although the purpose of this chapter is not to do the analysis of US size anthropometric national survey, it was important here to clarify that since different target markets are oftentimes associated with one specific ethnicity or age group, the size designation should be more valuable if it provides more information than the single one or two digits.

5.3.2 Summarizing the key elements

The results presented earlier clearly validate that some differences exist between sizes and shapes. They also demonstrate that these could also vary between ethnic groups and age groups. Without mentioning that ethnicity or age groups need to be written, we are convinced that length is an important variable and should be written/specified on garment size designations. We are also convinced that some girth measurements should also be written on the size designation. Although we discuss height and weight, we do not consider that these two variables would provide additional necessary information; on the contrary, they may have a negative impact on consumers' minds.

To summarize, we believe that for pants size designation, the waist, the hip, and in some cases the thigh girth and the inseams should be mentioned. In addition, since

Count of survey_ID — Waist Arr / Hip Arr

(A)

Hip Arr	25	26	27	28	29	30	31	32	33	34	35	36	37	38	39	40	41	42	(Blank)	Grand Total
32	0.09																			0.09
34	0.09	0.19																		0.28
35		0.09																		0.09
36	0.28	0.09	0.56	0.56	0.19	0.19	0.09													1.96
37	0.09	0.09	0.47	0.37	0.47	0.47	0.19													2.14
38	0.09	0.19	0.47	1.21	1.12	0.65	0.47	0.47	0.28											4.94
39		0.19	0.56	1.03	1.21	1.12	0.84	0.28	0.28	0.09	0.09	0.09								5.78
40			0.37	0.93	1.49	1.58	1.03	0.56	0.37	0.84	0.09	0.00	0.09							7.36
41		0.09		0.37	0.75	1.03	1.03	1.21	0.84	0.37	0.65	0.19	0.19							6.71
42			0.09	0.37	0.65	1.12	1.49	1.86	0.93	0.75	0.93	0.19	0.56	0.19	0.19					9.32
43				0.09	0.37	1.30	0.75	1.12	1.12	0.65	1.12	0.65	0.65	0.56	0.09					8.48
44				0.09	0.28	0.75	1.12	1.30	0.65	1.03	0.65	0.47	0.84	0.47			0.09			7.74
45			0.19			0.09	0.56	0.75	0.47	0.75	1.58	0.84	0.75	0.75	0.37	0.28	0.09			7.46
46						0.09	0.19	0.37	0.65	1.21	0.75	0.09	0.56	1.03	0.28	0.47				5.68
47						0.28	0.09	0.09	0.19	0.47	0.93	0.93	0.75	0.93	0.19	0.28	0.56			5.68
48								0.09	0.09	0.0.28	0.28	0.37	1.12	0.28	0.37	0.47	1.12			4.47
49								0.09	0.09	0.09	0.09	0.37	0.19	0.75	0.56	0.37	0.65			3.26
50								0.09				0.19	0.19	0.19	0.19	0.28	0.47	0.19		1.77
51								0.09				0.09	0.19	0.19	0.19	0.37	0.56	0.37		2.05
52								0.09	0.09			0.09				0.28	0.19	0.09		0.84
53															0.19	0.19	0.19	0.19		0.75
54																	0.09	0.28		0.37
55																	0.09			0.09
Grand Total	0.65	0.93	2.52	5.03	6.06	6.80	7.64	7.08	6.71	5.41	6.34	6.15	5.03	5.41	5.50	3.08	3.73	3.26	0.00	87

Count of survey_ID — Waist arr / Hip arr

(B)

Hip arr	25	26	27	28	29	30	31	32	33	34	35	36	37	38	39	40	41	42	Grand Total
32	0.20																		0.20
33	0.20	0.20																	0.40
34	0.20	0.61	1.21																2.02
35	0.20	0.81	0.81	0.81	0.61	0.20													3.44
36		0.81	1.82	3.04	1.21	0.81	1.01												8.70
37	0.20	1.42	2.02	2.63	2.02	2.02	2.02	0.81	0.20										13.36
38		0.61	0.40	2.23	3.85	3.44	3.24	2.83	0.81	0.20	0.20								17.81
39			0.40	2.23	2.02	2.63	2.43	2.83	1.21	0.40	0.40								14.57
40				0.20	0.40	0.20	0.61	1.42	3.04	1.01	2.02	0.81	0.20	0.40					10.32
41					0.20	0.40	1.21	1.21	2.23	1.21	1.42	0.81	1.82	0.20					10.73
42								1.42	0.61	1.82	1.62	0.61	0.40		0.40				6.88
43				0.20	0.20			0.20	0.61	0.40	1.42	0.81	0.20	0.20	0.81				5.06
44								0.20		0.20	0.40	1.42	0.20	0.20		0.20			2.83
45										0.20	0.40	0.20	0.40		0.61	0.20	0.61	0.20	2.83
46								0.20		0.20			0.20	0.20					0.81
Grand total	1.01	4.45	6.88	11.5	10.5	11.1	11.3	13.8	5.67	6.88	6.07	5.06	2.02	0.61	1.82	0.40	0.61	0.20	100.00

Fig. 5.3 Comparison of Afro-Americans (A) and Asian Americans (B) waist and hip measurement (inches) distribution.

previous studies have shown that people like the visual of a pictogram, we conclude that these measurements should be shown on a pictogram. At this point, many pictograms have been tested, and the perfect one has not yet been found. Having a pictogram with three or four key point measurements as described earlier should provide sufficient information so consumers could recognize themselves. It should serve the purpose to help consumers find the garment that suits them best. It would provide all of the essential information as it was suggested half a century ago such as the 14R – or 12T+, in a way that citizens from around the world could have a good understanding of the designated size. The only and biggest dilemma is which system should be used? The metric system, centimeters, or the use of the imperial system, inches? Thus this question is not solved, and neither is it clear which type of pictogram should be used.

5.4   Designing international size designations and methods of implementation

When the metric system became the new measurement system, its purpose was to facilitate all types of transactions that used measurements. In 1788, in the *Cahiers de Doleances*, people called for the reform of weights and measures. They were asking for "one law, one king, one weight, and one measure." As a result the metric system, which is the equivalent to the quarter of a meridian divided by 10 million, became mandatory starting on July 1, 1794. This modern system of measurement, equivalent to near the length of 3 ft or an *aune* (described later), allowed objects to be express in abstracted, commensurable units that relate to an absolute standard (Alder, 2002). It contrasted with the ancient system where measurements were inseparable from the object being measured and customs of the community that performed the measurement. At that time, not only did the physical standards differ from community to community, but also the technique of measurement depended on local custom.

5.4.1   A parallel with the apparel industry

A parallel could be done with the actual nonconformity and the absence of a specific system from the apparel industry. Nowadays, manufacturers/brands may have a same size designation although as mentioned earlier, their measuring points may differ from one another. As a result, one may use the waist at its narrow point to define its sizing, whereas another may use the waist where the garment is in position when worn, which could be 1 in. lower than the narrowest point of the waist. Or again, different manufacturers/brands may use the exact same measuring point, associated with the same size number, but this may be based on different underlying measurements. As a result, both could use the waist at the same position, both can use the same size designated number such as 12, but one could refer to a size 12 as being 28 in. (71 cm) waist, and the second could consider a size 12 for a waist of 32 in. (81 cm). Therefore the measurements themselves are different. Some manufacturers/brands use this to their advantage as a marketing strategy, and others just size the items as it was done by their predecessors. Although consumers seem to be unhappy with the actual size designation, making it mandatory to be installed on a garment may have a negative impact on retailers and manufacturers.

5.4.2 Introduction of a uniform size designation

The author believes that to enhance the consumer's shopping experience, the size designation should ensure a meaningful and comprehensible garment size (the key measuring points), its shape, and maybe its type of fit (loose, tighter, etc.). This size designation should bear a definitive relationship to a garment's key measurements and convey adequate information to consumers of any target market. It would eliminate this discrepancy that plagues the apparel industry. By instituting a new uniform size designation, this new system would provide a standardized way of communicating the size, the shape, and the fit of various garments without requiring a change in garment construction. Manufacturers, distributors, and retailers worldwide will use the same system (pictogram with common key points writing their own measurements) to ensure consistency. The question to raise now is how to get the best designed label? By initiating a design contest!

5.4.2.1 A design contest

To initiate this uniform way of communicating sizing, a design competition needs to be held. Design students and industry professionals worldwide could be invited to submit their designs for a new size designation. The winning design should then be tested in various countries before being used throughout the industry as the basis of an international campaign. If a new size designation is appealing, it would at first be adopted de facto by interested apparel manufacturers and all other interested parties. Then, if it becomes unbeaten, meaning it pleases consumers, retailers, and manufacturers/brands, and if it is suitable for worldwide use, it may become *de jure*.

5.4.3 Benefits of a uniform size designation

The whole apparel industry, designers, manufacturers, wholesalers, distributors, retailers, government and trade associations, academicians, and consumers should all benefit from it. For consumers, it would:

- help in finding and selecting the proper garment suitable for their size and shape,
- reduce wasted time when shopping (i.e., fitting room),
- improve the shopping experience,
- increase satisfaction,
- reduce returns/exchanges, and
- eliminate their confusion since they understand the information on the label.

For retailers, it would:

- facilitate ordering correct sizes and shapes for target market(s),
- increase consumers' loyalty,
- reduce returns/exchanges,
- reduce workload at the fitting rooms,
- reduce end-of-season leftover inventories (markdowns and garbage), and
- enable employees to work a priori with consumers and not a posteriori.

For designers and manufacturers, it would:

* improve target market selection decision,
* facilitate communication with retailers and consumers,
* enable better quality control, and
* improve communication/relations with subcontractors.

For government, trade associations, and academicians:

* for academicians, it would offer a common sizing and fit research language,
* for trade associations, *it would improve* communications with members and reduce legal/ quasi-legal problems, and
* for governments, it would facilitate consumer education, facilitate relations with the industry, and enable possible links with other areas of concern (i.e., population fitness and eating habits).

References

Alder, K., 2002. The Measure of All Things; The Seven-Year Odyssey That Transformed the World. Abacus, London.

Ashdown, S.P., 1998. An investigation of the structure of sizing systems: a comparison of three multidimensional optimized sizing systems generated from anthropometric data with the ASTM standard D5585-94. Int. J. Cloth. Sci. Technol. 10 (5), 324–341.

Beazley, A., 1997. Size and fit: procedures in undertaking a survey of body measurements. J. Fash. Mark. Manag. 2 (1), 55–85.

Bernier, B., Davignon, M.-F., Saint-Martin, J., 2003. Sur la piste : Géographie, histoire et éducation à la recherche de la citoyenneté. Éditions du Renouveau pédagogique inc., Bibliothèque nationale du Québec et Bibliothèque nationale du Canada.

Bryk, N.V., 1988. American Dress Pattern Catalogs 1873–1909. Dover, New York.

Burns, L.D., Bryant, N.O., 2002. The Business of Fashion Designing, Manufacturing and Marketing. Fairchild Publications, New York.

Chun-Yoon, J., Jasper, C.R., 1996. Key dimensions of women's ready-to-wear apparel: developing a consumer size-labelling system. Cloth. Text. Res. J. 14 (1), 89–95.

United States Department of Commerce Office of Technical Services (USDCOTS), 1958. Commercial Standard CS215-58 Body Measurements for the Sizing of Women's Patterns and Apparel, a Recorded Voluntary Standard of the Trade. United States Government Printing Office, Washington, DC.

Cooklin, G., 1990. Pattern Grading for Women's Clothes, the Technology of Sizing. Blackwell Publishing Professional Books, London.

Fan, J., Yu, W., Hunter, L., 2004. Clothing Appearance and Fit: Science and Technology. Woodhead Publishing Limited, Cambridge.

Faust, M.-E., Carrier, S., 2009. Women's wear sizing: a new labeling system. J. Fash. Mark. Manag. 14 (1), 88–126.

Goldsberry, E., Shim, S., Reich, N., 1996a. Women 55 years and older. Part I. Current body measurements as contrasted to the PS 42-70 data. Cloth. Text. Res. J. 14 (2), 108–120.

Goldsberry, E., Shim, S., Reich, N., 1996b. Women 55 years and older. Part II. Overall satisfaction and dissatisfaction with the fit of ready-to-wear. Cloth. Text. Res. J. 14 (2), 121–132.

Gould-Decauville, P., Bruere, C., Uhalde-Roux, C., Khatar, L., 1998. Guide pratique des tailles dans 36 pays, Études bimestrielles La Vigie internationale du vêtir-textile, second ed. Fédération de la maille, Clichy, France.

Kadolph, S.J., 1998. Quality Assurance for Textiles and Apparel. Fairchild Publications, New York.

Kidwell, C., 1979. Cutting a Fashionable Fit: Dress-Markers' Drafting Systems in the United States. Smithsonian Institution Press, Washington, DC.

Kidwell, C., 2001. Http://explore.cornell.edu/scene.cfm?scene=The%203D%20Body%20Scanner&=3D%20-D%20Custom%20Fit&view=3D%20-D%20Custom%20Fit%2001.

O'Brien, R., Shelton, W.C., 1941. Women's Measurements for Garment and Pattern Construction. Bureau of Home Economics, Textiles and Clothing Division, Miscellaneous Publication, No. 454 US Department of Agriculture and Work Projects Administration, Washington DC.

Patterson, C.J., 2012. Fashion tales. In: 13th International Conference of ModaCult, Milano, June 7, p. 2012.

Rasband, J.A., Lietchy, E.L., 2006. Fabulous Fit: Speed Fitting and Alteration, second ed. Fairchild, New York.

Schofield, N.A., Labat, K.L., 2005. Exploring the relationships of grading, sizing, and anthropometric data. Cloth. Text. Res. J. 23 (1), 13–27.

Sears Roebuck & Co. 1902 Catalogue, 2002. Princeton Imaging, CD.

United States Department of Commerce National Bureau of Standards (USDCNBS), 1971. Voluntary Product Standard PS42 70, Body Measurements for the Sizing of Women's Patterns and Apparel. United States Government Printing Office, Washington, DC.

Winks, J.M., 1997. Clothing Sizes International Standardization, The Textile Institute. Redwood Books, Manchester.

Workman, J.E., 1991. Body measurement specifications for fit models as a factor in clothing size variation. Cloth. Text. Res. J. 10 (1), 31–36.

Workman, J.E., Lentz, E.S., 2000. Measurement specifications for manufacturers' prototype bodies. Cloth. Text. Res. J. 18 (4), 251–259.

Yu, W., 2004. Human anthropometrics and sizing system. In: Clothing Appearance and Fit: Science and Technology. Woodhead Publishing Limited, Cambridge, pp. 170–195.

Further reading

Alexander, M., Connell, L.J., Presley, A.B., 2005. Clothing fit preferences of young female adult consumers. Int. J. Cloth. Sci. Technol. 17 (1), 52–64.

Anderson, L.J., Brannon, E.L., Ulrich, P.V., Presley, A.B., Worondka, D., Grasso, M., Stevenson, D., 2001. Understanding fitting preferences of female consumers: development an expert system to enhance accurate sizing selection. In: National Textile Center Annual Report, 1998-A08-1-10. National Textile Center, Spring House, PA, pp. 1–10.

Canadian General Standards Board, 1997. Canada Standard Sizes for Women's Apparel—Trade Sizes (CAN/CGSB-49.203-M87). Ottawa, Canada.

Eckman, M., Damhorst, M.L., Kadolph, S.J., 1990. Toward a model of the in-store purchase decision process: consumer use of criteria for evaluating women's apparel. Cloth. Text. Res. J. 8 (2), 13–22.

Faust, M.-E., Carrier, S., 2011. 3D Body Scanning: Generation Y Body Perception and Virtual Visualization. Woodhead Publishing Limited, Cambridge.

Faust, M.-E., Carrier, S., Baptiste, P., 2006. Variations in Canadian women's ready-to-wear standard sizes. J. Fash. Mark. Manag. 10 (1), 71–83.

LaBat, K.L., Delong, M.R., 1990. Body cathexis and satisfaction with fit apparel. Cloth. Text. Res. J. 8 (2), 43–48.

Mason, A.M., De Klerk, H.M., Sommervile, J., Ashdown, S.P., 2008. Consumers' knowledge on sizing and fit issues: a solution to successful apparel selection in developing countries. Int. J. Consum. Stud. 32 (3), 276–284.

Oldham-Kind, K., Hathcote, J.M., 2000. Speciality-size college females: satisfaction with retail outlets and apparel fit. J. Fash. Mark. Manag. 4 (4), 315–324.

Otieno, R., Harrow, C., Lea-Greenwood, G., 2005. The unhappy shopper, a retail experience: exploring fashion, fit and affordability. Int. J. Retail Distrib. Manag. 33 (4), 298–309.

Yoo, S., Khan, S., Rutherford-Black, C., 1999. Petite and Tall-sized consumer segmentation: comparison of fashion involvement, satisfaction and clothing needs. J. Fash. Mark. Manag. 3 (3), 219–235.

Part Two

Body measurement devices and techniques

Full body 3-D scanners

6

Susan P. Ashdown
Department of Fiber Science and Apparel Design, College of Human Ecology,
University of Cornell, Ithaca, NY, United States

6.1 Introduction

Commercially made full-body 3-D scanners have been available since the late 1990s. These scanners have driven new developments in the arts, entertainment, medicine, and apparel. The capability of this technology to capture a 3-D image of a full human body quickly and easily has contributed to the development of realistic animations for gaming applications and movies, advances in medical diagnostics and treatment, various technological initiatives in the apparel industry, and many anthropometric studies that increase our understanding of the anthropometric variation among and within different populations. Recent developments in this technology are making it more widely available and driving even more uses, for example, as a tool for tracking body changes with exercise. Software systems that can be used to measure, manipulate, deconstruct, and modify 3-D body scans are now widely available and user-friendly. Scanners themselves have changed on the one hand from large, expensive stationary installations to smaller affordable units on tripods or handheld devices and on the other hand to multiple sensor installations that can capture the full 3-D body in motion in high resolution and with great precision. The availability of affordable 3-D printing is also driving more uses of 3-D body scans, from the marketing of small 3-D portrait statuettes to the 3-D printing of prosthetics.

Three-dimensional scanning of the body has been used by apparel researchers and the automotive and other industries that require anthropometric data for the last two decades, since the technology was perfected and commercialized for capturing the whole body in the late 1990s and early 2000s. Early scanners were expensive and had a large footprint, which limited their availability for many uses. Although some scanners could be set up as mobile units, most were stationary. The early scanners mainly used white light or eye-safe lasers for their light sources, though scanners using microwaves were also developed. The computing requirements for these early scanners were sometimes met with multiple computers to collect and display the data. Some scanners only captured surface data; others also had cameras to record color and texture information, and the microwave scanner could scan through clothing. Early body scanners in industry were used for custom fit manufacturing on a small scale, automated size selection, the generation of virtual fit avatars, the manufacture of dress forms, and for anthropometric studies.

With the recent introduction of improved sensors, infrared light sources, and upgraded computing capability, 3-D body scanning can now be done very

Anthropometry, Apparel Sizing and Design. https://doi.org/10.1016/B978-0-08-102604-5.00006-8

inexpensively, increasing the usefulness of scanning for the apparel industry and making anthropometric studies less costly and more practical. Issues such as calibration, resolution, scan volume, balance between data capture and reconstruction of the data, format of the output, format/reliability/validity of the extracted data, and participant interaction with the technology must be understood to make best use of scanning systems.

6.2   History of full-body scanning, its use in apparel sizing and fit, and anthropometric studies

6.2.1   Early scanners

The first full-body scanner that was widely marketed in the late 1990s was made by Cyberware and was primarily sold to the entertainment industry and military researchers (Robinette and Whitestone, 1994) (see Fig. 6.1). It used stationary sensors and an eye-safe laser as a light source and provided a rotating platform for the person being scanned. The technology was effective, but the scanners were not portable, and their cost was prohibitive for most users. At this stage, neither the usefulness nor the specific procedures of scanning appropriate for apparel research or anthropometric studies were well understood. Cyberware, Inc. was dissolved in 2011.

The next generation of scanners available by the mid-1990s included brands such as Vitronic/Human Solutions, TC2, Hamamatsu, Telmat, Wicks and Wilson, and TecMath (Daanen and Jeroen van de Water, 1998). These scanners were costly but at a level that could be justified for a wider range of research opportunities, and their usefulness was becoming more apparent. They were still relatively large installations and were not optimal for scanning in multiple locations, although some systems could

Fig. 6.1 Early Cyberware scanner with a rotating platform (from Flickr user NIOSH).

be set up in a truck and moved from one site to another. This generation of scanners had multiple sensors placed around the body that allowed the subject being scanned to remain stationary instead of rotating on a platform. They were all similar in that they consisted of a light source, sensing devices, software to assemble the scan from the data from the sensing devices, and software to extract information (generally simple linear measurements) from the assembled scan.

Basic 3-D body scan data from these scanners and from 3-D scanners in general are expressed in an *XYZ* coordinate space. These scanners captured and plotted points from the surface of the body, generally in a grid density of 2 mm × 2 mm resulting in about 27 points per cm^2 or about 300,000 points in a scan of an average-sized person. The points could be triangulated and surfaced to create a statue-like rotatable image. The systems that had cameras to capture color and texture information generally "wrapped" this information as a separate layer onto the 3-D model.

The volume of the scanning area differed among scan companies but was generally in the range of 2.1 m (height) × 1.2 m (depth) × 1 m (width) and accommodated a standing or seated person. The available scanners differed greatly in the mode of representation of the scan data on the computer screen, from a simple point cloud to a model with triangulated and surfaced points highlighted with simulated lighting and providing different choices in perspective rules for viewing the model (see Fig. 6.2). The displays also varied in how much postprocessing was applied to the scan data, beyond simply assembling or merging the different camera views. Any scanner creates redundant points, either throughout the surface of the scan (white light scans) or in the overlapping camera views (laser light scans) (see Fig. 6.3). The task of the software developer who generates the program to assemble and justify these redundant data points is to create a digital model that will have the same measurements and shape of the complex, organic body. Reducing the scan to essential points will also reduce the size of the file, without losing needed information (see Fig. 6.4). This process can also improve scan visualization.

Generally, these early scanners were marketed as a sophisticated measuring device that could extract a large number of linear measurements and some angle measurements automatically from the scan. These measurements (which were based on data taken by manual measuring tools such as tape measures, anthropometers, goniometers, and calipers), and not the actual 3-D scan data themselves, were often the focus of the data display. As is frequently the case with new technologies, many researchers did not have a good concept of how to use the actual 3-D scan itself, as 3-D data were unfamiliar.

The measurements that were generated from the scans varied, with some modeled on standard anthropometric practice and some modeled on the measurements made by tailors and the apparel industry. Scanner companies also developed the capability to scan and derive measurements from subjects in seated positions to generate data for the automotive and airline industries. Multiple studies were conducted to verify the reliability and validity of the measurements of the different scanners, comparing automatically derived scan measurements with manually taken measurements of the same subjects (Bradtmiller and Gross, 1999; Paquette et al., 2000; Mckinnon and Istook, 2001; Choi and Ashdown, 2010). Studies were also conducted on issues relating to

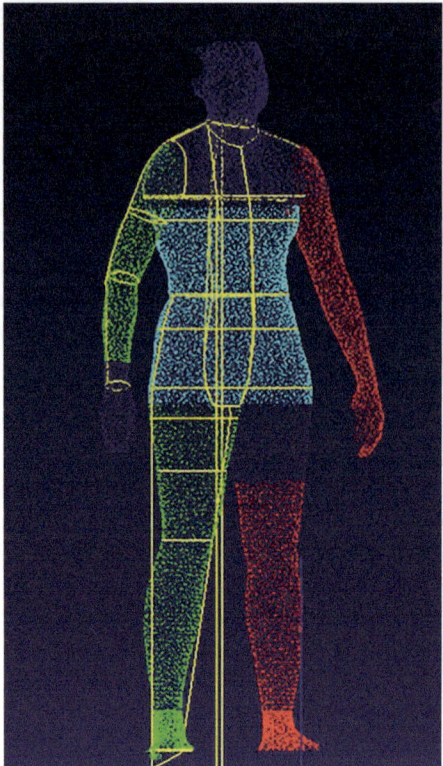

Fig. 6.2 Screen visualizations, one as a point cloud with body measurement locations indicated with lines and another as a triangulated and surfaced figure. Simulated lighting of the surfaced figure creates shadows that provide a visual reference to the 3-D shape.

the subject being scanned, addressing appropriate scan clothing, posture, breathing, and body sway (Brunsman et al., 1997; Mckinnon and Istook, 2002; Daanen et al., 1997a; Kim et al., 2015).

The output of the scanner was dependent on how the scanner and the scan software performed at many different stages of the process (appropriate calibration, reliable data capture by the individual sensing devices, number of sensing devices, reliable and valid integration of data from the multiple sensing devices into a 3-D model, valid location of measurement points [body landmarks], valid combination of on-the-surface and/or spanning of body prominences with the digital "tape measure," and valid choices on placing circumferential measurements that can either follow the body shape or follow the axis of the torso or limbs or are taken parallel to the floor). However, few studies reported or addressed these issues but only compared the final measurements from the scanner and their relationship with some version of corresponding manual measurements. It is understandable that researchers neglected to include this information, as the ability to examine many of these factors was not given to the user.

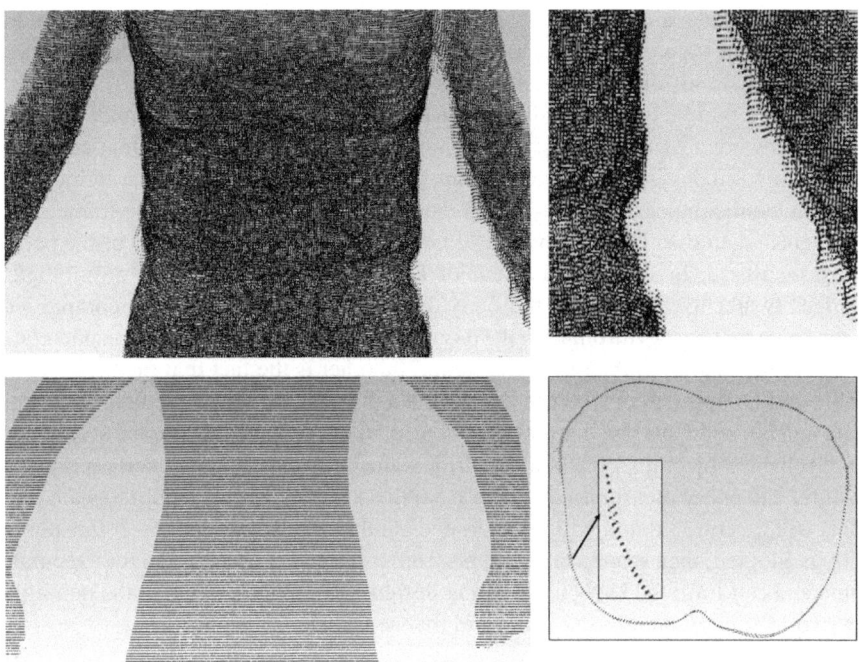

Fig. 6.3 Point clouds. Multiple points collected by a white light scanner (showing the wave effect) and organized points collected every 2 mm by a laser scanner (showing a slice from the scan with points from overlapping patches, captured by different cameras).

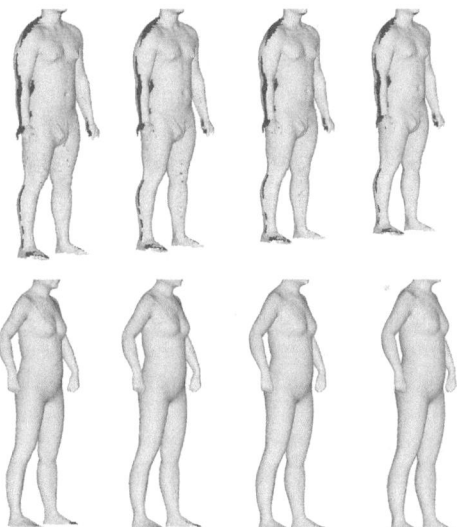

Fig. 6.4 Stages in merging and smoothing points, demonstrated on a white light scan image and on a laser scan image.

This lack of understanding of the scanning technology ultimately led to misunderstandings of the nature of scan measurements, which at times impeded effective development of methods of creating and using scan data.

There are two factors in the discussion of the "accuracy" of the scanner measurements that were not often discussed or resolved. One is the fact that the human body is an organic form that is in a state of constant change, so no method of measurement will result in identical measurements, even when taken within the same time frame. Measurements taken over time will vary even more. Changes in the posture of the person being measured, or in the placement of landmarks, combined with physiological changes (water loss or gain in the body driven by hormonal changes, compression of the spine that occurs throughout the day) result in measurement variation not related to the sophistication of the measuring tool. The other is the fact that the body is malleable and compressible, whereas the scan is a digital "statue" of the body. An argument can be made that the digital body created with a properly calibrated scanner is a more reliable object than the body itself; it is the body captured in a certain time and posture, and every measurement taken from the scan (as long as landmarks are identical) will be constant and will relate to one another in a valid manner. If this philosophy is adopted, then scan measurements from a calibrated scanner are the "accurate" value and can form the basis for clothing design and sizing systems in the nondigital real world if chosen and used in a valid manner.

6.2.2   Use of 3-D body scans and data from scans in apparel studies

Early use of scan data focused only on extracted linear measurements from the scans. Full-body 3-D scanning provided data for anthropometric studies of various populations, body shape analysis based on simple circumferences, extraction of measurements as the basis for custom-made clothing, automatic size selection, and the creation of avatars for size selection (Robinette et al., 1999; Bougourd and Treleaven, 2014; Devarajan and Istook, 2004; Hye, 1999; Corcoran, 2004; Lerch et al., 2007).

However, there were also early attempts to use the scans in ways that used more of the complex 3-D data from the scan. Robinette and Whitestone (1992) conducted an early study comparing 3-D head scans of army personnel with 3-D scans of helmets and proposed similar studies to align and merge clothed and unclothed body scans to directly compare clothing-to-body relationships in 3-D (Robinette and Whitestone, 1992). Circumferential slices of such scans provided precise data on the distance of the worn clothing from the body (see Fig. 6.5). Many studies of this nature have been conducted of firefighter gear for which the gap between clothing and body that captures still air can be thermally protective and for cooling vests that rely on contact of the vest with the body surface for conductive cooling (Park and Langseth-Schmidt, 2016; Deng et al., 2018; Branson et al., 2005).

Three-dimensional body scans are also used in the manufacture of dress forms that precisely duplicate the size, shape, proportions, and posture of an apparel firm's fit model. Most dress forms in the past (except those made for lingerie or swimwear) have

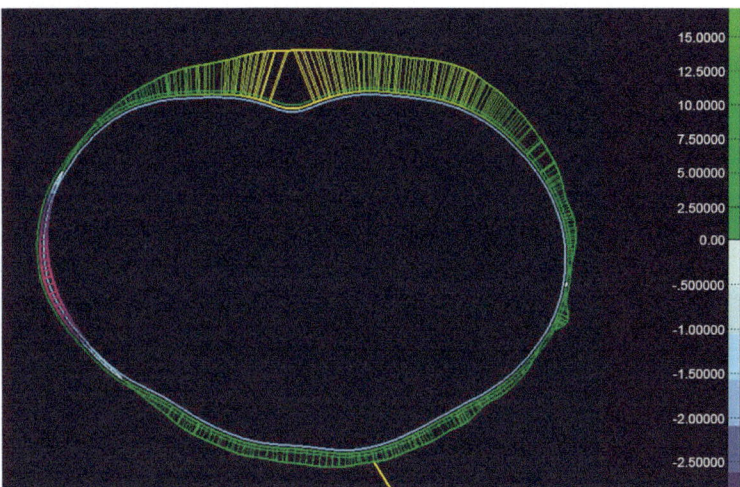

Fig. 6.5 Cross section of merged scans showing the relationship between body and clothing in two dimensions. The inner slice is from a seminude scan; the outer slice is from the same person in the same position wearing a cooling vest (Branson et al., 2005).

reflected garment shapes instead of body shapes, but this is changing with modern styles and materials that create clothing silhouettes that follow the curves of the body. Dress forms made from 3-D scans are used in product development and fit analysis (Haber, 2006) (see Fig. 6.6).

Body scans have also been used to examine body shape variation by assessing the 3-D shapes directly; for example, a lingerie company used 3-D scans to visually identify different bust shapes and configurations in the population. Using the method of subjective analysis of the different shapes, designers analyzed which shape variations had an impact on bra fit and made decisions about styles that are optimized for different women. Understanding bust shapes also made it possible to choose appropriate fit models for the development of specific styles. This shape analysis, making judgments from a visual assessment of the 3-D scan, can be very productive. Other means of sorting the population into shape groups have been developed based on data from 3-D scans using powerful statistical analysis methods (Song and Ashdown, 2011).

6.3   Different technologies used in full-body scanners

A variety of technologies have been used in the development of body scanners, and others are continually being introduced. However, the main distinctions among them can be categorized by the light source (or in some cases, the use of nonvisible wavelengths) that interact with the sensors used to collect the data (Daanen and Ter Haar, 2013; D'Apuzzo, 2009). ANIWAA, a technology analysis and comparison company, reviews and recommends scan technologies on their website (Lansard, 2018). Although there are some inaccuracies in their listings (i.e., Vitronic is listed as a structured light technology instead of a laser light technology), this is a useful compilation

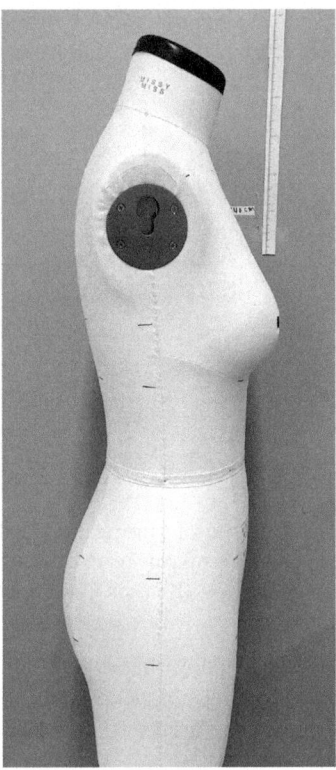

Fig. 6.6 Alvanon dress form modeled from a 3-D body scan. Unlike the traditional simplified dress form shapes used in clothing design, this form provides the anatomical curve of the buttocks and the undercut below the bust.

of some of the companies that produce body scanners in a rapidly changing market. They also list 3-D scanning apps and handheld scanners.

Laser: Laser scanners are considered the gold standard of 3-D body scanning. In booth form, these scanners are generally configured with sensors equally spaced around the body instead of focused on the front and back of the body. Laser scanners are somewhat sensitive to color and generally will not scan black objects well, though this can be adjusted to improve capture at the expense of the speed of the scan (normally about 12 s). Scanning can occur in the presence of white light, so the person being scanned does not need to be isolated from the person taking the scan. The major company currently providing laser scanning is Vitronic (Human Solutions).

Structured light: Structured light scanners take scans much more rapidly than laser scanners, but the data from white light structured scanners can be noisy and require more postprocessing. These scanners generally take multiple scans and then use computer algorithms to smooth the surface of the scan as they construct the 3-D digital model from the sensor data. The primary manufacturers of white light scanners acquire data from the front and back of the body and generate side data as part of the data assembly and smoothing process. White-light scanners can be very sensitive to color contrasts in the object being scanned. The scanning space must be isolated from the light from the environment, blocking the view of the person being scanned from the researcher. This can be a problem when a specific posture is required as the researcher cannot coach the subject into the desired position easily if the study participant is closed in a booth. TC2 is the primary vendor of structured light scanners.

Millimeter wave: Millimeter-wave scanners use nonvisible wavelengths that can travel through clothing and scan the surface of the body, unlike other scan technologies. The only scan company marketed to the apparel field using this technology is Intellifit. These scanners are used primarily for size selection in shopping malls. This technology is also used in the body scanners developed for security purposes in airports and other locations. Millimeter-wave scanning can occur in the presence of white light.

Infrared: Infrared scanners also use nonvisible wavelengths, but cannot scan through clothing. Infrared scanners are relatively recent and can be more affordable than other scan technologies. Infrared depth sensors are often used along with software that updates and refines the image continually based on new input, thereby developing effective handheld scanners. Though scan resolution cannot yet match stationary scanners, the technology is rapidly developing. Companies marketing infrared scanners include SizeStream and Styku. Most of the more recently developed scanner technologies use an infrared or near-infrared light source.

Four-dimensional scanning (3-D scans in motion): Full-body 3-D scans are taken with multiple cameras and LED lights in a stereophotogrammetry process, acquiring the data very quickly. This allows capture of body movement in 3-D over time. The scanner has a large footprint and high cost. Just as was the case in early body scanning, there is not yet a clear understanding of how these data can be best used in the apparel industry, resulting in great opportunities for researchers to develop new methodologies and data collection processes. 3dMD is the provider of full-body 4-D scanning capability.

6.4 Scanner issues relating to hardware and the subject being scanned

6.4.1 Calibration

Most scanners require an initial calibration process to align and coordinate sensors and periodic calibration to correct any drift or to ensure that the initial setup is being maintained. This process often makes use of a geometric object of known dimensions and can be simple or complex. The best designed systems have quick and easy calibration procedures with clear instructions and directions on what to do if the calibration fails.

6.4.2 Scan time

Most scanners are designed to collect the data for the scan in 3–12 s. Scan time is optimized for the amount of time that an unsupported person can stand still without starting to sway (Daanen et al., 1997b). Scans could be taken with higher resolution if the process were slowed down. However, current resolutions are appropriate for most anthropometric uses. The exceptions are scans of hands, feet, and faces, for the design of gloves, of shoes, and of facemasks. Foot scanners optimized to scan feet are marketed separately from body scanners and are available from many providers. Scanning hands and faces is another issue, and few scanners have been developed specifically for this purpose. A low-scan volume, high-resolution scanner developed for head scans by Human Solutions provides good hand data for glove design and face and head data for facemask and helmet design.

6.4.3   Posture, scanning apparel, and hair issues in scanning

Posture is a critical issue in scanning. For most measurement purposes a relaxed, natural posture is desirable in which the person's position is as close to bilaterally symmetrical as possible (weight equally distributed on both feet, shoulders, and hips square and balanced). The anthropometric position (head in the Frankfort horizontal plane, shoulders relaxed, arms at the sides, hands relaxed and facing forward, and feet with heels together and toes at a 60-degree angle) is ideal both to provide the most reliable measurement data and to maintain comparability with manually performed anthropometric studies from the past (see Fig. 6.7). However, this position will obscure sensors from parts of the body such as the underarm, inner thighs, and crotch, particularly for overweight or obese subjects. Therefore most scanners require that the subject being scanned stand with their feet, shoulder width apart (or further if needed to separate thighs), and with their arms abducted from the body.

Scanning a participant in the most common scan position may result in variation in stature and height measurements, and if care is not taken to coach the scanee into a relaxed position, he/she may brace his/her shoulders back, raise his/her shoulders, and/or lock his/her knees and elbows, all of which postures can affect the validity of the body measurements extracted from the scan. Defining and maintaining a standard position is important to conduct a study that is internally reliable. If it is important that the study be consistent with other anthropometric studies or if height measures are important, it may be necessary to take two scans of each subject, one optimized for circumferential measurements with arms and legs abducted and a second one optimized for height measurements, taken in the anthropometric position. In any case, it is important to coach the subject to assume a relaxed posture. The ISO standard on scanning metrics for anthropometrics specifies one anthropometric standing position and one standing position with limbs abducted. Also included in this standard is a standing pose with one arm extended forward and the other arm bent at a 90 degree at the elbow and a seated position (ISO International Standard, 2010).

If stature and crotch height are important measurements for the study, it may be preferable to take these two measurements manually. Scanners generally do not capture the top of the head well. Light sources can be scattered in hair, impacting the scan quality, and cameras are generally not aimed in a direction that will capture the top of the head. Also, even with the legs abducted, few scanners can actually capture the exact position of the crotch.

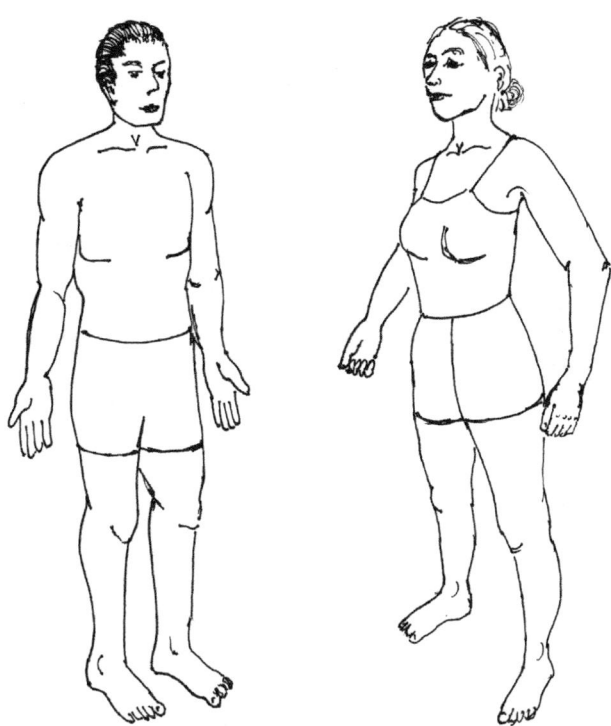

Fig. 6.7 Images of the anthropometric position and of the most common scan position with arms and legs abducted from the body. Fingers and thumbs are generally held together or in a soft fist for scanning as the resolution of the body scanner is not appropriate to capture individual fingers.

Best practice for getting a good scan of the head is to cover the hair with a close fitting cap, but of course if the hair is very long or thick then it can distort the shape of the head even if confined in a cap. This may not be a problem for most apparel studies for which it is only important to keep the hair away from the neck to get a good scan of the neck and shoulders. In this case, tying the hair up on the top of the head is the best solution.

In the past elaborate scan, clothing was deemed to be important to provide a sense of modesty and to provide consistency among the study subjects. Not only may this still be important for some populations, but also it may be acceptable, both for the scanees and the needs of the study, to scan in the study participants' ordinary underwear, only providing scan clothing for subjects whose underwear are unsuitable (many scanners would not scan black underwear). For apparel studies of women, there are advantages to scanning in their own most commonly worn bra, as this will provide a more valid breast shape for clothing than the more relaxed sports bras that have commonly been used in scan studies (Pei et al., 2019).

6.4.4 Scans in coordinate space

One other important issue is the origin point and the orientation of the scan in the coordinate space. Different scanners establish the origin (the 0,0,0 point in the *XYZ*-axis) at different points. Two popular choices for the origin are the center of the body being scanned, or centered on a plane at the feet of the scan. Most Cartesian coordinate systems are oriented so that the *X*- and *Z*-axes are on the transverse plane, and the *Y*-axis is up (Fig. 6.8), but other orientations are also used. Whatever the system, it is important that the person being scanned is not rotated in reference to the coordinate system, but is aligned with the frontal or sagittal plane of the body parallel with an axis. Often when taking scans from one software to another, the scan will need to be transposed into a different coordinate system. It is also important to know what the primary units are used in each software and that the scan is imported into the correct system and scale of measurements.

6.5 Reliability of scan generation from sensor views

The reliability of scan data is of course dependent on how reliably the *XYZ* coordinates are sensed and recorded. Proper calibration should ensure this for all systems, using a simple geometric object of known dimensions to check the system. The scanner software must be designed to take the scan data from each sensor or camera and assemble them into a 3-D model in a way that represents the body scanned accurately. This is not an insignificant task. The data points define a complex organic shape. There are many

Fig. 6.8 This image shows a Cartesian coordinate system with *Z* in the vertical position. Most scanners generate a scan with *Y* as the vertical direction and *X* and *Z* on the transverse plane.

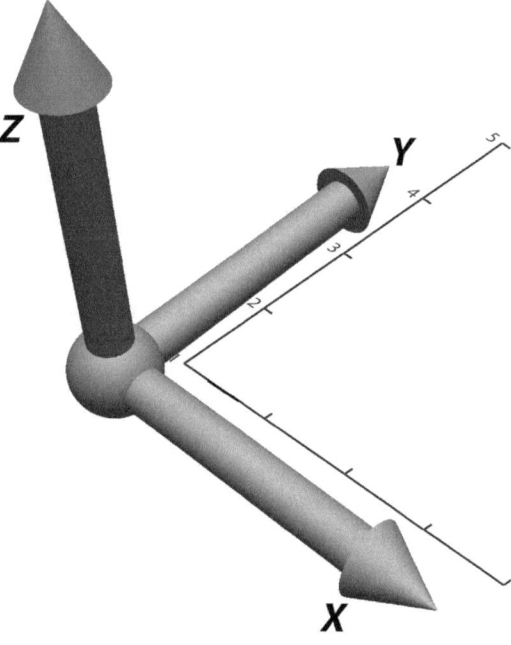

redundant points, and overlapping data will often not be aligned entirely with the surface. If only the outermost points are retained, the model is likely to be too big, and if only the inner points are retained, it is likely to be too small. Determining the optimal choice that retains only those points that represent the actual surface of the body is important. However, depending on the use to be made of the scan and the amount of variation of the scan data, the redundant and overlapping points may not be a problem. If the main goal of the study is to generate automated measurements then the measuring algorithm can be designed to take measurements appropriately in the data cloud (Fig. 6.9).

Some scanners only capture data from the front and back of the body and generate the missing parts of the scan on the relatively featureless side of the body. This can work well for a population that does not have much variation, but particularly for a population that has overweight or obese subjects, this can miss useful details of the side of the body. Whether this is a problem will again depend on what the goals of the research. It is not necessary to capture every detail of the body shape for some studies.

If the goal is to derive cross-sectional data, or to manually measure scans, or to 3-D print the scans, or to create a digital avatar, then it is often necessary to merge the data into a single shell, to reduce the number of redundant points and to make a "watertight" model by filling in missing data. In the early days of body scanning, this was done using 3-D software designed for engineering uses, such as PolyWorks, which was not very effective at manipulating organic shapes, but current software designed to manipulate 3-D data makes these tasks much easier. Some software systems for 3-D manipulation are only designed to work with NURBS models and therefore need to be converted from the XYZ coordinate points generated by the body

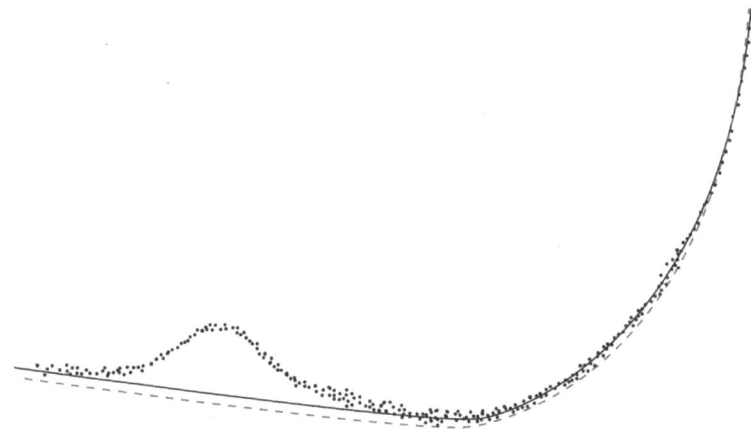

Fig. 6.9 Representations of circumferential measurements taken on a slice from a scan that has not been merged to delete redundant points. The measurement can be taken through the points on a path that averages the variation or outside all the points. The second method will give a slightly larger value.

scanner. Nonuniform rational B-spline (NURBS) models are based on mathematically generated curves and surfaces. Such models can be modified more easily than 3-D coordinate models by moving specific control points. NURBS models are more commonly used for geometric shapes; the complex organic shape of the human body is more difficult to represent and to modify using NURBS curves.

6.6   Automatic landmarking and data extraction software

The process of generating body dimensions automatically is distinct from the process of creating a 3-D digital body. Most scanner systems have the ability to automatically derive linear measurements from the scan, but the algorithms used by different scanner companies vary. The scanner that generates the best scan for a project may not have the best automatic measuring system for that project. Most scanners can accept a scan in a generic format made by a different scanner, so it is possible to scan in one system and derive measurements in another.

The first step in generating a measurement from a scan is to identify the points on the body to be measured, the body landmarks. These landmarks are well defined by anthropometrists (Hotzman et al., 2011), tailors (Roetzel and Fritz, 2014), and apparel practitioners (Joseph-Armstrong et al., 2010), though each group has different sets of landmarks, different placement of landmarks, and different methods of measuring between landmarks. Because there is great variation among body shapes, it can be very difficult to place landmarks reliably, and although a trained anthropometrist will generally place landmarks more reliably, some variation will always be present (Kouchi and Mochimaru, 2011). It is even more difficult to design software to reliably place landmarks on body scans. This is generally done by programming a search for certain geometric features on the body (Suikerbuik et al., 2004), but programming for every possible body configuration can be almost impossible, particularly given the variation of shape between underweight and overweight bodies.

Because clothing is designed to move with the body, many of the critical landmarks that define measurements are where the body articulates—the joints. Therefore landmarks are often placed by anthropometrists and apparel practitioners by palpating to feel the appropriate bony protuberances that define the center of joint movement (Fig. 6.10). It is not possible for the scan software to achieve the same level of precision and reliability by judging landmarks on joints based on the surface geometry of the body.

Other landmarks are established by common practice in the apparel industry or dictated by clothing styles. Placement of the landmark to measure waist circumference is a good example of this (Wren et al., 2014). The "natural" waist can be identified as the level of the spine of greatest articulation, between L4 and L5 of the lumbar vertebrae, or at the omphalion (belly button), or in normal weight women at the point where the body curves in the most at the side. In some body shapes, this area where the body curves in and makes a natural ledge is parallel to the floor, but in others, it tips down in front. On the other hand, most men have a small (or when less athletic, a substantial) bulge at the level of greatest articulation. However, none of this is helpful as currently

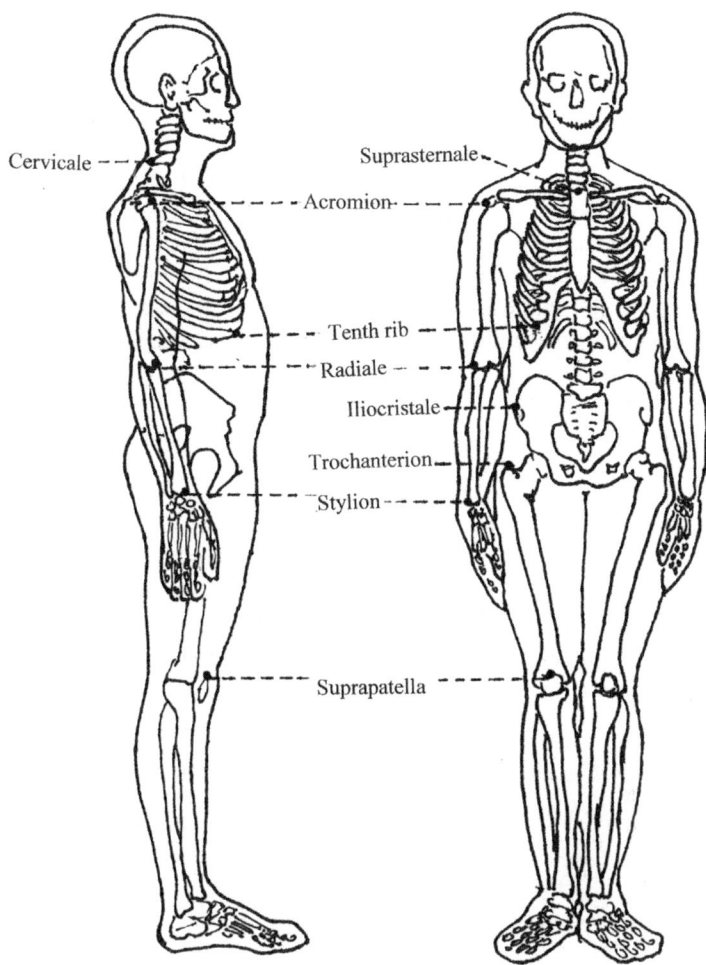

Fig. 6.10 Image showing skeletal landmarks, located by anthropometrists and apparel practitioners by palpating for the bony protuberances. Reliably finding the point on the bone closest to the center of the joint rotation requires training for some of the landmark locations.

most clothing is worn lower than the natural waist and has been since the 1950s. One way to identify the waist, depending on the goals of the study, is to have each individual being scanned identify their preferred waist position for themselves, by placing a snug elastic band at their own waist. Most scan software provides a variety of choices for waist placement, so another choice is to measure the waist at several different locations.

Of course the scanner software can quite easily identify the smallest circumference on the torso between the high hip and the tenth rib—but this is generally a point even higher than any preferred landmark for the natural waist. In a similar issue the landmark for the hip measurement is traditionally at the fullest protrusion of the buttocks; a circumference is taken parallel to the floor at this point. However, the largest circumference of the lower body is generally lower, where the bulge of the thighs adds to the circumference of the body (Fig. 6.11).

Another example of landmarking that is only determined by apparel practice is in the placement of the side seam, the line down the side of the body that divides the front of the body from the back of the body. This is an important landmark for apparel patternmaking because it is the dividing point to create arc measurements at the bust, hips, and waist that define the balance of the body. Most apparel practitioners can identify where the side seam should be placed to provide the best balance on different body types—but it is difficult to define this in terms that can be automatically located on a body scan reliably (Ashdown et al., 2008; Brownbridge et al., 2013).

Another area of the body that is complex and varied and therefore presents a problem for automated measurement extraction is the neckline measurement taken at the base of the neck (Huang et al., 2011). Though the cervicale at the back of the neck and the top of the sternum at the front of the neck are generally (but not always) easily identifiable on a scan, the actual placement of the neckline as it crosses the trapezius muscle on the side of the neck can be very difficult to determine. The precise curve of the neckline is also an apparel patternmaking construct. This measurement is therefore difficult when taken manually as well. Some apparel practitioners have devised a method for determining the desired curve using a chain that can be laid around the neck to determine the measurement more precisely than is possible with a tape measure.

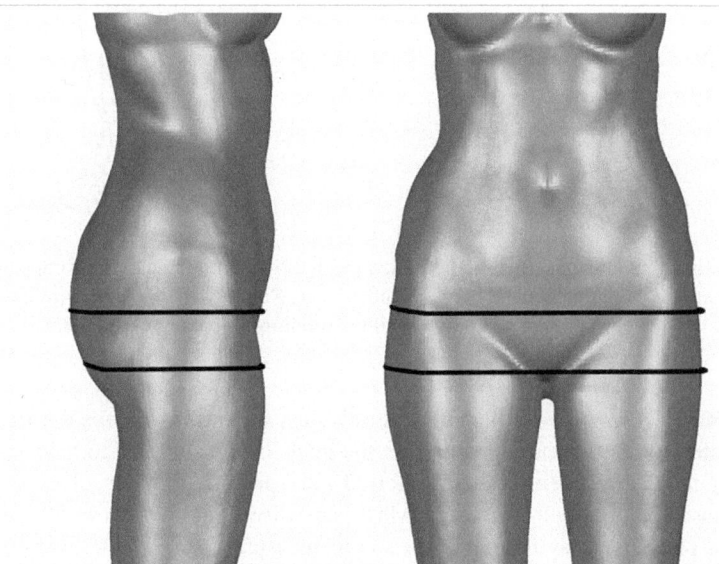

Fig. 6.11 Representations of the traditional hip measurement taken at the greatest protrusion of the buttocks (the higher measurement) and the largest circumference value that is generally lower on the body (where the bulge of the thigh adds to the circumference). The end-use requirements of the study will determine which measurement is most useful.

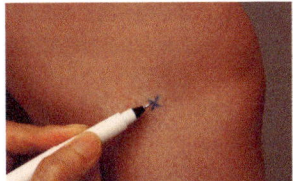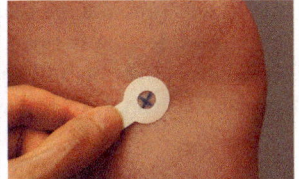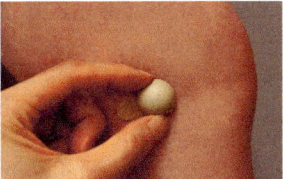

Fig. 6.12 Image of a physical landmark (a dome of a known dimension, highly engineered for a reliable size and shape) and a double-sided adhesive doughnut that ensures the dome can be precisely placed over a point landmark. This landmarking system is provided by Human Solutions.

Most scanner software developers recognize that the computer algorithms will not always place landmarks reliably and therefore make provision for easily indicating a different landmark point and retaking measurements. A trained operator can often improve the reliability and validity of a study by checking the measurements on a scan as they are taken and visually correcting landmark placement as necessary. In any case, scans should always be checked before the scanee leaves the booth to retake a scan if the person was not in the right position or if the scan is not optimized for some reason.

For optimal reliability of a study in which measurements will be derived, it is necessary to place landmarks manually, palpating for joint centers and marking the landmarks that are critical and cannot be reliably located automatically. If the scanner has capability to capture color or grayscale information along with the scan, then landmarks can be set with a washable marker or with an adhesive dot. If the scanner only takes *XYZ* coordinate data and no color information, a dimensional landmark can be used. Human Solutions makes a very precise dome, 20mm in diameter that can be captured in the scan (see Fig. 6.12). As the size of this dome is known, the precise landmark point can be identified at the top of the dome and transferred down to the surface of the scan. The bump of the landmark can then be removed from the scan.

Of course the advantages of scanning over manual measurements are the comparatively short amount of time needed with each study participant and the fact that highly trained assistants are not needed to conduct the study, as is true of a manually conducted anthropometric study. Manually placing landmarks adds time and requires skillful assistants, increasing the time and expense of an anthropometric study conducted with the scanner. However, if the number of manual landmarks is kept to a minimum, this training need not be overly extensive, and the study can be optimized for both reliability and cost.

6.7 How to assess scanner technologies for various end uses

With such a wide array of scanning systems available, with different capabilities, strengths, and weaknesses, scan systems exist for every end use. To decide the appropriate system for any one goal, many factors should be considered. Though it is convenient to use the same system to make the scans and to derive the desired data from the scans, this is sometimes not the optimal choice for the needs of the study. If one system is ideal for data collection and another for data extraction, it may be possible and desirable to purchase analysis software separately from the scan system.

If the primary goal is to derive linear measurements, for example, to conduct an anthropometric study to develop or improve a sizing system, or to collect measurements for a custom fit operation, the scan volume of the scanner can be limited. The most important parameters may be cost (to acquire multiple systems), size of the scanner, portability of the system, ease of setting up and taking down, and ease and reliability of calibration. The appeal or level of excitement generated by the experience of being scanned may be a factor, as the scanning experience itself, and the image generated on the screen are very different for different scanning systems. For example, for a younger demographic, a system that has the scanee drive the scanning process from an app on his/her own smartphone may be preferable. Depending on the needs of the study and access to trained operators or ability to train them, it may be desirable to acquire a system that will allow sensing of physical landmarks, which generally requires either a system that can collect color information or a system like the Human Solutions system for dimensional landmarks. One important factor in this case is ease of labeling the landmarks in the scan and access to an automated or semiautomated scan measuring system that allows selection and use of these points. Some systems will wrap the color image around the 3-D figure; it is preferable to look for a system that records color along with the 3-D points integrating the two datasets more effectively.

If the purpose of scanning is to conduct research requiring more interaction with the 3-D figure or 3-D printing, a system from which it is possible to create a highly detailed and watertight digital model may be preferable. Scan volume is an issue in this case, to allow capture of active body positions for research or groupings of people for 3-D printing (see Fig. 6.13).

For any scanning system, it is important to see an example of a raw scan and to understand how data are merged and cleaned to make a judgment on whether the system will provide the data desired for the particular end use. If linear measurements are the primary goal, then valid and reliable measurements can generally be acquired with fewer sensors, which can reduce both the footprint and the cost of the scanner. If more detailed 3-D information is desired, a system that captures more data will be preferable.

Another factor that is important with scanners is the technical support available for all issues in working with the scanner, from calibration, to acquisition of scans, to saving and merging of data, to output formats, to analysis, to methods of interfacing with other systems. The presence of a skilled and creative person with experience in computer systems, 3-D data, and data analysis on the research team and the use of generic software for viewing and manipulation of the scans can increase the range of possibilities for use of scan data.

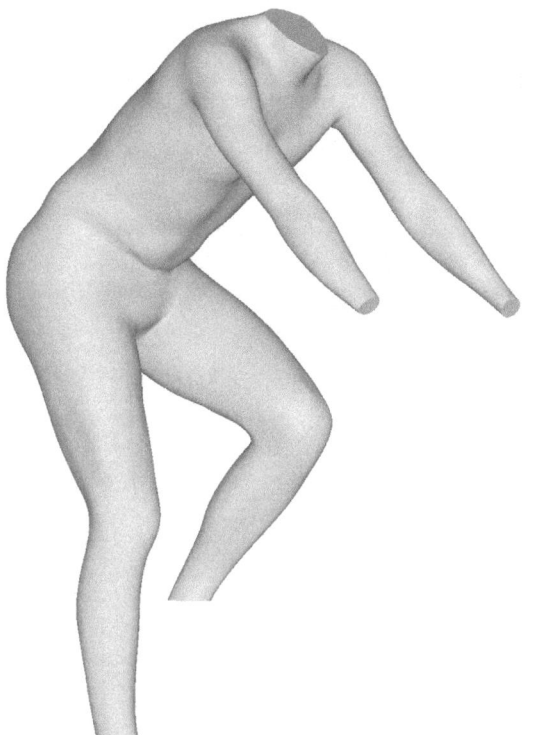

Fig. 6.13 Image of a body scan of a study participant on a bicycle. The ability to scan participants in active positions (which requires a large scan volume) allows researchers to investigate how body dimensions vary when the body is in motion.

6.8 Expanding the repertoire of measurements taken to make use of new concepts in the relationship of bodies to clothes

One of the primary benefits of 3-D body scanning is in the ability to keep a permanent digital model of study participants scanned, when they give permission in a signed consent form. This allows the researcher to return to the models when new knowledge or the need for additional measurements becomes apparent. As we work with the 3-D model, we have the opportunity to experiment with new ways to capture information about the body dimensions that can be difficult or impossible with a tape measure on the actual body. The "frozen in time and space" digital model makes it possible to measure in new ways, to ensure that adjacent measures are reliably related to one another (i.e., body movement or changes in posture have not impacted the relationship between measures), to take more measurements than can be tolerated if a person is required to stand stationary for a long period, to easily place points of reference around the figure such as planes from which to measure, to locate new measures such as largest or smallest circumferences that are difficult to find with a tape measure, and to capture actual body curves from the surface of the 3-D model.

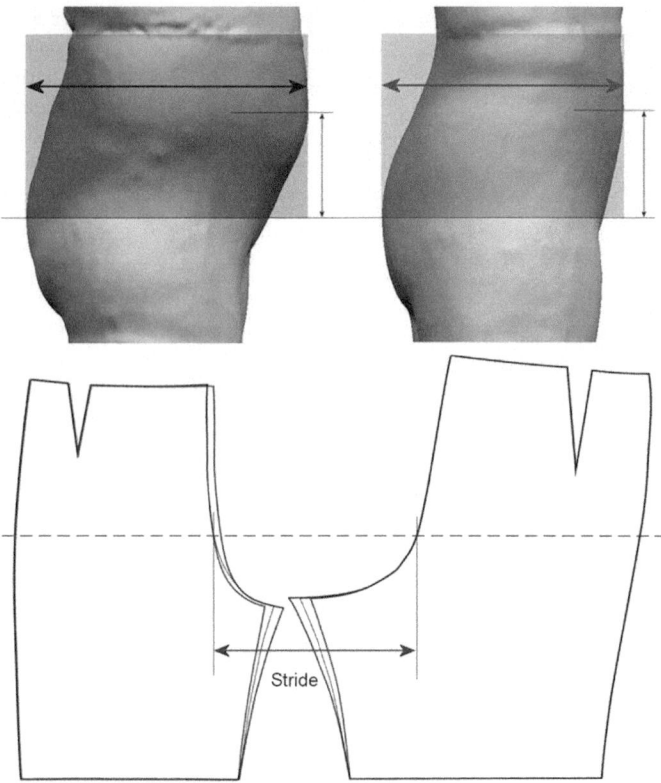

Fig. 6.14 Image showing a body depth measurement taken on two different study participants. This measurement is taken from the most posterior point on the buttocks to the most anterior point of the abdomen. This is not a traditional measurement for apparel production as it is difficult to take manually, but it directly relates to the two-dimensional pattern shown that is useful for fitting women's slacks.

All of these ways of manipulating the scan can add to our knowledge of actual body shapes, postures, and proportions. The next step is to find ways to incorporate this knowledge in the development of garment patterns (see Fig. 6.14).

6.9   Scans as avatars for fit analysis

In the place of fit avatars developed for virtual fit programs that are modified from linear body measurements, actual scans used as a fit avatar capture all body proportions, body configurations, and the unique posture of the individual. Virtual fit can be represented to more closely approximate the actual fit of the garment on the fit model or the individual, including the balance of the garment on the body. Software can identify joints within the scan and animate the scan to present different postures, or fit models can be scanned in multiple postures for analysis of the fit of clothing in active postures. Software to represent digital clothing on an avatar faithfully and in highly accurate detail is not yet developed. Fabric properties are not yet modeled in the detail needed to actually predict every wrinkle and stress fold in the garment when worn. However, some issues related to fit can be represented virtually at this stage, and further developments will improve these technologies over time.

6.10 Future trends

Future trends in body scanning for anthropometric studies will be driven by the recent and continual development of less expensive and more flexible ways of generating 3-D models of the body. Multiple technologies are being introduced now, some that use various sensed data to generate the model and others that depend on a large database of existing 3-D body scans and use more limited collected data to match and select an existing 3-D model from the database.

The introduction and improvement of infrared-based scanning combined with computing methods that interactively build a 3-D model from data generated as the sensor is moved around the body are making 3-D body scanning from handheld scanners practical. The cost of such scanners is a fraction of the cost of stationary scanners using laser light or white light. Studies show that the scans taken with these scanners, though lower in resolution than more traditional scanners, can be effective for the needs of an anthropometric study. Direct comparisons of the measurements taken by scanners using infrared light and scanners based on laser technologies indicate much promise in the new technologies (Soileau et al., 2016). Multiple sensors set up on tripods or incorporated into booth structures are also being marketed at much lower cost than earlier scanners. The portability of these systems makes them very appropriate for use in anthropometric studies.

It is difficult at this early stage to know which sensor devices will ultimately be successful in the market, but the Structure Sensor, designed to be connected to an iPad, is currently popular. This sensor company is also encouraging open-source development of software associated with the sensor, which opens up more possibilities for its ultimate use and success.

Systems to build a 3-D model from two-dimensional photographs have been under development for many years and are now reaching the stage of practical application. Models made from cell phone photographs are proliferating. One popular version uses a special bodysuit with graphics to assist in 3-D image generation. These systems are being introduced commercially, but few publications exist on studies of their effectiveness. However, once verified, these technologies can be revolutionary in their use in anthropometric studies. Their advantage is that they can be made available over the Internet, so that participants in anthropometric studies can be recruited remotely. Participants will be able to complete an online survey and then strip down and capture their body data in the privacy of their own home.

The same is true of systems that choose a scan from an existing database to match linear measurements taken of the body, but the inherent variability of the human form makes these systems less likely to be useful, because the number of variables necessary to find a perfect match is prohibitive.

Overall, testing is needed to establish reliability and comparability of the data from these new technologies to both manual measurements and more established scanners. Testing is also needed to establish protocols for scanning using these tools. For example, how quickly can a handheld scanner gather data on a whole body, and is it fast enough to prevent body sway? Even if a scan can be captured before sway in the body is introduced, stabilizing the arms will be necessary when scanning the body with a handheld scanner.

Overall, once this testing establishes the usefulness of these technologies, they have the potential to increase the number of anthropometric studies conducted, opening the door to studies on specific target markets, studies of children, studies of specific occupational groups, and longitudinal studies to track secular changes in a population. The reduced cost, portability, and accessibility of these technologies provide an impressive range of opportunities for anthropometric studies.

References

Ashdown, S.P., Sung Choi, M., Milke, E., 2008. Automated side-seam placement from 3D body scan data. Int. J. Cloth. Sci. Technol. 20 (4), 199–213.

Bougourd, J., Treleaven, P., 2014. National size and shape surveys for apparel design. Anthropometry, Apparel Sizing and Design. Woodhead Publishing, In, pp. 141–166.

Bradtmiller, B., Gross, M.E., 1999. 3D Whole Body Scans: Measurement Extraction Software Validation (No. 1999-01-1892). SAE Technical Paper.

Branson, D.H., Cao, H., Jin, B., Peksoz, S., Farr, C., Ashdown, S., 2005. Fit analysis of liquid cooled vest prototypes using 3D body scanning technology. J. Text. Apparel Technol. Manag. 4 (3) Available from: http://www.researchgate.net/publication/237462999_Fit_analysis_of_liquid_cooled_vest_prototypes_using_3D_body_scanning_technology.

Brownbridge, K., Gill, S., Ashdown, S.P., 2013. Effectiveness of 3D scanning in establishing sideseam placement for pattern design. In: Paper Presented at the 4th International Conference on 3D Body Scanning Technologies, Long Beach, CA e-space.mmu.ac.uk.

Brunsman, M.A., Daanen, H.M., Robinette, K.M., 1997. Optimal postures and positioning for human body scanning. In: Proceedings, International Conference on Recent Advances in 3-D Digital Imaging and Modeling, 1997. IEEE, pp. 266–273.

Choi, S.Y., Ashdown, S.P., 2010. 3D body scan analysis of dimensional change in lower body measurements for active body positions. Text. Res. J. 81 (1), 81–93. https://doi.org/10.1177/0040517510377822.

Corcoran, C.T., 2004. Technology Report: Mass Retailers Find Custom Clothing Fits Them Just Fine: Mass Retailers Embrace Custom. Women's Wear Daily, New York. 10, 12′.

D'Apuzzo, N., 2009. Recent advances in 3D full body scanning with applications to fashion and apparel. In: Gruen, A., Kahmen, H. (Eds.), Optical 3-D Measurement Techniques IX. Austria, Vienna.

Daanen, H.A.M., Jeroen van de Water, G., 1998. Whole body scanners. Displays 19, 111–120.

Daanen, H.A., Ter Haar, F.B., 2013. 3D whole body scanners revisited. Displays 34 (4), 270–275.

Daanen, H.A., Brunsman, M.A., Robinette, K.M., 1997a. Reducing movement artifacts in whole body scanning. In: 3dim. IEEE, p. 262.

Daanen, H.A., Brunsman, M.A., Robinette, K.M., 1997b. Reducing movement artifacts in whole body scanning. In: 3dim. IEEE, p. 262.

Deng, M., Wang, Y., Li, P., 2018. Effect of air gaps characteristics on thermal protective performance of firefighters' clothing: a review. Int. J. Cloth. Sci. Technol. 30 (2), 246–267.

Devarajan, P., Istook, C.L., 2004. Validation of female figure identification technique (FFIT) for apparel software. J. Text. Apparel Technol. Manag. 4 (1), 1–23.

Haber, H., 2006. WWDexectech: Better Dress Forms through Scan Data. Vol. 191(72). Women's Wear Daily, New York, p. 9.

Hotzman, J., Gordon, C.C., Bradtmiller, B., Corner, B.D., Mucher, M., Kristensen, S., Blackwell, C.L., 2011. Measurer's Handbook: US Army and Marine Corps Anthropometric Surveys, 2010–2011. No. NATICK/TR-11/017), ARMY NATICK SOLDIER RESEARCH DEVELOPMENT AND ENGINEERING CENTER MA.

Huang, H.Q., Mok, P.Y., Kwok, Y.L., Au, J.S., 2011. Determination of 3D necklines from scanned human bodies. Text. Res. J. 81 (7), 746–756.

Hye, J., 1999. Body Scanning has Immediate Applications. Vol. 178(39). Women's Wear Daily, New York, p. 15.

ISO International Standard, 2010. 3-D Scanning Methodologies for Internationally Compatible Anthropometric Databases, second ed. Reference number ISO 20685:2010(E).

Joseph-Armstrong, H., Hagen, K., Maruzzi, V.J., 2010. Patternmaking for Fashion Design. Pearson Education/Prentice Hall, Upper Saddle River, NJ.

Kim, D.E., LaBat, K., Bye, E., Sohn, M., Ryan, K., 2015. A study of scan garment accuracy and reliability. J. Text. Inst. 106 (8), 853–861.

Kouchi, M., Mochimaru, M., 2011. Errors in landmarking and the evaluation of the accuracy of traditional and 3D anthropometry. Appl. Ergon. 42 (3), 518–527.

Lansard, M., 2018. The Best 3D Scanners. ANAWII. https:/www.aniwaa.com/best-3d-body-scanners/ (Accessed 1 February 2019).

Lerch, T., MacGillivray, M., Domina, T., 2007. 3D laser scanning: a model of multidisciplinary research. J. Text. Apparel Technol. Manag. 5 (4), 1–8.

Mckinnon, L., Istook, C., 2001. Comparative analysis of the image twin system and the 3T6 body scanner. J. Text. Apparel Technol. Manag. 1 (2), 1–7.

Mckinnon, L., Istook, C.L., 2002. Body scanning: the effects of subject respiration and foot positioning on the data integrity of scanned measurements. J. Fash. Market. Manag. 6 (2), 103–121.

Paquette, S., Brantley, J.D., Corner, B.D., Li, P., Oliver, T., 2000, July. Automated extraction of anthropometric data from 3D images. In: Proceedings of the Human Factors and Ergonomics Society Annual Meeting. Vol. 44(38). SAGE Publications, Sage CA: Los Angeles, CA, pp. 727–730.

Park, J., Langseth-Schmidt, K., 2016. Anthropometric fit evaluation of firefighters' uniform pants: a sex comparison. Int. J. Ind. Ergon. 56, 1–8.

Pei, J., Fan, J., Ashdown, S.P., 2019. Detection and comparison of breast shape variation among different 3D body scan conditions: nude, with a structured bra, and with a soft bra. Text. Res. J. 1–12. https://doi.org/10.1177/0040517519839398.

Robinette, K.M., Whitestone, J.J., 1992. Methods for Characterizing the Human Head for the Design of Helmets. No. AL-TR-1992-0061)ARMSTRONG LAB WRIGHT-PATTERSON AFB OH.

Robinette, K.M., Whitestone, J.J., 1994. The need for improved anthropometric methods for the development of helmet systems. Aviat. Space Environ. Med. 65 (5 Suppl), A95–A99 (0095–6562).

Robinette, K.M., Daanen, H., Paquet, E., 1999. The CAESAR project: a 3-D surface anthropometry survey. In: Proceedings. Second International Conference on 3-D Digital Imaging and Modeling, 1999. IEEE, pp. 380–386.

Roetzel, B., Fritz, E., 2014. Bespoke Menswear: Tailoring for Gentlemen. H.F. Ullmann Publishing, Berlin.

Soileau, L., Bautista, D., Johnson, C., Gao, C., Zhang, K., Li, X., et al., 2016. Automated anthropometric phenotyping with novel Kinect-based three-dimensional imaging method: comparison with a reference laser imaging system. Eur. J. Clin. Nutr. 70 (4), 475.

Song, H.K., Ashdown, S.P., 2011. Categorization of lower body shapes for adult females based on multiple view analysis. Text. Res. J. 81 (9), 914–931.

Suikerbuik, R., Tangelder, H., Daanen, H., Oudenhuijzen, A., 2004. Automatic Feature Detection in 3D Human Body Scans (No. 2004-01-2193). SAE Technical Paper.

Wren, P., Gill, S., Apeagyei, P., Hayes, S., 2014. Establishing a pre and post-3D body scanning survey process for able-bodied UK women aged 55 years+ to determine an appropriate waist position for garment development. In: 5th International Conference on 3D Body Scanning Technologies, 21 October 2014–22 October. Lugano, Switzerland, p. 2014.

Functional measurements and mobility restriction (from 3D to 4D scanning)

7

Anke Klepser, Simone Morlock, Christine Loercher, Andreas Schenk
Hohenstein Institute for Textilinnovation gGmbH, Boennigheim, Germany

7.1 Introduction

Clothing has various functions. Alongside the culturally conditioned covering of nakedness, it should also protect the human body from environmental conditions. At the same time, clothing cannot be a rigid shell, as it is worn during movement. In fashionable everyday clothing, relatively moderate position changes of the body are carried out, but in functional sport clothing or protective work clothing, highly dynamic movements are carried out in some instances. To make sure clothing products can guarantee an optimal fit, a specific protective function, and an ergonomic comfort, substantiated analysis of the body is necessary, because clothing can only protect well when it fits well, especially in movement. For the development of standard body measurement tables, in part, laborious series of measurements were designed, which, in the last few years, have been mostly carried out with 3-D body scanners (Kouchi, 2014; Bougourd, 2014; Hohenstein Institute and Human Solutions GMBH, 2008; Morlock, 2015b; Morlock et al., 2009). Manually, digitally, or otherwise, body measurements are recorded standardized in a standing posture (ISO—International Organisation for Standardization, 1989).

Body measurements are the basis for the development of clothing products. In movement, various muscles are activated. The body surface changes, and therefore so do the body dimensions. These movement-related changes of the body measurements had been neither studied nor implemented in size charts until now. Thus a disparity occurs between the basics of standard size charts and the application requirements of the user. In the area of fashion, this can lead to fit problems. However, in sport and work clothing as well as personal protective clothing, the consequences are far more serious. Here, it can lead to mobility restriction, reduced performance, or even limited protection for the user. Therefore it is of great importance that the movement-related dimensional changes in measurements (functional measurements) and their influence on the interaction between the wearer and the clothing are scientifically researched. Through the use of 3-D scanning technology, studies in this topic area are much easier and quicker to carry out, especially as large groups of test subjects can be very effectively measured. For the first time, Morlock et al. (2018) applied 3-D body scanners to establish functional measurements. They have dealt in detail

Anthropometry, Apparel Sizing and Design. https://doi.org/10.1016/B978-0-08-102604-5.00007-X

with the systematic recording of functional measurements the analysis of differing measurements, which result from the comparison of measurements in different motion postures with the standard standing position, and the practical preparation of data. The study and its results are presented in Section 7.2.

Three-dimensional body scanners, in addition to recording body surfaces to analyze body measurements, are being increasingly used to scientifically examine the fit of clothing products. This means initial studies do analyze the fit and function of the products not only when standing but also in typical practical positions (Baytar et al., 2012; Ashdown et al., 2005; Ashdown, 2011). Thus restrictions of mobility and movement caused by clothing can be not only visually evaluated but also quantified. Section 7.3 of the chapter describes the research situation of the assessment of mobility restrictions due to clothing starting from the first analog approaches.

The recording of movement and its influence on body dimensions or ergonomic wearing comfort with a 3-D scanner is only possible step by step. Everything must be done position by position, with each one being held during the recording time. Therefore it is unclear whether the results can illustrate the real dynamic changes of the body. The technical development in the area of 4-D scanning systems (scanning during movement) opens a whole new promising research method for detailed requirement analysis. A look at this technology concludes the chapter with Section 7.4.

7.2   Functional measurements

Body measurements, on which the development of clothing is based, sometimes differ seriously from the measurements of size charts in the practice of movement (standing, sitting, kneeling, bending, etc.). Length and circumference measurements sometimes change considerably, as shown in Fig. 7.1. The distance between the seventh cervical vertebrae and forth lumbar vertebrae (length of the back contour) of a male subject with the German size 50 (chest girth: 100 cm) has a difference of 12.3 cm between

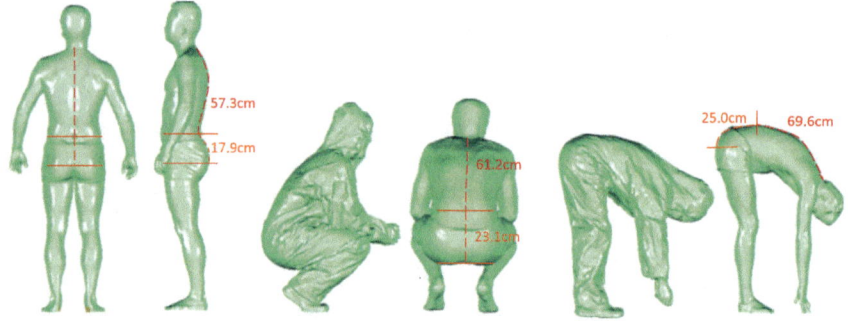

Fig. 7.1 Change of body measurements in different body positions.

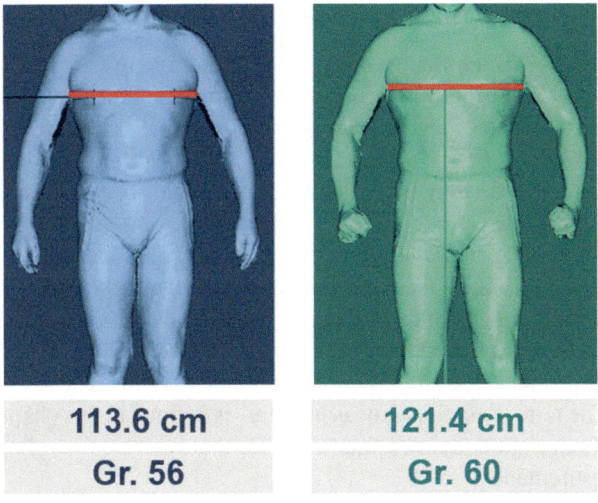

113.6 cm   **121.4 cm**
Gr. 56   **Gr. 60**

Fig. 7.2 Change in the chest girth with a slight lifting movement of the arms.

standing and a bent over forward body position. This means an ease of 21.5% is caused by the different body position. The small measurement "hip depth" is extended by 7.1 cm, which is an increase of 39.7%.

Furthermore, movement-related differences in measurements have a significant effect on the size allocation for the renting, shipping, and online retailing of work clothing and protective clothing. This is because the chest girth, which, as the primary measurement, determines the assignment of clothing sizes, changes considerably as shown in Fig. 7.2. The chest girth of a test subject varies between the German sizes 56 (chest girth: 112 cm) and 60 (chest girth: 120 cm) only on the basis of an easily simulated light lifting motion. This is because every clothing size covers a scope of 4 cm (e.g., size 56 covers the chest girth of 110–114 cm). The range, therefore, includes three clothing sizes or a difference in the chest girth of about 8 cm.

Obviously dimensional changes like these affect the fit and the ergonomic wearing comfort of the clothing. However, it is not yet known what the average degree of extremity of movement-related changes is nor to what extent the fit and comfort are affected by it. This blank space should be filled by the "functional measurements" study (Morlock, 2015a, b). The aim of the project was the foundational research of movement-related body changes and the deduction of the respective dimensional differences (functional measurements). For the efficient application of functional measurements in clothing, the results were implemented into both ergonomic- and movement-oriented size system. The connection to traditional clothing sizes and known size designations ensures the acceptance and application of the new size system in practice. The focus was on the ergonomic design of work clothing and protective clothing and thus the analysis of typical postures and movements.

7.2.1 Definition of functional measurements

Body measurements, which change during movement, and research into them have, to this day, no clear definition. In various scientific works, names such as dynamic or functional anthropometry and dynamic or functional measurements are used (Gersak, 2014; Gupta, 2014; Todd and Norton, 1996). So far, this describes in most cases the manual or three-dimensional static recording of body measurements in different posture positions.

A dynamic recording of the body surface is already possible through the technical development of 4-D scanners (see Section 7.4) (Guan et al., 2009, 2012). A derivation of functional measurements of these has not yet been scientifically implemented. But this should be seen as the next step in the research of dimensional changes through movement. Against this background a differentiation of terms is necessary to delineate the methods. The term "functional measurement" is defined in this chapter as the static recording of changes in body measurements: for the dynamic recording, accordingly "dynamic measurements."

7.2.2 Distinction between standard measurements, functional measurements, and ergonomic measurements

For the design of clothing, personal protective equipment (PPE), workplaces, and man-machine interfaces, anthropometric data are used. For these in general, two different measuring systems are used: size charts (Hohenstein Institute and Human Solutions GmbH, 2008) and ergonomic standards (DIN Deutsches Institut für Normung e.V., 2005; Scheffler and Schüler, 2013; Jürgens et al., 1998; Hsiao et al., 2014; ISO—International Organisation for Standardization, 2011). In manufacturing, size charts serve as a basis. The variation of body measurements in the population is taken into account by the definition of clothing sizes for different body types. Each clothing size is assigned production-relevant dimensions such as body height and waist girth. The recording of clothing-technical measurements, whether for manufacture or individual production, is carried out using the standard anthropometric posture. In this the person being measured is upright, legs hip width apart, and arms slightly spread laterally (Ashdown, 2011; Kouchi, 2014; Morlock, 2015a, b; ISO—International Organisation for Standardization, 2010).

The movement-related variation in body measurements is partially reproduced in ergonomic standards (DIN Deutsches Institut für Normung e.V., 2005; Scheffler and Schüler, 2013; Jürgens et al., 1998; Hsiao et al., 2014). The focus of these ergonomic standards lies on the design of workplaces, machines, and safety products. Ergonomic measures are usually represented as the 5th, 50th, and 95th percentile and are differentiated according to sex and age group. However, these only indicate the percentage distribution of a measurement within a sample of people. The size reference necessary for the construction of clothing does not exist, so the ergonomic data cannot be implemented in manufacturing in accordance with the size. Functional measurements bind the two questions. On the one hand, they produce the size reference; on the other hand, they show the body changes due to movement. This is considered an important foundation for development of the production of clothing products with high ergonomic wearing comfort.

7.2.3   Analysis of functional measurements

To derive and analyze functional measurements on a larger sample size for the first time, a measurement study was initiated (Morlock, 2015a, b). For this, 93 subjects (men and women) between the ages of 17 and 65 were recorded and measured with a 3-D body scanner. The three-dimensional recording of the subjects was done with the Vitus Smart XXL. They were scanned at Hohenstein in the south of Germany. The participants were from the local area. Nevertheless the random sample shows a good regional distribution.

To cover the variety of body shapes of the population, the clothing size system of the last German sizing survey SizeGERMANY is divided into body height rows and figure types (Hohenstein Institute and Human Solutions GMBH, 2008). The types of men's figures are determined by the difference between the chest girth and waist girth and in women by the difference between the bust girth and the hip girth. For men the body heights are differentiated between extra short, short, normal, tall, and extra tall. The types of figures are divided into extra slim, slim, normal, heavy, and extra heavy. The women accordingly have the differentiated body height rows of short, normal, and tall and the figure types narrow hips, normal hips, and wide hips.

The sample encompasses the size ranges: 42–64 (chest girth: 84–128 cm) for men and 36–52 (bust girth: 72–104 cm) for women. The body heights and figure types could be largely covered. Although the subjects were randomly selected, a good distribution of body shapes could be achieved.

7.2.3.1   Landmarking

The prerequisite for recording reproducible body measurements in measuring studies is working with anthropometric landmarks. Many landmarks are identified in anthropometry by palpation of bony structures under the skin, for example, the spinous process of a vertebra. Three-dimensional laser scanners only capture the surface of the body. Bony structures are only visible, if at all, on slim people, who are assigned to the small clothing sizes. To be able to reproducibly measure and analyze the length and circumference measurements, physical markers must be applied to the subjects' bodies prior to the scanning process. So the changes can be traced and contrasted by the movement.

In the area of motion capturing, retroreflective markers are used to perform motion analysis. These enable the identification and tracking of anthropometric points by camera systems. In principle, these markers would also be suitable for use in 3D scanners, but they are very expensive due to the retroreflective surface in the purchase price. A cost-effective alternative is commercially available polystyrene balls. In the project, various sizes were tested.

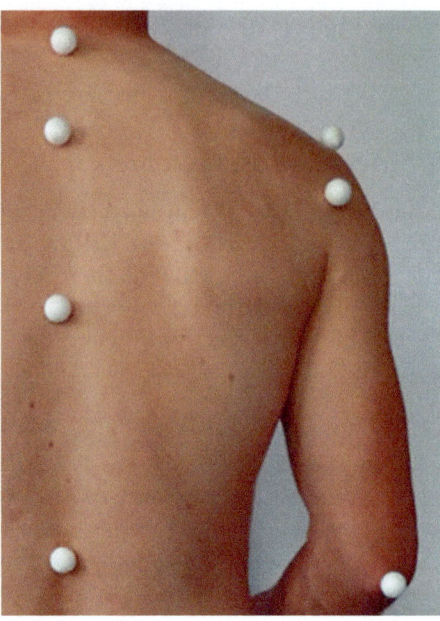

Fig. 7.3 Example of physical markers on the human body according to anthropometric landmarks.

Here, there were two requirements: on the one hand, the smallest possible size, so that measuring sections are not negatively affected, and, on the other hand, that the scanner system can detect these reliably and the visibility is given on the 3-D scan.

Polystyrene balls with a circumference of around two centimeters fulfilled these conditions. The markers were attached to predefined anthropometric landmarks (e.g., the seventh cervical vertebrae and acromion) using a double-sided adhesive tape on the skin (see Fig. 7.3).

In the study, 16 markers on anthropometric landmarks and auxiliary points (such as arm crease) were adhered per test person prior to the scan (see Fig. 7.4).

7.2.3.2 Posture

The analysis of the movement-related change of body measurements is the focal point of the functional measurements. The challenge was to define positions that can be captured with a 3-D scanner and put into practice by subjects in a reasonable time window (see Section 7.2.3.3). Therefore a limit of 10 scan positions was required. Whole-body scans can be carried out in a few seconds. Nevertheless, it is scientifically proven that postures of subjects exhibit a large variety (Han et al., 2010; Lashawnda and Istook, 2002; Lu and Wang, 2010; Schwarz-Müller et al., 2018). The postures to be taken for the scan are usually communicated verbally by the scan personnel to the test person. To implement the instructions the subjects need the ability of proprioception, that is,

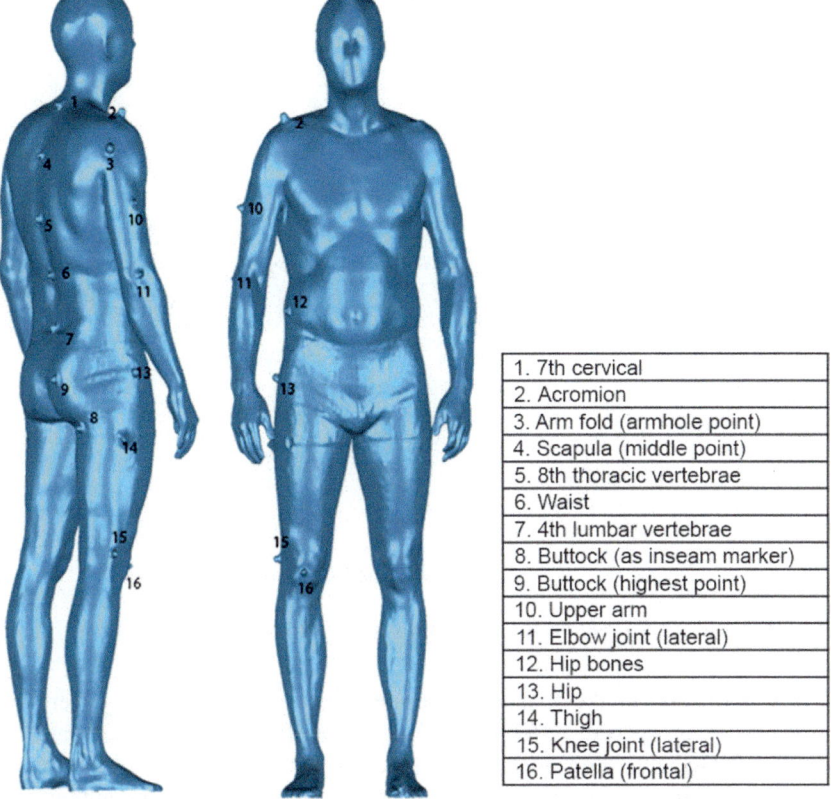

1. 7th cervical
2. Acromion
3. Arm fold (armhole point)
4. Scapula (middle point)
5. 8th thoracic vertebrae
6. Waist
7. 4th lumbar vertebrae
8. Buttock (as inseam marker)
9. Buttock (highest point)
10. Upper arm
11. Elbow joint (lateral)
12. Hip bones
13. Hip
14. Thigh
15. Knee joint (lateral)
16. Patella (frontal)

Fig. 7.4 Position of the marker.

the awareness of body position and body movement in space. This is not equally distinct in all subjects. The reproducibility of body postures by the subjects represents a significant challenge in the investigation of functional measurements. Even to perform simple postures, mobility and holding strength are necessary. For example, in the case of one posture in the project, the subjects had to lift the right arm sideways to shoulder height (see position number 5 in Fig. 7.5). This could not be performed equally by all the subjects. Some did not have the shoulder joint mobility to bring their arm to shoulder height. Others, on the other hand, found it hard to develop the power to keep their arms steady for the period of recording. Similar differences in performance were found in the "Bend" and "Squat" positions. The problems that arose particularly in these positions are described in more detail hereafter (see Section 7.2.5). In general a special challenge can be seen in the fact that the static attitudes in the 3-D detection relate to movements. It is certainly much easier to perform movements in natural speed. Four-dimensional scanning systems will provide a solution to this challenge in the future. Nevertheless, future research will also have to deal with the problem of reproducible forms of movement.

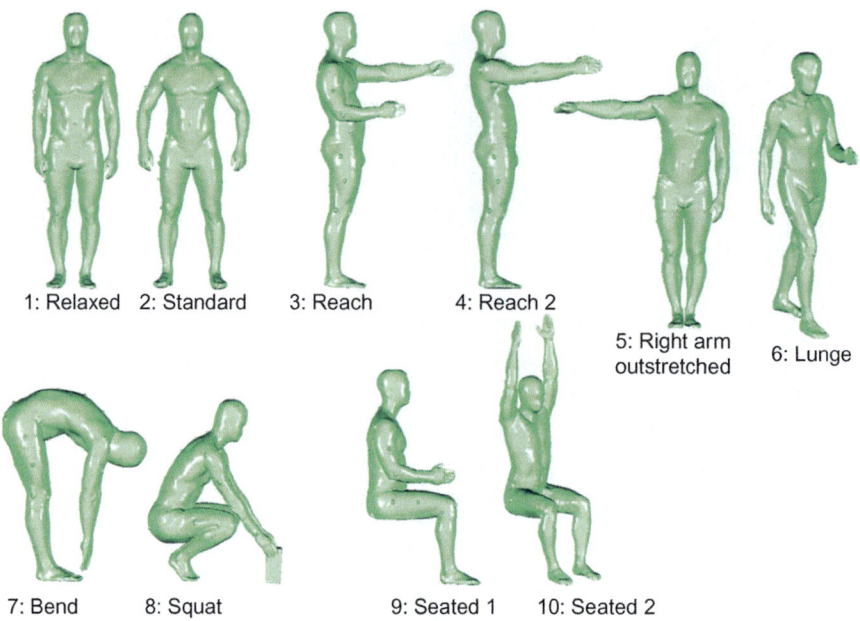

1: Relaxed   2: Standard   3: Reach   4: Reach 2

5: Right arm   6: Lunge
outstretched

7: Bend   8: Squat   9: Seated 1   10: Seated 2

Fig. 7.5 The defined positions for the analysis of the range of movement (ROM).

The derivation of the functional measurements is based on the difference that results when comparing different body positions, for example, the "Bend" to the standing ISO standard posture, which is used to determine the body measurements. The standing standard posture has been called "Relaxed" since the implementation of the German sizing survey SizeGERMANY (Hohenstein Institute and Human Solutions GMBH, 2008; ISO—International Organisation for Standardization, 2010). It serves as a reference for all measuring sections (Kouchi, 2014; ISO—International Organisation for Standardization, 1989). In addition, the "Standard," "Reach," and "Seated 1" positions were recorded in the same way as the SizeGERMANY sizing survey. Additionally, six further positions were defined. These positions are "Reach 2," "Right Arm Outstretched," "Lunge," "Bend," "Squat," and "Seated 2" (see Fig. 7.5). All 93 subjects were scanned in these positions.

7.2.3.3   Recording of the movement-oriented positions

The 3-D recording of the positions was done with the Vitus Smart XXL 3-D body scanner (Vitronic, n.d.). The scanner works with the optical triangulation measurement principle and provides a reliable and accurate system that has been tested in numerous previous measuring studies and allows accurate, noncontact, three-dimensional imaging of the human body. The advantage of this method is that, in contrast to the manual measurement, 3-D data on posture can be extracted and the scanned data are still available for later evaluations. It is part of the category of laser scanners, which scan the surface by means of light and thus create a 3-D scan. This has the advantage that the scanning process is relatively fast (duration: approx. 10s/scan) and the physical markers on the subject are easily recognizable.

The disadvantage of using a 3-D body scanner for such a measurement survey is that the scan area is limited and, depending on the position, it can lead to severe shadowing of individual parts of the body (Kouchi, 2014). The detectable area amounts to $2100 \times 750 \times 750$ mm. Four lasers and eight cameras are positioned in the four columns. The rigid setting makes it difficult to detect horizontal areas and areas such as the armpit or crotch. Furthermore, for positions like "Bend" or "Squat," body parts conceal other areas. This also leads to unrecorded points. The limited recording space of the scanner means that not all positions can be scanned. Therefore positions had to be adjusted and referred to the examined areas.

To reduce shadows the use of handheld 3-D scanners were discussed. Compared with 3-D body scanners, they can also be used to detect areas that are concealed by other parts of the body. This is because the device is guided manually around the body. The method requires experience in operation. Even with adequate skill a full-body scan takes several minutes. As a result the subject has to hold the individual positions much longer in comparison with the 3-D body scanner. This is neither feasible nor reasonable for many subjects in terms of physical condition. Due to the long recording time, individuals may be scanned using handheld systems. For larger groups of subjects, these scanners are not the method of choice.

For future studies in the field of functional measurements, it must be examined whether body scanners with structured light are better suited for this usage. These enable even shorter recording times and less shadowing than a laser scanner, since significantly more cameras with different viewing angles can be used.

7.2.3.4 Dimensions

In total, 21 measurements were taken from each of the 93 persons in the 10 different positions (see Table 7.1), which were taken on the 3-D scans in 10 different positions with the aid of the measurement software ScanWorx (Human Solutions Gmbh, n.d.). These include, alongside the primary measurements, chest girth, other girth, length, and distance measurements. The definition of the measuring sections is basically based on the ISO standard (ISO—International Organisation for Standardization, 1989). However, some measurement definitions had to be adjusted. The arm length was not measured according to the standard from the acromion to the wrist bone, but from the marker position of the "arm crease" (see Section 7.2.3.1) to the wrist bones. This allowed the determination of the maximum arm reach in the movement that, in connection with the change in back width, must be covered by a clothing product. The measuring section of the upper arm circumference was also recorded differently. In accordance with standards, measurements are taken at the strongest point directly at the transition from the shoulder to the arm. However, this rule cannot be applied to the measurement taken by 3-D body scans. This is because of the fact that large shadowing occurs at the transition from the shoulder to the arm, and so the measurement cannot be taken comprehensively. A standard-compliant registration would lead to measurement errors. The measurement values would be too small

Table 7.1 Defined measurements

	Measurements
1.	Body height
2.	Chest circumference
3.	Underbust circumference
4.	Waist circumference
5.	Distance—waist to apex point
6.	Hip circumference
7.	Distance—waist to seat (in seating position)
8.	Leg length
9.	Shoulder width
10.	Across back width
11.	Upper arm circumference
12.	Back waist length
13.	Distance—seventh cervical to fourth lumbar vertebrae
14.	Distance—waist to crotch level
15.	Knee height
16.	Calf circumference
17.	Arm length—standard
18.	Arm length—arm fold to finger tips
19.	Arm length—center-back-neck-point-to-wrist length
20.	Upper arm length
21.	Thigh circumference

(Kirchdörfer, 2009). When examining functional measurements, it is crucial to identify areas of the body that are subject to major changes. This was presumed with regard to the upper arm girth not directly at the transition from the shoulder to the arm. For this reason the marker determining the measurement was set on the most pronounced form of the upper arm, when viewed laterally.

In addition to the lengths, girths, and distances, new measuring sections were recorded in the project, such as the distance between seventh cervical and fourth lumbar vertebrae, to be able to determine the dimensional differences relevant for clothing technology. All defined and recorded measurements are listed in Table 7.1.

The registration of individual measurements is not necessary in every position. Because, in the different positions, depending on girth, proportional shadowing and overlapping of different body areas can appear on the scan. This can lead to individual measurements not being able to be exactly determined. Here the reliability of the respective measurement had to be evaluated individually for each scan. Nevertheless the aim was to register as many dimensions as possible for as many postures as possible to ensure the comparability of the individual measurements.

From 93 scans in 10 positions, a total of around 8000 body measurements were taken. This demonstrates the very high processing effort that is necessary to research the functional measurements. Even though the sample size of 93 subjects seems to be

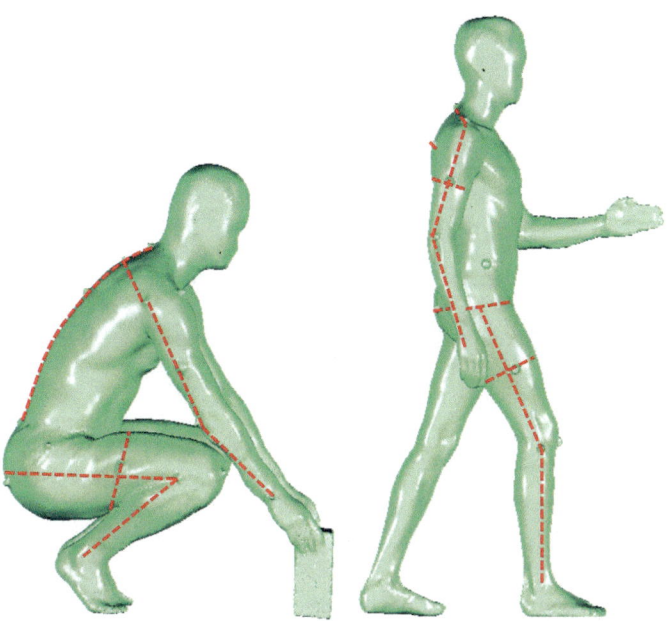

Fig. 7.6 Example representation of the functional measurements.

rather small compared to other series measurements, the number of measurements to be taken multiplies due to the different forms of movement to be investigated.

Fig. 7.6 shows an example of the registration of individual body measurements in two scanning positions. The markings illustrate the measurement positions. Comparability and reproducibility are important points not only for the definition of the measurements but also for the acquisition of the scans.

7.2.4 Analysis of the functional measurements

The main question before starting the analysis was in what kind of form the functional measurements should be presented. It made more sense only to represent the measurement differences that arise between the different positions and the ISO standard posture, instead of the complete measurements (see Section 7.2.3.2). This makes it possible to use the functional measurements independently of a specific size system.

As the first step the complete range of the measurement forms in the different positions was shown broken down. Fig. 7.7 shows this approach with the example of the back width. From left to right, five different positions can be seen ("Reach 2" to "Right Arm Outstretched"). For each position the line graph visualizes the minimum (dashed line), average (continuous line), and maximum (dotted line) difference of the measurement. For example, the back width of the subjects in the "Seated 2" position shows a variance of 16.2 cm, a minimum of −15.5 cm, an average of −8.7 cm, and a maximum of 0.7 cm. The range between the smallest and the largest measurement in the

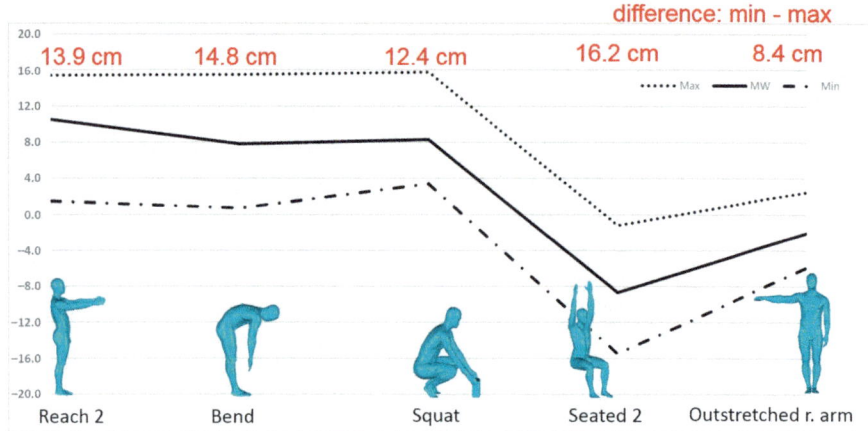

Fig. 7.7 Evaluation of the back width measurement.

positions is significant. This must be able to cover a clothing product to ensure ergonomic wearing comfort. However, only the positive changes are relevant for the clothing construction. Thus the increase in the back width in the "Reach 2" position must be taken into account when developing the pattern. The reduction of the back width in "Seated 2," on the other hand, does not matter, because in this case the clothing product cannot cause any restriction in movement.

In a further step, these results were considered separately gender-specific according to clothing sizes. The aim was to provide pattern makers and designers with measurement parameters for clothing development, especially for the pattern making, which enable the development of products that support, not limit, the human ROM.

The evaluation of the measurement differences showed a large variance of individual values. This may be due not only to the sample size but also to the fact that the subjects were not able to reproduce the defined positions with 100% accuracy. Thus no significant differences between the sizes could be derived. To achieve a higher statistical certainty, three size groups were formed: small men's size, German size 44–48 (chest girth: 69–88 cm) or small women's size, German size 36–40 (bust girth: 72–80 cm), medium men's size 50–54 (chest girth: 100–108 cm) or medium women's size 42–46 (bust girth: 84–92 cm), and large men's size 56–60 (chest girth: 112–120 cm) or large women's size 48–52 (bust girth: 96–104 cm). The advantage is that a larger number of values are found in each group.

For each of these size groups, the average value, as well as the minimum and maximum values, was recorded. Furthermore, minimal differences between the body height ranges could be identified. These were also accounted for. In regard to the figure types, no differences could be determined. This could be due to the small sample size but possibly also to the fact that figure types in general have a smaller influence on the functional measurement. This must be specified in further research.

7.2.5   Results

The results of the analysis can only be presented in part in this section. The amount of postures and recorded body measurements does not allow for a full-scale presentation. Therefore, particularly striking results will be discussed here. The back width showed great differences in the individual positions and is presented here as an example for the other 20 measurements. The individual execution of the positions by each subject had a large influence on the collected values. Especially in the "Bend" and "Squat" positions, large variances arose with a significant influence on the back length. These two positions are also considered in the succeeding text. The influence of muscle work on the body geometry, using the example of the upper arm, is found at the end of the result analysis. But first the general structure of the functional measurement tables is presented.

7.2.5.1   Structure of the functional measurement tables

In the amalgamation of the investigated measurement differences, the conception of a multidimensional matrix was paramount. In contrast to the conventional size charts, a complete representation of the functional measurements in relation to clothing size and size range for different types of movement for men and women can be implemented holistically. The size system was coupled with the traditional German sizing charts for men and women in regards to primary dimensions and size designation but was simplified by the formation of size groups. The measurement differences divided into small, medium, and large sizes can easily be transferred to other systems.

As in the SizeGERMANY sizing survey, a differentiation according to the body height series of short, normal, and tall was also carried out here. In addition to the average values of the individual measurement differences, the new tables also show the maximum extent of the individual measurements. The variance of the individual values is immense, and therefore the average sometimes shows a rather restrained dimensional change in the movement. The maximum value, however, shows the maximum required ROM that a clothing product should cover.

Fig. 7.8 shows, in part, a functional measurement table resulting from the findings of the measurement survey. This table shows the average value, that is, the average deviation or average degree of severity of the different measurements for the reference position (see Section 7.2.4) and the maximum value. Additionally, the data were structured according to both positions 4–10 (see Section 7.2.5) and size groups. The advantage of this type of presentation lies in the comparability. Thus, in the construction, the measure that changes the most, for example, can be directly picked out.

Due to the complexity of the tables, the results were transcribed into a pivot table. This allows the application-specific selection of results that are to be displayed using the filter function. Positions or measurements can be displayed individually or in comparison. Thus the large amount of data can be put together in a user-friendly and individual manner.

Men size

Measurement	44-48 (chest circumference: 88-96 cm)						50-54 (chest circumference: 100-108 cm)						56-60 (chest circumference: 112-120 cm)					
	4	5	6	7	8	10	4	5	6	7	8	10	4	5	6	7	8	10
Chest circumference [cm]	-1.0		-1.5	-1.2	-0.8	-3.0	-1.5		-2.5	-1.4	-0.8	-4.5	-2.5		-3.0	-1.6	-0.8	-7.5
MAX	2.9		0.3	2.7	2.8	-2.0	2.5		0.4	2.5	3.0	-2.5	1.5		0.3	4.0	3.0	-4.0
Waist circumference [cm]	0.5			5.0		0.0	0.5			5.0		0.0	0.5			5.0		0.0
MAX	1.7			10.7		3.4	2.2			7.9		2.4	1.8			9.2		2.4
Hip cirumference [cm]	0.5			3.0			0.5			3.5			0.5			4.0		
MAX	0.9			5.9			1.0			6.9			0.4			6.9		
Underbust circumference [cm]	0.5			5.0		0.0	0.5											

Fig. 7.8 Extract of the functional measurement table (men).

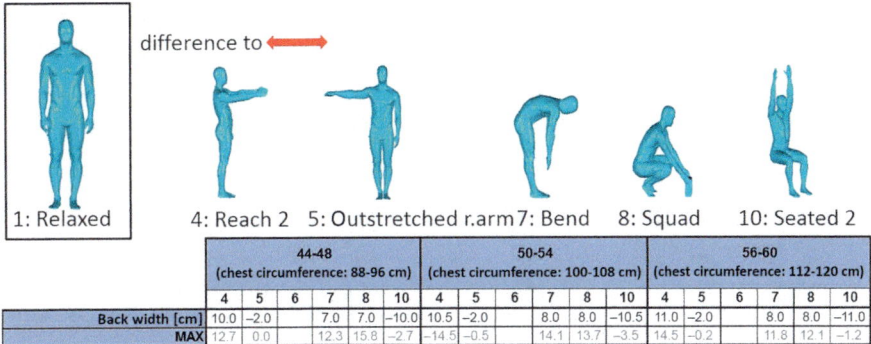

difference to ←——→

1: Relaxed   4: Reach 2   5: Outstretched r.arm   7: Bend   8: Squad   10: Seated 2

	44-48 (chest circumference: 88-96 cm)						50-54 (chest circumference: 100-108 cm)						56-60 (chest circumference: 112-120 cm)					
	4	5	6	7	8	10	4	5	6	7	8	10	4	5	6	7	8	10
Back width [cm]	10.0	−2.0		7.0	7.0	−10.0	10.5	−2.0		8.0	8.0	−10.5	11.0	−2.0		8.0	8.0	−11.0
MAX	12.7	0.0		12.3	15.8	−2.7	−14.5	−0.5		14.1	13.7	−3.5	14.5	−0.2		11.8	12.1	−1.2

Fig. 7.9 Extract from the table of differences in measurements (back width).

7.2.5.2 Measurement example: Back width

In addition to the length of the back, the distance "waist to lower torso mark," and the arm reach, significant dimensional changes in back width were determined. The average difference of the "Relaxed" to "Reach 2" position in men was 10.0 cm on average. For the average difference of the "Relaxed" positions to the "Bend" position, the measurement was 3.0 cm smaller (see Fig. 7.9). As justification, it can be stated that the arms extended forward in the "Reach 2" position make the "arm crease" marker positions on the left and right move away from each other. The back width is significantly responsible for the wearing comfort and also affects the primary measurement of the chest girth, which is responsible for the clothing size allocation. Therefore these significant changes must be taken into account in the pattern development to prevent the clothing causing mobility restrictions.

7.2.5.3 Individual variation of positions

The execution of individual positions can vary greatly due to the individual mobility of the subjects. This, in part, has a strong impact on the measurement results. The taken body measurements can only be meaningfully examined when the positions are taken comparably. Especially the "Bend" and "Squat" positions were executed varyingly. In Fig. 7.10, different subjects can be seen in the "Bend" position. The curvature of the back and the extension of the leg differ significantly. As a result the distances between fingertips and ground have a great variability. The difference between the subjects varied up to 50 cm.

To counteract this, a solution approach was found. The "Bend" position was carried out twice by every subject. Both variations were scanned to be able to evaluate the result later with regard to body height (see Fig. 7.11):

• Version 1: Bend forward as far as possible, according to physical condition.
• Version 2: Bend to the reference point, at the height of the middle knee point.

Version 1 establishes the maximum range of movement, and version 2 makes the value fundamentally comparable.

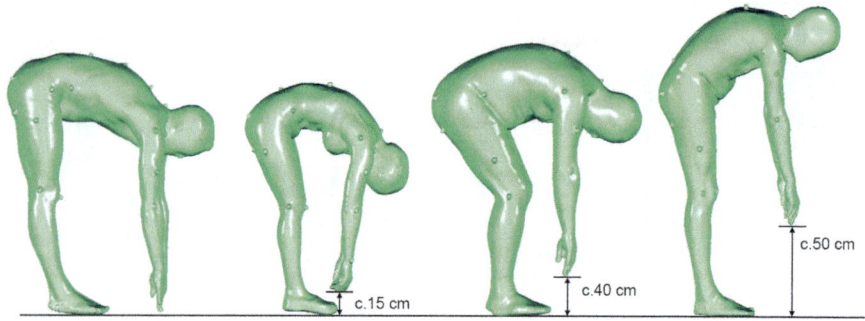

Fig. 7.10 Examples of different executions of the "Bend" posture.

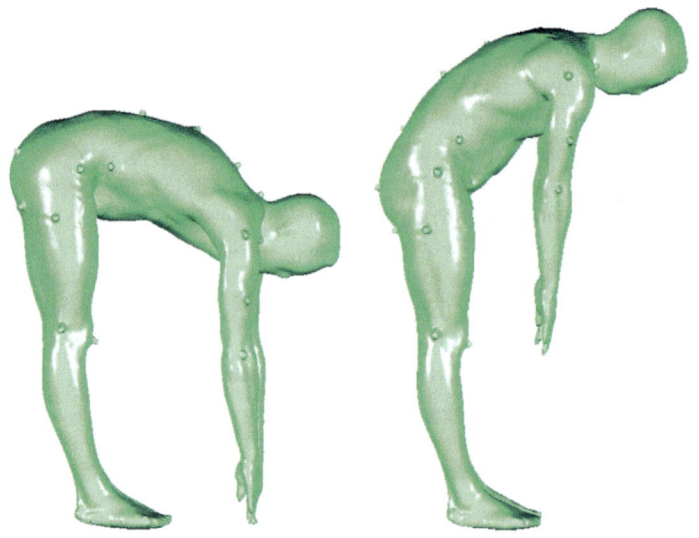

Fig. 7.11 Versions 1 and 2 of the "Bend" posture.

In Fig. 7.12, it is clear that the "Squat" position, just like the "Bend" position, was not, and was not able to be, carried out equally by all subjects. Different executions have an effect on the position of the markers on the back and equivalently influence the measurement distances.

In general the back length in the "Squat" position is longer than in the "Relaxed" position. This is because through the flexion of the spine, the spinous processes of the vertebrae move away from each other and thus so does the corresponding marker (see Fig. 7.12). Subject 3 (Fig. 7.12, third from the left), however, performs the position with a very straight thoracic spine. Therefore the marker points converge in this area. This leads to a difference of -3.7 cm. The shortening of the back length was only seen in a few subjects. The examples of the "Bend" and "Squat" positions suggest that it would be useful in future studies to form posture groups. These would allow a

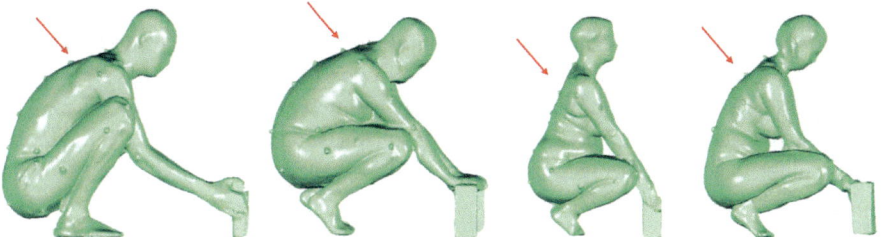

Fig. 7.12 Examples of different executions of the "Squat" posture.

differentiated analysis and evaluation of the functional measurements. However, a larger sample size would be necessary for this.

7.2.5.4 Muscular influences on the body geometry

Some measurements did not show the expected dimensional changes. For example, the upper arm girth (measured by the most pronounced form; see Section 7.2.3.4) of an athletic subject only changes up to 2 cm from the straight to the bent arm. This equals 4.9% (see Fig. 7.13). Body analyses revealed that the triceps shows its most pronounced form when the arm is stretched. The biceps, on the other hand, clearly shows when the arm is bent. By this counteractive work of agonist and antagonist, only the shown small dimensional change of the upper arm girth occurs. It is striking that the position of the measuring section changes in the two positions following the most pronounced form. A marking of the measuring section is therefore not possible. Anthropometrically speaking the measurement is difficult to measure reproducibly. For implementation in clothing products, however, the maximum shape is essential.

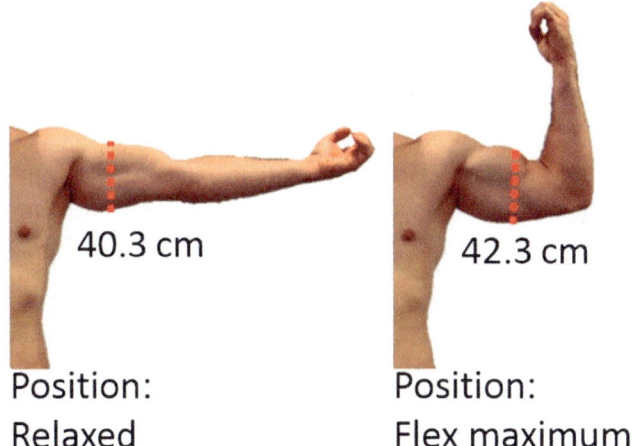

Fig. 7.13 Shape of the upper arm.

It should be noted that, as already mentioned, the static positions recreate movements. The muscle work is static rather than dynamic work. Even for trained subjects, it is difficult to precisely replicate the muscular work of a movement in a static position. It can be assumed throughout that the body geometry will vary in both cases. As a result the measurements could also be different. This is fundamental to investigate. As 3-D scanners are unable to record movement, 4-D scanners should be seen as a solution for these analyses.

7.2.6 Outlook

A further challenge, in addition to measuring the dimensions and deriving the dimensional differences, is implementing them in optimized apparel products for sports, work, and PPE. The correlation to the textile material and its respective elasticity presents a central topic (see Section 7.3.1). Fig. 7.14 shows an example of the representation of the evaluated dimensional differences (back width and arm reach) between the "Relaxed" and "Reach 2" positions, as they are to be considered in a product. When the arms are stretched forward, the back width increases by 26% in the large-size range (chest girth 112–120 cm) and the arm reach by 7%. This corresponds to an increase of 11 cm in the back width and 5 cm in the arm reach. Due to the increase in back width and arm reach, there is a required extra length of about 16 cm that must be available in a shirt, for example, in the movement of the user. This extra length must be taken into account not only in the production of patterns by the construction of special clothing elements such as movement folds or adjustment options but also when using textile materials with defined properties of stretch and/or elasticity. The implementation of the functional measurements in pattern development is a new field of research.

7.3 Examinations of mobility restrictions caused by clothing

Due to a lack of information about the changes of the human body surface during movement, necessary comforts were not always sufficiently considered in the development of clothing. However, a lack of extra width or length hinders the wearer when executing movement. With the analysis of functional measurement and the development of a movement-oriented measuring system, important foundations were laid to develop clothing with high ergonomic wearing comfort for different sizes and figure types. Yet, as scientific studies have shown, the static recording of functional measurements has set some limits (see Section 7.2.5). With the further development of scanning systems from 3-D to 4-D, static positions can be replaced by movement in the future. This would allow for the analysis of more realistic, if not real, circumstances. This opens up the possibility to investigate completely new research issues. But before arriving at the new 4-D scanners in Section 7.4, the research work in the field of mobility restriction should be looked at.

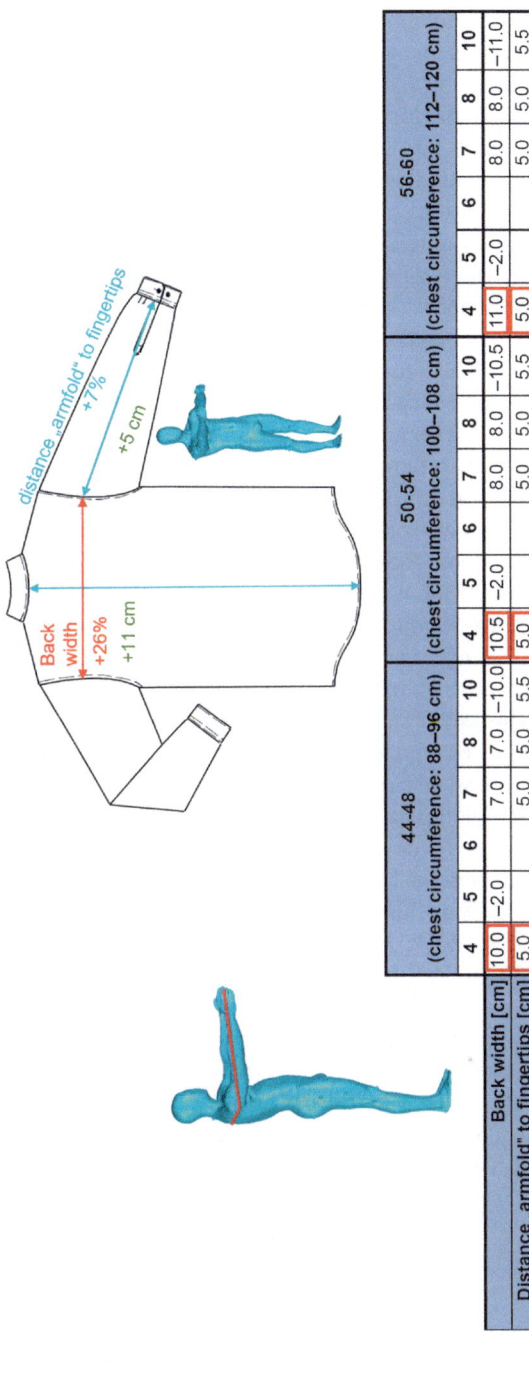

		44-48 (chest circumference: 88–96 cm)						50-54 (chest circumference: 100–108 cm)						56-60 (chest circumference: 112–120 cm)					
		4	5	6	7	8	10	4	5	6	7	8	10	4	5	6	7	8	10
Back width [cm]		10.0	−2.0		7.0	7.0	−10.0	10.5	−2.0		8.0	8.0	−10.5	11.0	−2.0		8.0	8.0	−11.0
Distance „armfold" to fingertips [cm]		5.0			5.0	5.0	5.5	5.0			5.0	5.0	5.5	5.0			5.0	5.0	5.5

Fig. 7.14 Exemplary presentation of implementation of differences in clothing.

7.3.1 Mobility restrictions caused by clothing

Clothing restricts the movement of the wearer to different extents according to the flexibility and elasticity of the material. In the area of fashion, good ergonomics are not part of the target values. Here, visual appearance takes center stage. Mobility restrictions, that is, the reduction of the ROM, are accepted by the customers (Gill, 2018). Critical movements are avoided if possible. The tight pencil skirt is not intended for stepping over hurdles, and you cannot mount an engine overhead wearing a sport coat. The combinations of material stiffness and pattern reduce mobility; in these examples, either the step size or the shoulder rotation is restricted. For example, excessive and overly dynamic movements in fashionable clothing can lead to seam damage (Schmid and Mecheels, 1981).

Such damage remains for the most part without serious consequences for the wearer in everyday life. The situation is very different for PPE, military equipment, and sport or outdoor wear. Movements here are repeated multiple times and are often highly dynamic. This stresses seams and materials maximally. If areas are stressed too much, the protective effect or functionality can be reduced (Baytar et al., 2012). Ergonomics are as just as important requirement as the protective function for these products. Mobility restrictions at work or during sport represent a further physical, and therefore also mental, strain (Akbar-Khanzadeh et al., 1995). Clothing can severely restrict the performance of the wearer, if they permanently have to act against resistance. Impairments of this kind result in longer processing times (in a work environment), earlier muscle fatigue, and higher stress (Adams and Keyserling, 1993). In a physically demanding job (e.g., in the fire department), clothing should be as supportive as possible (Watkins and Dunne, 2015). The goal of clothing development in this context must, therefore, be maximum protection with minimal impairment of the mobility of the wearer. In some cases, even clothing-technical solutions for the passive support of the users can be realized.

Most clothing products are developed on the basis of the standard anthropometric posture defined in standards and size charts (see Section 7.2.2). The persons to be recorded stand upright, legs hip width apart, and arms slightly spread laterally (Ashdown, 2011; Kouchi, 2014; Morlock, 2015a, b; ISO—International Organisation for Standardization, 2010). People are, however, in this static position for very short moments throughout the day, as they are predominantly in movement. They walk, sit, bend the torso forward, or stretch the arms out upwards. Above all at work and during sport, the body posture changes regularly and dynamically. With each movement the body dimensions and geometry vary (see Section 7.2). The variations in body proportions have a direct effect on the fabric and on the whole clothing product. The material is stretched or deformed. This also influences the interaction between body and clothing. An increase in material stretch usually also means an increase in clothing pressure on the body (Lim et al., 2006). If the pressure is too great, the wearer feels the clothing is uncomfortable and can be limited in his mobility. To counter this, clothing must accommodate for the changes in the body and expand or reduce analogously to it. This can be achieved on the one hand by material characteristics or by the tailoring. Woven fabrics are limited in their flexibility due to their binding and thus also the associated distortion resistance. In general, knitted fabrics are more elastic and stretchable than woven fabric.

They have a high elasticity, which can adapt to the changes in girth and length of the body. However, not every material can be used in every field of application. Not every ROM can be compensated for solely by material properties. In many cases, design adjustments are necessary.

Clothing-technical solutions for the reduction of mobility restrictions so far have been scientifically considered in individual examples (Boorady, 2011; Ashdown, 2011). Concrete examples in PPE are protective work parts with preformed knee areas or jackets for overhead work with specific armhole construction and elastic elements (see Fig. 7.15). In sports, results from application analysis lead to, among others, bicycle or motorcycle parts with preformed fit for the riding position. The patterns were adjusted so that the parts fit perfectly with a 90 degree angle in the hip and knee joints.

Targeted material positioning or application-specific pattern designs require basic body shape analysis. The investigation of the body regions that are subject to change processes leads to clear functional descriptions. Thus the corresponding adaptations can be made in the product development process (Watkins and Dunne, 2015; Morlock, 2015a, b). In addition, it is important to quantify the interaction between clothing and wearer, as well as the mobility restrictions experienced through clothing. Clothing products with high ergonomic wearing comfort can only be developed through evaluating the limitations of movement first (Saul and Jaffe, 1955; Adams and Keyserling, 1993; Watkins and Dunne, 2015; Lockhart and Bensel, 1977; Gregoire et al., 1985; Alexander and Laubach, 1973; Son et al., 2013, 2014; Huck, 1988, 1991; Adams, 2000; Graveling and Hanson, 2000; Coca et al., 2008, 2010; Son and Xia, 2010).

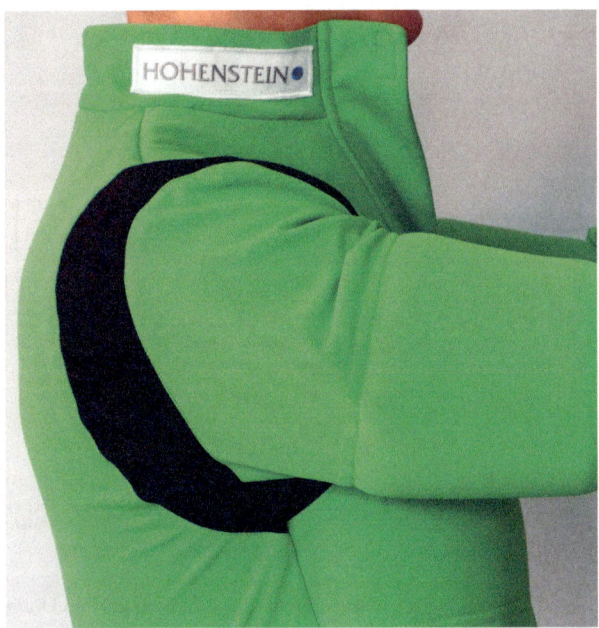

Fig. 7.15 Jacket for overhead work with adapted armhole construction and elastic zone.

7.3.2 Methods to analyze the human-garment interface

The development of an optimal fit for mass production is a challenge due to a variety of factors. A decisive criterion is the heterogeneity of the clothing sizes and body shapes of the customers (Morlock and Schenk, 2016; Boorady, 2011). What influence do they have on the interaction between user and product? This should be considered for workwear, particularly PPE. Especially here the fit during movement is of vital importance. Initial research approaches to the analysis of the human-garment interface aimed for an objective evaluation of the products in terms of range of movement and reach. Thus Saul and Jaffe (1955) developed 28 performance tests to research the impairment of gross motor skills through military clothing; an example is the subject sits on the floor with their feet on the test structure. He bends forward as far as possible, and with a bar on the structure, the hip mobility can be quantified (see Fig. 7.16 on the left).

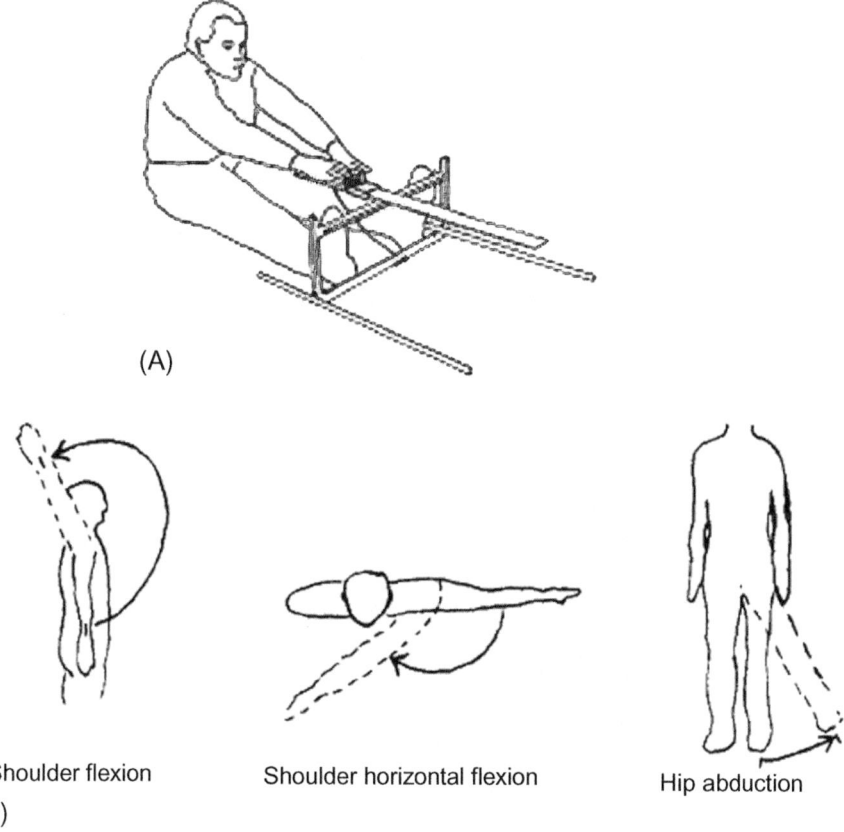

(A)

(B)

Shoulder flexion   Shoulder horizontal flexion   Hip abduction

Fig. 7.16 Examples for the analysis of the range of motion (Watkins and Dunne, 2015; Adams and Keyserling, 1993).

Lockhart and Bensel (1977) carried out similar studies for military clothing protective against cold climate and Gregoire et al. (1985) for protective suits against chemical and biological warfare agents. The variables in the studies, among others, were mobility, speed of movement, and coordination. These were examined in general performance tests: For example, how deep the subjects can bend forward, and what distance they have covered after five steps. Alexander and Laubach (1973) examined the arm reach of air force pilots in clothing protective against cold climate. The test setup was modeled on the working environment of the pilots. They were sitting and had to complete different reaches in a modified pegboard test. Alongside military clothing, such analyses are especially carried out in the area of fire-resistant clothing (see Fig. 7.16 on the right) (Son et al., 2013, 2014; Huck, 1988, 1991; Adams, 2000; Graveling and Hanson, 2000; Coca et al., 2008, 2010; Son and Xia, 2010). Beginning with Huck (1988) up until Son et al. (2014), the methods are comparable. The majority of the mobility was quantified by analog devices such as goniometer, flexometer and meter. Son and Xia (2010) used a motion capture system for the analysis of the mobility of the subjects. In principle, quality standards must be complied with for all ergonomic analyses. Hsiao and Halperin (1998) established a six step program:
Determination of the relevant body.dimensions:

1. Determination of the target group (sex, age, and profession)
2. Selection of the number of subjects (cost-benefit)
3. Data collection as a basis for the statistical evaluation
4. Calculation of specific dimensional changes
5. Implementation of the necessary adjustments to the clothing products

This guarantees solid data for the development of ergonomic products and a practice-oriented use of anthropometric data. The measurement of the range of movement is considered as a recognized method, and the results are considered as objective and quantifiable.

Analog methods, however, do have the disadvantage of being very laborious and prone to mistakes. Two persons are necessary for anthropometric measurement. One measures the subject, and the other notes the results. The accuracy of the measurements depends highly on the measurer (Kouchi, 2014). The manual recording of data contains the risk for transfer errors. Landmarks that are used for different measurements must be marked consistently. The time required, depending on the number of measurements, is not insignificant and can possibly be stressful for the subjects. This is because the survey is carried out on an unclothed body (in underwear) and requires a close proximity to the subject. Especially dimensions such as crotch length are usually perceived as rather uncomfortable.

Partially in studies, large differences were seen between the results of measurements of different measurers (interobserver error) (Goodwin et al., 1992; Kouchi, 2014). Here, reproducible methods must be developed so that the performers achieve a consistent identification of the anatomical landmarks and use of the anthropometric devices. Three-dimensional scanner systems in combination with semiautomatic measuring software reduce said sources of errors. For this reason, 3-D body scanners have been used in various scientific works (see Chapter 6).

7.4   Dynamic recording of body surfaces

So far, movement has been digitalized with motion capture technology (Watkins and Dunne, 2015). Defined points are tracked on the body or on the silhouette and turned into data that a computer can read. Animation and control of applications such as video games and computer-animated films or biomechanical analyses in medicine and sports science are among the purposes.

The output is an animated, sometimes very abstracted skeleton, as the focus in bio-mechanics lies on the range of movement of individual joints. Information about changes in body surface area cannot yet be reliably derived using motion capture tech-nology, although scientific research has already been carried out in this area. Dynamic recording of the whole 3-D surface using motion capture technology was presented by Park and Hodgins (2006). With 350 markers the geometry of the skin and its deforma-tion was only recorded limitedly but was then transferred to a static 3-D scan and inter-polated. This achieves a higher resolution of the surface mesh but still not an exact representation accurate of the skin surface. For a better resolution, more markers have to be used, which is limited for the currently available motion capture technology. The reflective markers are relatively large, and the effort of calibration grows with each marker. The automatic tracking when there are too many markers also has to be com-plexly corrected by hand. Meanwhile, there are further research approaches. However, the majority of studies work with interpolation of points and are also coupled with great effort (Aguiar, 2014; Wu et al., 2011; Rose and Murphy, 2015).

Four-dimensional technologies have been mostly used in medicine so far. By means of computed tomography, magnetic resonance tomography, or ultrasound, the interior of the body (e.g., organs, skeleton, and embryos) is visualized in three dimensions and in stop motion (Li et al., 2008; Rivera-Rivera et al., 2015; Schreibmann et al., 2014; Smith et al., 2005). In recent years, scanner systems have also been developed to record the body surface that are able to display areas of the body faster and more precisely, such as the face, feet, or back, in both three-dimensional and motion (Vialux, n.d.; 3dMD, 2019; Lane, 2013; kle-point, n.d.; Sinfomed, n.d.). They mainly work with structured light. The three-dimensional recording by means of structured light enables the digitization of recorded surfaces in the form of point clouds. Here, projectors project a defined pattern (e.g., stripes and checks) onto the area to be detected, which is captured by cameras from defined angular positions. Using various algorithms and measuring principles, such as the tri-angulation method, light-section method, coded light approach, and phase shift method, the detected areas can then be reconstructed in three dimensions. On this basis, series recordings allow the scanning of movement. The research group led by Michael Black at the Max Planck Institute for Intelligent Systems (MPI-IS) works, according to its own statement, with the world's first 4-D body scanner. This is based on a system of 3dMD with 22 stereo cameras, 22 color cameras, and spot pattern pro-jectors. The system enables movement recording with 60 pictures per second (pps). With good image quality, it can be assumed that the more pps made, the better and more detailed the movement can be detected. The scanner was individually developed for the MPI-IS. The challenge with four-dimensional recording is that multiple sensors are required simultaneously for complete body recording to synchronize the individual scans and display them from any angle needed. However, the purchasing price of the whole system rises significantly with the number of sensors. Therefore the detection of movement has been limited so far to small body parts such as the feet and face. Furthermore, it must be taken into account that the more pps recorded, the greater the storage capacity required and the greater the effort involved in the data processing. However, as technology advances, this hurdle is likely to be overcome quickly.

For many software developers in the field of scanning, 4-D is the next logical step in development (Heindl et al., 2015). Research work in which 4-D body scanners were used is so far exclusively from MPI-IS. It is concerned with the simulation of people in motion (Guan et al., 2009, 2012). Studies on the analysis of body measurement and proportion changes, range of movement, or mobility restrictions in clothing as a foundation for clothing development and optimization still do not exist.

In addition to structured light scanners, the use of photogrammetric systems is applicable for the four-dimensional recording of whole bodies. In photogrammetry, images are created from different spatial positions and perspectives and calculated using appropriate software to form a 3-D model of the desired area. So far the systems have been used for the digitalization of people and objects in static positions (Botspot, 2019; 3Dcopysystems, 2019). According to the manufacturer, however, series recordings of 10 pps are possible. Moderate movements can also be depicted with this technology. This could be a solution, at least for some areas of application and fit analysis as well as research of mobility restrictions in clothing development.

Four-dimensional technology has not yet been verified for use in body dimension and proportion analysis as a basis for clothing design or to study mobility restrictions. There are no methods for fit assurance with this technology yet. Four-dimensional scanning technology is still in its infancy and can grow by being tested in a variety of applications, such as in the clothing industry. Due to the further development of hardware and software, it can be assumed that all systems will become much cheaper in the medium term. Future research work will seek to gain access to this technology for clothing development. At Hohenstein a publicly funded research project began in 2018. First, mobility restrictions will be researched with 4-D scanning (photogrammetry) (Klepser, 2017).

7.5   Conclusion and future trends

The three-dimensional recording of people and their clothes with the help of scanning technology allows extensive scientific analysis of body changes and the interaction between body and textile shell. The fast recording in just seconds makes the digital archiving of large numbers of subjects possible. Measurements can be ascertained in a significantly higher number and multiple times, without burdening the subjects. Body forms and measurements can be analyzed in a more differentiated manner. This is because individual measuring sections can be subdivided into sections and the entire geometry of the body region can be examined. Databases with 3-D scans thus form the basis for new scientific questions.

In the presented "functional measurements" project, for the first time, body measurements were also scientifically analyzed with respect to movement and derived from a larger sample. There are tables available for the clothing industry that describe the average changes in body size in movement, related to clothing size, and that are a valuable complement to the standard size tables. However, the research has revealed many other challenges, such as the reproducibility of the different positions, the reproducibility of the landmarks and measurements on the body, the limitations of the 3-D body scanner used in the analysis of the functional measurements, and the significantly greater processing effort compared to standard sizing surveys. Future works should aim to examine these aspects in a more detailed manner.

Thus the development of clothing products with high wearing comfort can be significantly supported. But the challenge, in addition to recording the kinematic standards of people, is their implementation in the clothing product. The change in the body surface and the change in length of the extremities must be able to be reproduced by the textile, for example, in the buttock area. Especially in functional clothing the material and the implementation of the ROM in the pattern present a particular challenge. There the required extra length must be provided in such a way that the wearer of the clothing product can bend, kneel, or sit optimally. All the layers of a product must be designed flexibly enough that the movement can be compensated for by either constructive ease, stretch, or special clothing elements. Therefore value-based material recommendations are to be considered mandatory for the development of the pattern. For this purpose the textile materials must be tested for their flexibility and/or elasticity and correlated with the functional measurements.

In the development process, 3-D scanners can be a further foundation for the optimization of products. If these are used for fit analysis, the virtual three-dimensional representation will allow for individual adjustments of the view by rotation or zoom. In addition, it is possible not only to view scans side by side but also to place them inside one another, whereby a direct comparison can be implemented, for example, of the unclothed body with the clothed body. Three-dimensional software can be used to identify and quantify compacted or bridged areas. Mobility restrictions can also be clearly identified by supplementing the fit analyses with application of typical positions. The restriction of ranges of movement through clothing can be impressively visualized using 3-D scanner technology and also quantified using appropriate surveying software. Here, too, there is still a great need for research, since so far only a few studies have been carried out. There are still missing some basic methods that allow the subjects to occupy an identical position with and without clothing. The presented research shows first approaches that need to be continued.

The recording of the human body in static positions with 3-D body scanners has its limitations. Dynamic movements are only recreated. This means that the subjects have to tighten muscle groups, which are not needed for just holding static positions, a requirement that can be met only by a few and sometimes not at all. In addition, it can be assumed that the expression of the musculature in the static position differs from that of dynamic movements. The technical development from 3-D to 4-D scanning systems could be a medium term answer to this problem. Although only moderate movement speeds are currently recordable, the further development of hardware and software will soon allow more dynamic movements to be recorded. Technically, this is already possible but only at a considerable financial expense. Although 4-D technology is still in its infancy, it is critical to connect it to clothing development through research. It is important to examine the potential that scanning in motion offers clothing manufacturers.

7.6 Source of further information and advice

Some further information and advices of project "functional dimensions" can be found in these proceedings and articles:

- Motion-oriented 3D analysis of body measurements. Proceedings of AUTEX World Textile Conference 2017, Korfu, Greece.
- Design of a motion-oriented size system. Proceedings of 12th Joint International Conference CLOTECH 2017, Lodz, Poland.
- Design of a motion-oriented size system for optimizing professional clothing and personal protective equipment. Proceedings of ITMC 2017—International Conference on Intelligent Textiles and Mass Customization, Gent, Belgium
- Design of a motion-oriented size system for optimizing professional clothing and personal protective equipment. J. Fash. Technol. Text. Eng. S4, 014. https://doi.org/10.4172/2329-9568.S4-014.

The project "mobility restrictions" is currently in the processing state (duration: from 01.05.2018 to 30.04.2020). Further information about 4-D scanner systems and scanning of movement is given in the proceedings of the last 3DBODY.TECH conference http://www.3dbody.tech/cap/home.html.

Acknowledgments

The IGF projects 18993 N and 20163 N under the auspices of the Research Association Forschungskuratorium Textil e.V., Reinhardtstraße 12-14 10117 Berlin, were sponsored via the AIF as part of the program to support "Industrial Community Research and Development" (IGF), with funds from the Federal Ministry of Economics and Energy (BMWi) following an order by the German Federal Parliament.

References

3Dcopysystems, 2019. 3DCOPYSYSTEMS [Online]. Available from:https://3dcopysystems. com/ (Accessed 15 June 2018).

3dMd, 2019. Temporal-3dMD Systems (4D) [Online]. Available from:http://www.3dmd.com/ 3dmd-systems/dynamic-surface-motion-capture-4d/ (Accessed 30 July 2018).

Adams, D.R., 2000. Protective Ensembles for Firefighting Challenge the Balance and Stability of Firefighters. National Fire Academy, Osceola County, FL, USA.

Adams, P.S., Keyserling, W.M., 1993. Three methods for measuring range of motion while wearing protective clothing: a comparative study. Int. J. Ind. Ergon. 12, 177–191.

Aguiar, E.D., 2014. Performance Capture Methods. European Conference on Computer Vision (ECCV), Zürich.

Akbar-Khanzadeh, F., Bisesi, M.S., Rivas, R.D., 1995. Comfort of personal protective equipment. Appl. Ergon. 26, 195–198.

Alexander, M., Laubach, L., 1973. The effects of personal protective equipment upon the arm-reach capability of USAF pilots. In: Interagency Conference on Management and Technology in the Crew System Process.

Ashdown, S.P., 2011. Improving body movement comfort in apparel. In: Song, G. (Ed.), Improving Comfort in Clothing. Woodhead Publishing, Cambridge (GB).

Ashdown, S.P., Slocum, A.C., Lee, Y.-A., 2005. The third dimension for apparel designers: visual assessment of hat designs for sun protection using 3-D body scanning. Cloth. Text. Res. J. 23, 151–164.

Baytar, F., Aultman, J., Han, J., 2012. 3D body scanning for examining active body positions: an exploratory study for re-designing scrubs. In: 3rd International Conference on 3D Body Scanning Technologies, Lugano (CH).

Boorady, L.M., 2011. Functional clothing—principles of fit. Indian J. Fibre Text. Res. 36, 344–347.

Botspot, 2019. Available from: https://www.botspot.de/ (Accessed 15.06.2018).

Bougourd, J., 2014. National size and shape surveys for apparel design. In: Gupta, D., Zakaria, N. (Eds.), Anthropometry, Apparel Sizing and Design. Woodhead Publishing, Cambridge (GB).

Coca, A., Roberge, R., Shepherd, A., POWELL, J.B., Stull, J.O., Williams, W.J., 2008. Ergonomic comparison of a chem/bio prototype firefighter ensemble and a standard ensemble. Eur. J. Appl. Physiol. 104, 351–359.

Coca, A., Williams, W.J., Roberge, R.J., Powell, J.B., 2010. Effects of fire fighter protective ensembles on mobility and performance. Appl. Ergon. 41, 636–641.

DIN Deutsches Institut für Normung e.V, 2005. Ergonomie—Körpermaße des Menschen—Teil 2: Werte. DIN Deutsches Institut für Normung e.V. DIN 33402–2.

Gersak, J., 2014. Wearing comfort using body motion analysis. In: Gupta, D., Zakaria, N. (Eds.), Anthropometry, Apparel Sizing and Desing. Woodhead Publishing, Cambridge (GB).

Gill, S., 2018. Human measurement and product development for high-performance apparel. In: Mclaughlin, J., Sabir, T. (Eds.), High-Performance Apparel. Woodhead Publishing, Cambridge, GB.

Goodwin, J., Clark, C., Deakes, J., Burdon, D., Lawrence, C., 1992. Clinical methods of goniometry: a comparative study. Disabil. Rehabil. 14, 10–15.

Graveling, R., Hanson, M., 2000. Design of UK firefighter clothing. In: Kuklane, K., Holmer, I. (Eds.), Ergonomics of Protective Clothing. National Institute for Working Life, Stockholm.

Gregoire, H., Call, D.W., Omlie, C.R., Spicuzza, R.J., 1985. An automated methodology for conducting human factors evaluations of protective garments. In: Proceedings of the Human Factors and Ergonomics Society Annual Meeting, Vol. 29, pp. 916–919.

Guan, P., Reiss, L., Hirshber, D.A., Weiss, A., Black, M.J., 2012. Drape: DRessing Any PErson. ACM Trans. Graph. (Proc. SIGGRAPH). 31 (4) 35:1–35:10.

Guan, P., Weiss, A., Balan, A.O., Black, M.J., 2009. Estimating human shape and pose from a single image. In: International Conference on Computer Vision, pp. 1381–1388.

Gupta, D., 2014. Anthropometry and teh design and production of apparel: an overview. In: Gupta, D., Zakaria, N. (Eds.), Anthropometry, Apparel Sizing and Desing. Woodhead Publishing, Cambridge (GB).

Han, H., Nam, Y., Choi, K., 2010. Comparative analysis of 3D body scan measurements and manual measurements of size Korea adult females. Int. J. Ind. Ergon. 40, 530–540.

Heindl, C., Bauer, H., Ankerl, M., 2015. ReconstructMe SDK: a C API for real-time 3D scanning. In: 6th International Conference on 3D Body Scanning Technologies, Lugano (CH).

Hohenstein Institute and Human Solutions GMBH, 2008. Size GERMANY Homepage [Online]. Available from: www.sizegermany.de (Accessed 17.07.2018).

Hsiao, H., Halperin, W., 1998. Occupational safety and human factors. In: ROM, W.N. (Ed.), Environmental and Occupational Medicine. Lippincott-Raven, Philadelphia, PA, USA.

Hsiao, H., Whitestone, J., Kau, T.-Y., Whisler, R., Routley, J.G., Wilbur, M., 2014. Sizing firefighters: method and implications. Hum. Factors 56, 873–910.

Huck, J., 1988. Protective clothing systems: a technique for evaluating restriction of wearer mobility. Appl. Ergon. 19, 185–190.

Huck, J., 1991. Restriction to movement in fire-fighter protective clothing: evaluation of alternative sleeves and liners. Appl. Ergon. 22, 91–100.

Human Solutions Gmbh, n.d., ANTHROSCAN [Online]. Available from: http://www.humansolutions.com/fashion/front_content.php?idcat=141&lang=5 (Accessed 31.07.2018).

ISO—International Organisation Für Standardization, 1989. ISO 8559:1989: Garment Construction and Anthropometric Surveys—Body Dimensions. First Edition 1989-07-01 ed.

ISO—International Organisation Für Standardization, 2010. ISO 20685:2010 3-D Scanning Methodologies for Internationally Compatible Anthropometric Databases. ISO 20685:2010. ISO, Genf (CH).

ISO—International Organisation Für Standardization, 2011. ISO 26800:2011 Ergonomics—General Approach, Principles and Concepts. ISO 26800:2011. ISO, Genf (CH).

Jürgens, H.W., Mathdorff, I., Windberg, J., 1998. Internationale anthropometrische Daten als Voraussetzung für die Gestaltung von Arbeitsplätzen und Maschinen. Bundesanstalt für Arbeitsschutz und Arbeitsmedizin.

Kirchdörfer, E., 2009. SizeGERMANY—Erläuterungen zu den Größentabellen—Frauen. unveröffentlicht.

kle-point. n.d., Studie in Uedem: Tübinger Forscher analysiert Füße mit 3D-Scanner in der Bewegung [Online]. Available from: http://www.kle-point.de/aktuell/neuigkeiten/eintrag.php?eintrag_id=54227 (Accessed 31.01.2017).

Klepser, A., 2017. Grundlagenuntersuchung zur Erschließung der 4D-BodyScanner-Technologie für die Analyse bekleidungsbedingter Mobilitätsrestriktionen (Unpublished Project Proposal) IGF Nr. 20163N. Hohenstein Institut fuer Textilinnovationen gGmbH, Bönnigheim.

Kouchi, M., 2014. Anthropometric methods for apparel design: body measurement devices and techniques. In: Gupta, D., Zakaria, N. (Eds.), Anthropometry, Apparel Sizing and Design. Woodhead Publishing, Cambridge (GB).

Lane, C., 2013. The potential for dense dynamic 4D surface capture illustrated with actual case studies. In: 4th International Conference on 3D Body Scanning Technologies, Long Beach (USA).

Lashawnda, M., Istook, C., 2002. Body scanning: the effects of subject respiration and foot positioning on the data integrity of scanned measurements. J. Fash. Mark. Manag. 6, 103–121.

Li, G., Citrin, D., Camphausen, K., Mueller, B., Burman, C., Mychalczak, B., Miller, R.W., Song, Y., 2008. Advances in 4D medical imaging and 4D radiation therapy. Technol. Cancer Res. Treat. 7, 67–81.

Lim, N.Y., Yu, W., Fan, J., Yip, J., 2006. Innovation of girdles. In: Yu, W. (Ed.), Innovation and Technology of Women's Intimate Apparel. Woodhead Publishing, Cambridge (GB).

Lockhart, J.M., Bensel, C.K., 1977. The Effects of Layers of Cold Weather Clothing and Type of Liner on the Psychomotor Performance of Men. Army Natick Research and Development Labs.

Lu, J.M., Wang, M.J.J., 2010. The evaluation of scan-derived anthropometric measurements. IEEE Trans. Instrum. Meas. 59, 2048–2054.

Morlock, S., 2015a. Entwicklung eines ergonomisch- und bewegungsorientierten Größensystems für Funktionsmaße zur optimierten Gestaltung von Berufs- und Schutzbekleidung. IGF Nr. 18993N (unpublished project proposal). Hohenstein Institut für Textilinnovation gGmbH.

Morlock, S., 2015b. Passformgerechte u. bekleidungsphysiologisch optimierte Bekleidungskonstruktion für Männer mit großen Größen unterschiedl. Körpermorphologien, Hohenstein: HIT.

Morlock, S., Schenk, A., 2016. 3D-basierte Entwicklung eines innovatven Verfahrens zur Passformdiagnose von Bekleidung. Hohenstein Institut für Textilinnovation gGmbH.

Morlock, S., Wendt, E., Kirchdörfer, E., Rupp, M., Krzywinski, S., Rödel, H., 2009. Grundsatzuntersuchung zur Konstruktion passformgerechter Bekleidung für Frauen mit starken Figuren. Bekleidungsphysiologisches Institut Hohenstein; Technische Universität Dresden, Hohenstein.

Morlock, S., Lörcher, C., Schenk, A., 2018. Entwicklung eines ergonomisch- und bewegungsorientierten Größensystems für Funktionsmaße zur optimierten Gestaltung von Berufs- und Schutzbekleidung. IGF Nr. 18993N. Hohenstein Institut für Textilinnovation gGmbH.

Park, S.I., Hodgins, J.K., 2006. Capturing and animating skin deformation in human Motio. ACM Trans. Graph. 25 (3), 881–889.

Rivera-Rivera, L.A., Turski, P., Johnson, K.M., Hoffman, C., Berman, S.E., Kilgas, P., Rowley, H.A., Carlsson, C.M., Johnson, S.C., Wieben, O., 2015. 4D flow MRI for intracranial hemodynamics assessment in Alzheimer's disease. J. Cereb. Blood Flow Metab 36 (10), 1718–1730.

Rose, B., Murphy, M., 2015. Mapping Wearer Mobility for Clothing Design. European Publication Server, Cambridge.

Saul, E.V., Jaffe, J., 1955. Effects of Clothing on Gross Motor Performance. U. S. Army Natick Laboratories, Natick, MA (USA).

Scheffler, C., Schüler, G., 2013. KAN-Studie 51: Rohfassung eines Leitfadens für die richtige Auswahl und Anwendung anthropometrischer Daten Sankt Augustin. KAN Kommission Arbeitsschutzund Normung.

Schmid, U., Mecheels, J., 1981. Kräfte an Textilien und Nähten der Kleidung in Abhängigkeit von Körperbewegungen und Kleidungsschnitt. Bekleidung Wäsche 2, 77–82.

Schreibmann, E., Crocker, I., Schuster, D.M., Curran, W.J., Fox, T., 2014. Four-dimensional (4D) motion detection to correct respiratory effects in treatment response assessment using molecular imaging biomarkers. Technol. Cancer Res. Treat. 13, 571–582.

Schwarz-Müller, F., Marshall, R., Summerskill, S., 2018. Development of a positioning aid to reduce postural variability and errors in 3D whole body scan measurements. Appl. Ergon. 68, 90–100.

Sinfomed. n.d., AWB-Mapper 4D [Online]. Available from: http://www.sinfomed.de/4D-Wirbelsaeulenvermessung/ (Accessed 30.07.2018).

Smith, A., Chudleigh, T., Maxwell, D., 2005. Incorporating 3D and 4D ultrasound into clinical practice. Ultrasound 13, 4–11.

Son, S.Y., Bakri, I., Muraki, S., Tochihara, Y., 2014. Comparison of firefighters and non-firefighters and the test methods used regarding the effects of personal protective equipment on individual mobility. Appl. Ergon. 45, 1019–1027.

Son, S.Y., Lee, J.Y., Tochihara, Y., 2013. Occupational stress and strain in relation to personal protective equipment of Japanese firefighters assessed by a questionnaire. Ind. Health 51, 214–222.

Son, S.Y., Xia, Y., 2010. Evaluation of the effects of various clothing conditions on firefighter mobility and the validity of those measurements made. J. Hum. Environ. Syst 13, 15–24.

Todd, W.L., Norton, M.J.T., 1996. Garment-Doffing kinematic analysis. Cloth. Text. Res. J. 14, 63–72.

Vialux. n.d., Vialux 4D Scanner [Online]. Available from: https://www.vialux.de/de/4d-video. html (Accessed 14.06.2019).

Vitronic. n.d., 3D Body Scanner [Online]. Available from: https://www.vitronic.com/industrial-and-logistics-automation/sectors/3d-body-scanner.html (Accessed 31.07.2018).

Watkins, S.M., Dunne, L.E., 2015. Functional Clothing Design. Bloomsbury, New York (USA).

Wu, C., Varanasi, K., Liu, Y., Seidel, H.-P., Theobalt, C., 2011. Shading-based dynamic shape refinement from multi-view video under general illumination. In: IEEE International Conference on Computer Vision (ICCV), Barcelona.

Further reading

Ashdown, S., Loker, S., Schoenfelder, K., Lyman-Clarke, L., 2004. Using 3D scans for fit analysis. J. Text. Apparel Technol. Manag. 4, 1–12.

Ernst, M., Detering-Koll, U., Güntzel, D., 2012. Investigation on body shaping garments using 3D-body scanning technology and 3D-simultion tools. In: 3rd International Conference on 3D Body Scanning Technologies. Hometrica Consulting, Lugano (CH).

Klepser, A., 2018. Entwicklung einer Methode zur Quantifizierung von Shaping Effekten, IGF19442N. Hohenstein Institut für Textilinnovation gGmbH.

Nam, J., Barnson, D.H., Asdown, S.P., Cao, H., Jin, B.P., 2005. Fit analysis of liquid cooled vest prototypes using 3D body scanning technology. J. Text. apparel Technol. Manag. 4, 1–15.

Nawaz, N., Troynikov, O., Kennedy, K., 2012. Investigation into fit, distribution and size of air gaps in fire-fighter jackets to female body form. In: 3rd International Conference on 3D Body Scanning Technologies, Lugano (CH).

Psikuta, A., Frackiewicz-Kaczmarek, J., Frydrych, I., Rossi, R., 2012. Quantitative evaluation of air gap thickness and contact area between body and garment. Text. Res. J. 82, 1405–1413.

Part Three

Testing and evaluation of fit

Fabric, seam and design applications in developing body-contouring jeans for better size and fit

8

Aisyah Mohd Yasim, Rosita Mohd Tajuddin
Faculty of Art & Design, Jalan Kreatif, University Technology Mara (UiTM), Shah Alam, Selangor, Malaysia

8.1 Research background

Jeans have undergone different types of modification in the manufacturing industry to suit customers' demand such as yoga jeans, jeggings, and body-contouring jeans (Rahman, 2011). These modifications were made as women in today's market pay more attention when buying clothes, as they want to satisfy their individual style and taste and at the same time feel and look more attractive by hiding imperfections and accentuating their beauty. This has led to the development of various clothing types that were made to emphasize physical appearance while satisfying their needs and wants.

Body contouring jeans that emphasizes types of fabric, seam and design would help to resolve problems encounter by wearers especially at stomach, hips and thigh areas. This will also aesthetically improve shaping of the seat, hips, and flattened stomach for a more natural and sophisticated appearance.

8.1.1 Body-contouring jeans

A body-contouring system for jeans is based on the concept of actively shaping the body's silhouette that contours and accentuates the stomach, hips, and buttocks of the wearer (Tulin and Laney, 2015). As jeans are said to be the most fashionable item that will last throughout the years (Wong and James, 2014), the innovation of body-contouring jeans is to create a garment that acts like a shape wear while being trendy at the same time. Besides that, it also provides comfort and ease of wear and smoothens the stomach, thus creating a slimming effect in a nonbinding, nonconstricting manner.

The shaping fit system includes the manipulation of design, seam, and fabric of the traditional jeans. According to Gonzalez, the founder of "IVIDO Push-Up Jeans," the idea of push-up jeans is to create pants with body-shaping function. It uses no padding but rather a combination of strategically stitching/seaming, pocket placement, and

Anthropometry, Apparel Sizing and Design. https://doi.org/10.1016/B978-0-08-102604-5.00008-1

special fabric to enhance a woman's shape. Designers need to identify suitable fabrics as well as types of seams and design in designing jeans that can enhance woman's shape yet providing them comfort and ease of wearing.

8.2   Methodology

8.2.1   Fabric test

A body-contouring system for jeans was based on the concept of actively shaping the body's silhouette (Tulin and Laney, 2015). Several methods were used by previous research in producing jeans with body-shaping function. According to Tulin and Laney (2015) a stretch denim was used when producing form-fitting jeans. Besides that, Carl Chiara, a designer of Levi's Jeans, stated that the new "Curve ID" line was developed using super-stretch denim fabric. At the same time, based on the questionnaire data that were compiled by the researcher, it can be summarized that the majority of the respondents preferred the mixture of denim and denim Lycra fabric when buying form-fitting jeans. Texas jeans and denim Lycra were selected for the experimentation as these fabrics have stretch properties that would serve as an important attribute of body contouring jeans.

8.2.1.1   Testometric test/tensile strength test

Upon identifying the type of fabric that was most preferred by Malaysian women, the researcher conducted the Testometric test. This method was crucial to examine both of the fabric structure on their elongation break (stretch ability) and also force break (fabric strength). This step was important as the design of the jeans by placing the right fabric panel with different stretch properties would lead to comfort while giving a lifting effect to the rear area of the wearer. Hence, in this study, several experimentations had been conducted to test out the fabric tensile strength. Three different fabrics were tested, and it was found that Texas jeans gave the most elongation break (millimeter), while denim Lycra gave the most force break (Newton).

The general design of the body-contouring jeans also played a huge role in the manufacturing part. According to Tulin and Laney (2015), the usage of different levels of fabric strength and stretch ability on the design in making jeans panels will give full support and comfort. By taking advantage of the characteristics of each panel, the researcher can make full use of the panels by putting them in places where the shapes and dimension of the panels can be altered to suit the design.

8.2.1.2   Seam strength test

Seam strength test was later conducted to identify the right type of seam to be applied on the jeans design. A curved riser seam was applied to physically lift the seat and visually created a shapely seat. Also, slightly forward-placed side seams were used to draw the eye inwards to create a narrower silhouette (Laney, 2015). Therefore, two types of seams were tested, which included straight seam and curved seam.

Implementation of the correct seam gave different results on the elongation break and force break. Hence the researcher had conducted multiple experimentations on seam strength. Based on the experiment conducted by the researcher, it was found that the curve seam and zigzag seam gave the highest elongation break (millimeters) and force break (Newton), thus making them suitable to be applied in constructing body-contouring jeans.

8.2.1.3 Fabric experimentation on fabric tensile strength

This process is crucial to indicate which fabrics were most suitable to be applied and to determine the type of seam to be used. Three types of fabric that were experimented were denim Lycra, Texas jeans (hard), and Texas jeans (soft). All of these fabrics will be tested using tensile strength machine on their warp and weft direction. After the fabric tensile strength test, the right fabric to be used on the body-contouring jeans was found. The seams that were selected for this study were straight seam and curve seam.

Testometric machine (STM566) in Fig. 8.1 was used to conduct tensile strength test. Three types of fabrics involved denim Lycra, Texas jeans (hard), and Texas jeans (soft) were tested. The main purpose in conducting this experiment was to determine the fabric elongation break, millimeters (stretch ability), and fabric force break, Newton (strength). Fig. 8.2 shows how the fabric was affected when force is applied onto the fabric during the experiment. The experiment was repeated five times for each sample to get the mean reading. Data were collected based on the reading of the graph and tabulated in a systematic table to make comparisons.

8.2.1.4 Seam strength test

Upon getting the result of the fabric elongation break, millimeter (stretch ability), and force break (strength), the chosen fabric was tested out again to test the application of seam on the chosen fabric. The two types of seams that were tested were straight seam and curve seam. Fig. 8.2 shows how the seam was affected when force was being applied.

Fig. 8.1 Testometric machine.

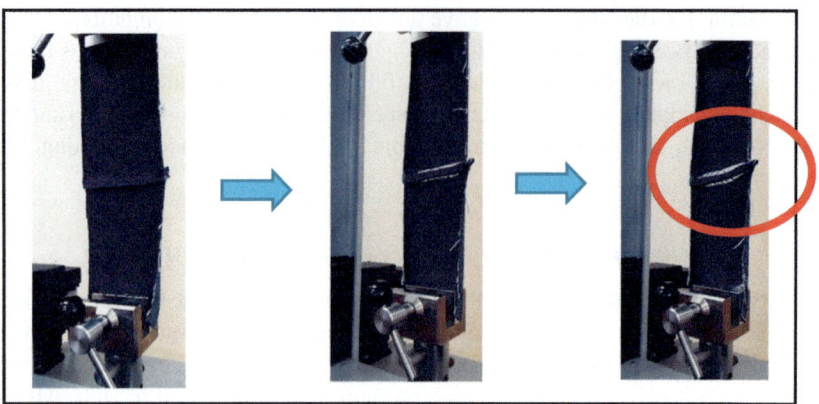

Fig. 8.2 Force being applied on seam.

8.3 Results of the tensile strength test

It can be seen in Fig. 8.3 that the sample cut in the weft direction for denim Lycra did
not break after the application of force but broke in the warp direction. Texas jeans
(soft) were the least affected fabric after the application of force, while Texas jeans
(hard) broke in both direction.

TYPE	WARP	WEFT
Lycra denim		
Texas jeans (soft)	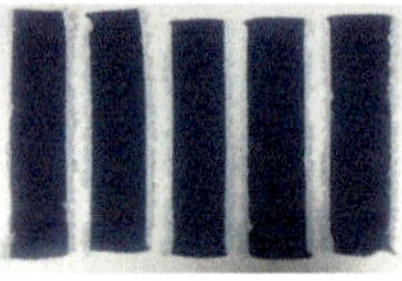	
Texas jeans (hard)	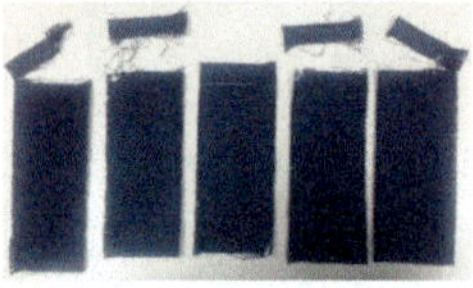	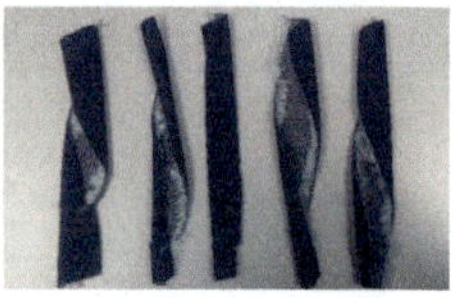

Fig. 8.3 Fabric condition after applying force (tensile strength test).

8.4   Findings and discussion

8.4.1   Findings of the tensile strength test

The graph generated during the tensile strength test is showed in Tables 8.1–8.3, to make comparisons on the fabric and seam strengths. The test displayed that the two materials selected to be used in the final product design were Texas jeans (soft) and denim Lycra, while Texas jeans (hard) was eliminated as it broke in both directions when force was applied.

Based on Tables 8.1 and 8.2, Texas jeans (soft) was applied in the weft direction for most parts of the garments as it has the highest mean reading on the elongation break of 82.99 mm compared with the warp direction of mean reading 54.38 mm. In a specific implementation, Texas jeans (soft) is placed at the inner thigh area, all the way down to the ankle as this part encountered the highest elongation break, millimeter (stretch ability), with 100% elastic recovery. Hence, this would enhance the comfort elements on the proposed design and would not sag after several uses.

On the other hand, based on Tables 8.1 and 8.2, the Testometric results showed that denim Lycra offered the highest force break in the weft direction with a mean reading of 514.45 mm compared with the warp direction with a mean reading of 1388.52 mm. The fabric was chosen in the weft direction as this provided a 100% recovery state. Hence, denim Lycra fabric was the most suitable and appropriate to be applied at the hips, front rise, and back yoke areas as this fabric had the highest force break, Newton (strength), whereby the hardness and strength of the fabric helped to contour the problematic area.

Meanwhile, based on Table 8.3, curve seam was tested on both denim fabrics, namely, denim Lycra and Texas jeans (soft). Texas jeans (hard) was eliminated from the seam strength test as it broke in both warp and weft directions during the fabric Testometric test. Based on the results of the seam Testometric test on denim fabric, it was confirmed that curve seam demonstrated a higher elongation break and force break when compared with a straight seam. Hence, to allow ease of movement, a curve seam was suitable to be applied on the design. By applying a curve seam, this would not only increase comfort to the wearers but also prevent the seam from breaking and at the same time reduce force.

8.5   Design implementation and development

Results from the preliminary study showed that Malaysian women prefer jeans design that was simple, relaxed, and suitable for everyday wear. Thus, after implementing the components of body-shaping wear such as curve seam into the design, it was found that the design of conventional push-up jeans and body-shaping wear at the rear area was too extravagant for Malaysian consumer.

Fig. 8.4 showed a flow of diagram for body-contouring jeans. Components includes incorporating two denim fabrics with different stretch and strength properties as panels, cutting the fabric according to the panels for body-contouring jeans, placing

Table 8.1 Fabric stretch ability (elongation break, mm) versus strength (force break, N)—weft.

Types	Thickness (mm)	(Min)		(Mean)		(Max)		Elastic recovery (in.)
		Elongation break (mm)	Force break (N)	Elongation break (mm)	Force break (N)	Elongation break (mm)	Force break (N)	
Denim Lycra	0.8	64.73	439.59	70.55	514.45	77.23	628.50	100% recovery
Texas jeans (1)	0.7	70.11	83.52	82.09	321.84	93.29	454.29	100% recovery
Texas jeans (2)	0.6	66.16	931.20	71.62	984.68	75.95	1076.80	Break

Table 8.2 Fabric stretch ability (elongation break, mm) versus strength (force break, N)—warp.

Types	Thickness (mm)	(Min)		(Mean)		(Max)		Elastic recovery (in.)
		Elongation break (mm)	Force break (N)	Elongation break (mm)	Force break (N)	Elongation break (mm)	Force break (N)	
Denim Lycra	0.8	50.27	1190.10	53.44	1388.52	57.57	1518.20	+0.5
Texas jeans (1)	0.7	50.68	477.64	54.38	482.63	62.57	487.89	100% recovery
Texas jeans (2)	0.6	59.11	1494.50	62.17	1751.24	67.44	1869.50	Break

Table 8.3 Fabric stretch ability (elongation break, mm) versus strength (force break, N) seam.

Types	Thickness (mm)	Straight		Curve		Zigzag		Elastic recovery (in.)
		Elongation break (mm)	Force break (N)	Elongation break (mm)	Force break (N)	Elongation break (mm)	Force break (N)	
Denim Lycra	0.8	75.75	484.90	81.36	491.54	N/A	N/A	Break
Texas jeans (1)	0.7	84.80	322.17	85.13	339.90	N/A	N/A	100% recovery

Fig. 8.4 Implementing different characteristics of fabric panel on the design to act as shape-contouring panel (front).

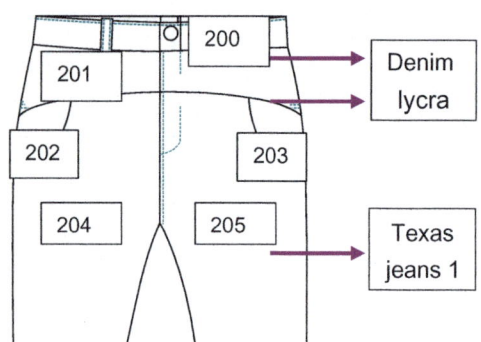

the panels in specific positions of the jeans, and forming the panels into jeans. Mixing two different fabrics that have different stretch and strength properties to targeted areas would control towards the shaping and flexibility of body-contouring jeans.

Fig. 8.4 showed a specific implementation of patterns for body-contouring panels that are patterned for a pair of jeans. There were six panels each of which was applied to a different area of the surface of the jeans. The most top panel 200 was positioned around the waistline. Meanwhile, panel 201 was placed at the lower abdomen. The bottom left and right panels, which were panel 202 and panel 203, were placed at the upper hip area. In a finished pair of jeans, as in Fig. 8.4, panels 204 and 205, which formed most parts of the garment, met at the crotch point all the way down to the ankle.

In this implementation, denim Lycra was applied at panel 200 around the waistband and panel 201 implemented at the lower abdomen area. In implementations, since denim Lycra had lower elasticity and higher strength, once applied to the design, the panels tend to limit the natural stretch of the fabric. These areas of the jeans with panels having less stretch will provide firm support for the wearer in the areas where the panels were positioned. As a result the abdominal area will be pushed in by the hardness of the fabric, creating a flatter-looking stomach. At the same time, denim Lycra was applied at panels 202 and 203 creating a fake pocket design. The strength and the hardness of the fabric helped to give support and smooth the outer hips.

Fig. 8.4 showed specific implementation of panels 204 and 205 where the panels were joined together at the crotch to form the garment. Texas jeans (soft) were applied at panels 204 and 205. The panels started from the crotch all the way down to the ankles. The incorporated panels 204 and 205 provided a high stretch effect in which the panel could stretch along the weft direction. The high stretch properties allowed the panels to mold and shape the natural curves of the wearer's body in a comfortable manner. Besides, Texas jeans (soft) will be applied in the weft direction for most parts of the garments specifically at the inner thigh area, as this has the highest elongation break, millimeter (stretch ability), with 100% elastic recovery, making the final product design comfortable and will not sag after several uses (Fig. 8.5).

In this implementation, denim Lycra is applied at panel 206 around the waistband and at the back yoke panel 207. These areas of the garments with panels having less

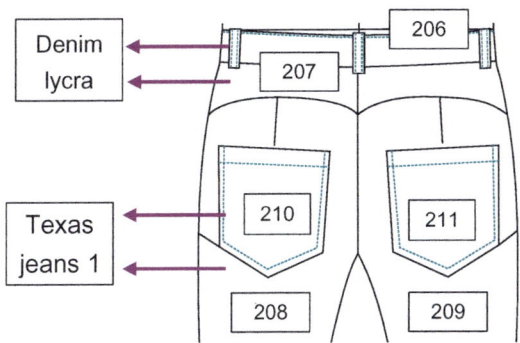

Denim
lycra

Texas
jeans 1

206
207
210
211
208
209

Fig. 8.5 Implementing different characteristics of fabric panel on the design to act as shape-contouring panel (back).

stretch (lower elasticity) can provide firm support for the wearer in the areas where the panels were positioned.

Meanwhile, back panels 208 and 209 formed the entire garment, which joined at the crotch, all the way to the midthigh, and towards the ankle. Texas jeans (soft) is implement on these panels and joined with panel 207 (back yoke) using curve seam. This cooperation of design allowed for higher elasticity with 100% recovery state that was designed to lift the seat of the wearer as it curves around and cups the natural shape of the seat. Meanwhile, panels 210 and 211, which were positioned as the back pockets, were cut in bigger design placed closer to each other to give a fuller look at the rear area.

8.6   Size and fit evaluation

Fit is one the most crucial attributes when buying jeans and pants (Zhang et al., 2002). A pair of good-fitting jeans will be able to conform the body, providing well-being, health, comfort, and ease of wearing to the wearer. Hence, to create a garment to fit the lower body, different body sizes and shapes are taken into consideration.

The results of the posttest were as shown in Figs. 8.6–8.8, whereby the researcher made a side-by-side comparison of conventional jeans and body-contouring jeans to test the different body size and shape to fit body-contouring jeans.

8.6.1   Respondent 01 in size M having wide rear area

Based on Fig. 8.6, body-contouring jeans were tested on respondent 01 with a wide rear area with rectangle body shape. In this case the respondent wished to have smaller dimensions of the rear area and to have curvier shapes at the rear and hips. When wearing conventional jeans, it can be seen that the problematic area protrudes to the sides creating a bulge at the bottom part or the rear. This eventually makes the rear area of respondent 01 looking wide and square. Not only that, the tightness of the conventional jeans caused respondent 01 to have a wedge at the crotch, making uncomfortable and creating an unattractive look.

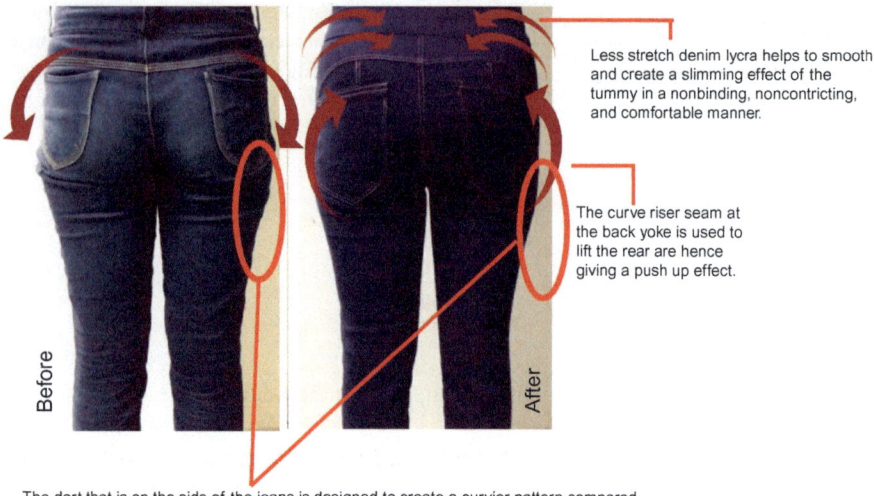

Less stretch denim lycra helps to smooth and create a slimming effect of the tummy in a nonbinding, noncontricting, and comfortable manner.

The curve riser seam at the back yoke is used to lift the rear are hence giving a push up effect.

Before

After

The dart that is on the side of the jeans is designed to create a curvier pattern compared to the conventional pattern of jeans thus reshaping the curves of the user's hips.

Fig. 8.6 The effect of body-contouring jeans on respondent 01 with size M with wide rear area.

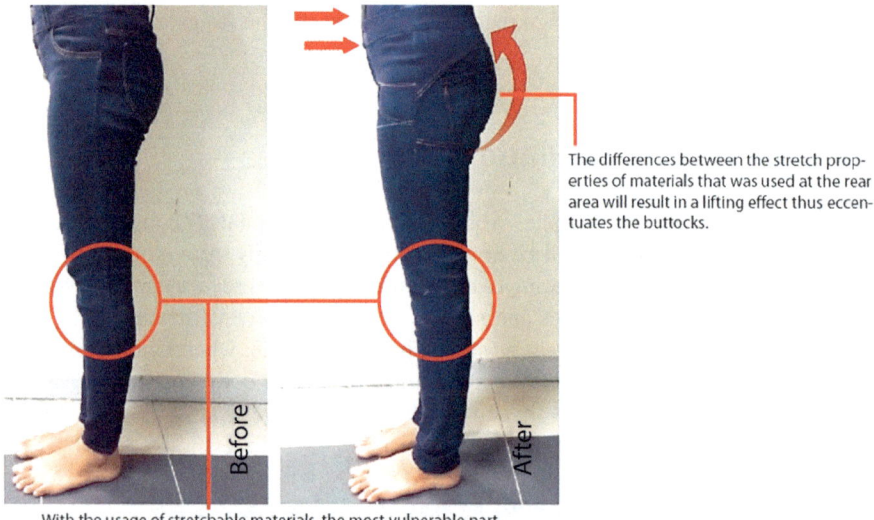

The differences between the stretch properties of materials that was used at the rear area will result in a lifting effect thus eccentuates the buttocks.

Before

After

With the usage of stretchable materials, the most vulnerable part, which is the knee would not sag or overstretch after repeated wear.

Fig. 8.7 The effect of body-contouring jeans on respondent 02 size S having rectangle body shape with flat rear area.

Therefore the usage of different fabric panels in body-contouring jeans would help to solve the problem. A less stretch denim Lycra is placed at the back yoke to push in the problematic area, which would also create a slimming effect on the waist in a nonbinding, nonconstricting, and comfortable manner. Meanwhile a high stretch fabric

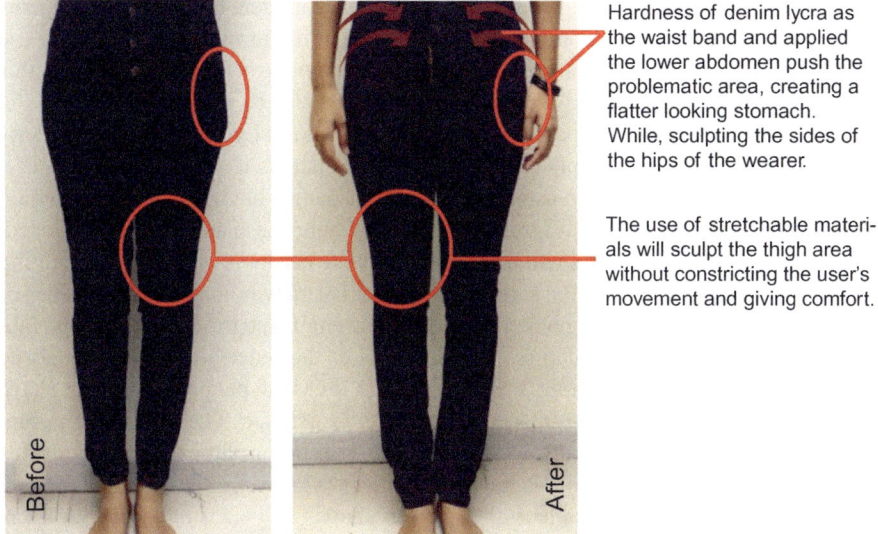

Hardness of denim lycra as the waist band and applied the lower abdomen push the problematic area, creating a flatter looking stomach. While, sculpting the sides of the hips of the wearer.

The use of stretchable materials will sculpt the thigh area without constricting the user's movement and giving comfort.

Fig. 8.8 The effect of body-contouring jeans on respondent 03 (front view).

was used at the rear area to give comfort to the wearer, thus eliminating the bulge at the sides and wedge at the center. In addition, a curve seam was used at the back yoke replacing the straight seam to lift the rear area, hence giving a push-up effect. Four additional darts were added to create a curvier pattern compared with the conventional pattern, which helps to make the wide and rectangle rear appear curvier as the rear area was pushed outwards instead of sideways.

8.6.2 Respondent 02 in size S having rectangle body shape with flat rear area

Fig. 8.7 showed respondent 02 in size S having rectangle body shape with flat rear, wearing conventional jeans on the left side versus wearing body-contouring jeans on the right side. Rectangle-shaped women tend to have a straight body with no significant difference between the sizes of their hips, waists, and shoulders and wish to have an appearance of fuller rear area. In this design the back yoke has less stretch properties compared with the rear area. The different stretching properties of each fabric leads towards lifting effect that would accentuates the rear area, making the body shape looks fuller and curvier.

8.6.3 Respondent 03 in size S having pear body shape

Fig. 8.8 showed respondent 03 in size S having a pear-shaped body type. Women with a pear-shaped body type have larger lower torso and smaller upper torso. They usually have small chest and flat stomach, whereby the hips are slightly wider than the shoulders. In this case, wearing conventional jeans, respondent 03 often has problems in tightness around the thigh and rear area but loose or ill-fitting at the waist. The use of only one type of fabric usually tends to create a muffin top at the tummy area. In this implementation a hard fabric (denim Lycra) with less stretch properties is applied at the waistband and at the lower abdominal area. Since denim Lycra has lower elasticity and higher strength, once applied to the design, the panels tend to limit the natural stretch of the fabric. These areas of the jeans with panels having less stretch would provide firm support for the wearer in the areas where the panels were positioned. As a result the abdominal area will be pushed in by the hardness of the fabric, creating a flatter-looking stomach. At the same time, denim Lycra was applied at both sides of the upper hips creating a fake pocket design. The strength and the hardness of the fabric helped in giving smooth and contour supports at the outer hips. These proposed designs not only were targeting the problematic areas of people with pear shape but also helping to contour the hips of people with larger size. The high stretch properties allowed the panels to mold and shape the natural curves of the wearer's body in a comfortable manner specifically at the inner thigh area to ensure that the final product design is comfortable for people with different shapes and sizes and will not break after several uses.

8.7 Conclusion and implications

In this study, based on the data that have been analyzed, it can be concluded that there was a strong association between fabric, seam, and design when designing body-contouring jeans. Body-contouring jeans are essential in catering to different sizes especially to those women who were large sizes as this jean could fit well at the lower body. Body-contouring jeans were tested on women with different body shapes and sizes to find the significance of size and fit for lower-body garment. Using fabric with different stretch properties as panels helps to contour, mold, and give support at the targeted area of the body while creating a better fit towards women with different shapes and sizes.

Meanwhile the implementations of the curve seam design and additional darts were recommended as these efforts help to accentuate and create an appearance of a fuller rear towards women with flat rear area besides lifting and contouring the rear area of women with a wider rear area. These three components of body-contouring jeans marked hand in hand to bring out the best of their characteristics, which showed a significance effect towards the fitting of body-contouring jeans despite of different body shapes and sizes. As women are typically categorized into four types of body shape, which include hourglass, pear, apple, and rectangle, designers and manufacturers need to understand their target market niche when designing a good-fitting jeans wear. The design should be able to fulfill the ever-changing customers' needs and demands.

A body-shaping fit jean affects the silhouette of the garment and this would enhance the overall shape of the body. The fit system includes the use of different fabrics with stretch and strength properties as the panel, application of darts to create allowance, curve seaming, and right construction techniques for pants. The construction of body-contouring jeans concept is also applicable to pants, shorts, capris, and other clothing in which shaping and support are desirable in the waist, buttock, hip, and thigh areas. Since Malaysia is composed of a dynamic population from different ethnic backgrounds with unique body sizes and shapes, hence, it is crucial for apparel manufacturers especially who are involved in jeans wear production to use this technique as a way to provide clothing satisfaction that meets diverse customer's expectations.

Each body shape would exclusively experience unique fit issues (Saeidi, 2018). Consumers nowadays preferred fit, comfort, and performance as the main attributes when buying apparel (Hendriksz, 2016). Therefore understanding apparel fit from consumer's perspective is vital. Different body sizes and shapes give different impact on jeans' design in terms of silhouette, form, performance, and look. Findings from this study suggested that choosing the right fabric with different properties is important while using the right type of seams and formulating good pattern construction would enhance the performance of jeans wear among wearers in the future.

References

Hendriksz, V., 2016. How to Solve the Denim's Industry Problem With Fit. Retrieved January 5 2019 from: https://fashionunited.uk/news/fashion/how-to-solve-the-denim-s-industrys-problem-with-fit/2016110422373.

Rahman, O., 2011. Understanding consumers perception and behaviours: implication for denims jeans design. J. Text. Apparel Technol. Manag. 7 (1), 47–58.

Saeidi, E., 2018. Men's Jeans Fit Based on Body Shape Categorization. LSU Doctoral Dissertationsp. 4784.

Tulin, K., Laney, D., 2015. Body Shaping Fit System. Journal of US Patent, United States.

Wong, C., James, G., 2014. Push Up Jeans and Related Production Process. Journal of US Patent, United States.

Zhang, Z., Li, Y., Gong, C., 2002. Casual wear product attributes: a Chinese consumer' perspective. J. Fash. Mark. Manag. 6 (1), 53–62.

Further reading

9 Jeans That Will Flatter Your Butt. Retrieved November 9, 2016 from: http://www.whowhatwear.co.uk/jeansflatter-your-butt/.

Alexander, M., Connell, L., Presley, A., 2005. Clothing fit preferences of young female adult consumers. Int. J. Cloth. Sci. Technol. 17 (1), 52–64.

Bougourd, J., 2007. Sizing in Clothing: Developing Effective Sizing Systems for Ready-to-Wear Clothing. Woodhead, New York.

Bryson, D., McCAnn, J., 2009. Smart Clothes and Wearable Technology. Woodhead Publishing, United Kingdom.

Darliana, M., Baba Md, D., 2010. Development of Malaysian anthropometric database. In: Conference on Manufacturing Technology and Management, Kuching, Malaysia.

Engel, C., Bell, R., Meier, R., Martin, M., Rumpel, J., 2011. Young consumers in new marketing ecosystem: analysis of their usage of interactive technologies. Acad. Market. Stud. J. 11 (2), 23–44.

Grace, I., Ruth, E., 2005. Apparel Manufacturing: Sewn Product Analysis. Pearson Education, New Jersey.

Helena, M., Letsiwe, L., Adamski, A., 2014. Body shape versus body form: a comparison of the body shapes of female Swazi consumers with those of body forms used in apparel manufacturing. J. Fam. Ecol. Consum. Sci. 42, 87–101.

Women Shape Wear. Retrieved November 9, 2016 from: http://www.hourglassangel.com/blog/ discover-acts-shapewear/.

Denim Trends Now. Retrieved November 28, 2016 from: http://www.popsugar.com/fashion/- 2017-Denim-Trends-42237711#photo-42237711.

Know Your Body Shape. Retrieved November 11, 2016 from: https://www.pinterest.com/ brandilamb14/wardrobe-for-your-body-type/.

Karen, M., Rosalie, J., 2006. Perry's Department Store: A Product Development Simulation. Fairchild Publication, United States.

Lee, S., 2011. Pants Having Body Shaping Function. Journal of US Patent, United States.

Mariane, C., Kotsipulos, A., Dallas, M., Eckman, M., 1995. Fit of Women's Jeans: An Exploratory Study Using Disconfirmation Paradigm. Vol. 8. pp. 208–213.

Panero, J., Zelnik, M., 1979. Human Dimensions and Interior Designs: A Source Book of Design Reference Standard. Whitney Library of Design, p. 30.

Paul, R., 2015. Denim and Jeans: An Overview, Denim: Manufacture, Finishing and Applications. Woodhead Publishing, pp. 1–5.

Rahman, M., Rahman, M., 2015. Investigation of the bulk, surface and transfer properties of chlorine bleached denim apparel at different condition. Eur. Sci. J. 11, 213–227.

Wu, J., Delong, M., 2006. Chinese perceptions of western branded denim jeans: a Shanghai case study. J. Fash. Mark. Manag. 10 (2), 238–250.

Evaluation of pattern block for fit testing

V.E. Kuzmichev
Ivanovo State Polytechnic University, Ivanovo, Russian Federation

9.1 Introduction

Among all the factors that influence consumers' demands, habits, and satisfaction, the fit is the most essential in determining the quality and sales volume of the product. Consequently, the achievement of good clothing fit is the pivotal criterion for pattern makers. Clothing fit is the indicator by which to judge design, construction, tailoring, and production, which are integrated with comfort, clothing appearance, pattern, etc. Generally, a good fit of clothing signifies objectively acceptable numeric indicators established by the producers, on the one side, and subjectively satisfactory appearance and wearing comfort, such as clothing pressure, ease allowance, etc., demanded by the consumers, on the other side. Two terms are used to describe the quality of clothes: fit and balance.

The first term, *fit*, is described psychologically and physiologically by a number of aspects: thermophysiological comfort, sensorial comfort, body movement comfort, and aesthetic appeal (Li, 2001). Fit plays an important role in clothes design, as it significantly affects the appearance and comfort of clothes and describes the concordance between body measurements and clothes dimensions and the total look of the system evaluated (LaBat and DeLong, 1990).

Acceptable values of ease allowances which are designing in pattern block of different garment styles should prevent the misfit arising (Surikova et al., 2017).

Many scholars have extensively researched the fit of clothing and have found that a well-designed pattern block, using 3D simulation and bodyscanning technologies, is the foundation of many different kinds of properly fitting clothes (Naglic et al., 2016; Lin and Wang, 2016; Wang et al., 2014). To obtain clothing fit in accordance with body morphology, the processes of body classification and pattern block shaping should both be explored more carefully. Fit should be considered as an important factor to evaluate the rationality of body classifications and pattern block making (Beazley, 1997). Well-fitted garments mainly rely on accurate pattern making.

The second term, *balance*, expresses the best situation, which occurs when the weight, size, and location of clothing and its parts are spread in such a way around the body that they do not destroy the harmonic view of the system in total. The term balance is used to find the numeric indicators to predict the fit before clothes production.

Anthropometry, Apparel Sizing and Design. https://doi.org/10.1016/B978-0-08-102604-5.00009-3

The boundaries between the two score terms of fit and balance are not very obvious sometimes. They disappear when differences between the volume of the clothes and the body indicated by air gaps become smaller or completely disappear. For loose clothing designed by means of pattern blocks with positive ease allowances, the indicators of fit and balance can be calculated without problems, but for tight-fitting clothes, such as compression garments, which have negative ease allowances, the numeric indicators of balance are never used (see Section 13.6 in Chapter 13).

By means of balance indicators, which can be calculated after pattern block making, the fit of clothes can be predicted before production.

The main problems with ill-fitting clothes are hiding in the pattern block. For fit prediction, the pattern block should be checked to find the basic indicators of balance. The balanced pattern block is the combination of its parts (front, back, sleeve), in which each part should be designed in accordance with the desirable level of clothes fit in total. Bad-fitting clothing can be produced only by an imbalanced pattern block, but properly fitting clothes need only a balanced pattern block.

So, the schedule of special indicators can help to evaluate the balance between the body and clothes construction, on the one side, and predict the fit, on the other side.

9.2   Approaches to fit criteria

Clothing appearance is evaluated within the scope of aesthetic (harmonic) comfort which is a complex perception contributing to overall image and well-being of the wearer. This appearance is judged by front, profile, and back silhouettes, balance between the shapes of body and clothes, proportion and position of pieces, distribution of folds, etc. To achieve a satisfactory fit, it is inadequate to study only one aspect without consideration of others. The synergy effects dependent on body morphology, construction of pattern block, clothes type and style, and properties of textile materials should be experimented with synchronously to observe the variation of fit and balance and to establish the practical criteria of properly fitting clothes.

To observe fit objectively, numeric indicators were proposed to describe the features of clothes evaluated in the system "body-clothes." The clothing compression is the most frequently used indicator to reveal the interactive force applied to the soft tissue of the human body, which will influence the movement and comfort accordingly (see Chapter 13). On the one hand, a good fit of clothes is determined only when acceptable wearing comfort, satisfactory appearance, and appropriate clothing pressure are obtained. On the other hand, a misfit of clothing is determined from uncomfortable feelings of the wearer, defective appearance, and clothing compression that is too high. The reasons for misfit are mainly related to the absence of anthropometric information, defective methods of pattern block making, and an ignorance of textile material properties.

In fit investigations, the clothing itself and the wearer's body are regarded as a concomitant system, because the wearing effects vary tremendously from one wearer to another due to the diverse morphology of human bodies. To meet the demands of different body types, ready-to-wear mass production (RTW) clothes are produced according to sizing systems that provide adequate sizes with suggested intervals of different body measurements. In accordance with the sizing system, the consumers can select compatible RTW clothes with sizes that are very close to their morphology. However, a standard size cannot provide a good fit every time (Turner, 1994). After all, similar body measurements taken from different people do not equal to a chosen body type (Hu et al., 2018).

Due to ongoing investigations of body morphology and the differences between different body types, new information is available to achieve a more effective sizing system and to improve clothing fit and comfort (Pei et al., 2018). For example, new devices can be used to obtain additional body measurements of slope, shape, and thrust of the shoulder area of individual consumers, which are then used in adjustments to an RTW shirt (Kim et al., 2017).

To determine whether a garment fits well or not, a criteria of assessment should be created. Well-fitting clothes should be judged against the following criteria (Gill, 2015; Erwin et al., 1979):

(1) the grain direction relating to fabric structure (mainly woven) in accordance with principles of grain alignment;
(2) the relationship between silhouette, construction, and styling lines of clothes;
(3) the acceptable appearance of clothes without stress folds or unnecessary creases;
(4) the balance, relating to concordance between the human body and clothes;
(5) the ease allowance or air gap, which shows linear or volumetric differences between the human body and clothes.

These principles are expected to meet different criteria in regard to the definition of clothes fit (Gill, 2015). By means of a combination of subjective and objective evaluations, the fit criteria can be established. But due to the huge number of styles and silhouettes and the constant changing of people's understanding of the fit and clothes performance, universal scores aren't stable.

Subjective evaluations are principally conducted with sensory analysis using questionnaires (participants) after a training session of scoring criteria. The participants are asked to answer the questions with a range of scores, which accordingly reveals their opinions toward a clothing fit. The sensory evaluation is used to find the influence of pattern blocks and textile materials (Fujii et al., 2017) and to investigate the suitable range of ease allowance in accordance with fit, beauty, comfort perception, and purchase intention (Monobe et al., 2017). The sensory evaluation method can obtain participants' assessments by a scale from worst to best (Xue et al., 2016) or from "very uncomfortable" to "very comfortable" (Gu et al., 2016).

Objective evaluations of clothes fit are executed with measuring devices and algorithms to estimate fit through numerical indicators and values, such as the combination of the lengths from front waist line to back waist line across shoulder neck points (SNPs) and shoulder points (SPs) measured on pattern blocks and the angle between the front edges of ready clothes (Yan et al., 2017).

With the development of virtual technology, many researchers have leveraged visualization tools to measure compression and ease allowances (air gap, volumetric, projection, etc.) during virtual try-on. The color gradient is used to visualize the distribution of air gap in the "clothes—avatar" system (Lage and Ancutiene, 2017). Summarily, the fit evaluation is conducted objectively and subjectively in real and virtual environments, from which the criteria of good fit and pattern adapting approaches can be achieved.

9.3 Indicators of fit and balance

There are three main factors that influence the fit of RTW clothes:

(1) construction of the pattern block;
(2) clothes manufacturing (cutting, sewing, heat-moisture treatment, etc.);
(3) fullness of consideration of the textile material properties.

To evaluate fit and balance, the following indicators are used:

(1) the visual destruction of the outline shape, not in accordance with traditional styles;
(2) the appearance of stress folds or unnecessary creases that were not planned or designed;
(3) the differences between the 3D shape of the body and clothes parts of shoulder area;
(4) the bottom line isn't parallel to the floor (if the designed bottom line should be parallel);
(5) the edges and seams don't follow the anthropometrical levels or lines in accordance with the design;
(6) the bending of edges and seams under forces arising in clothes;
(7) nonconcordance between the clothes parts in a dynamic state; usually, this evaluation needs expert opinions and some special devices to record and parameterize the changes in the clothes due to movement.

The indicators listed are used for balance evaluation in the static state (Aldrich, 2008). The number of indicators that can be used for evaluation of fit and balance depends on function and the planned level of clothes quality.

The most serious issue is to find the cause of any defects that appear. Sometimes the defect has several causes, which can reflect many small mistakes arising step-by-step from the pattern-block making to the clothes production. Fig. 9.1 shows how two main elements of the system "male-garment" and "female-garment" could be attached one to another with imbalances arising from two different causes (see Chapter 10).

As a result of the nonconcordance shown in Fig. 9.1, the side seam isn't perpendicular to the floor (the angle between side seam and vertical line is not zero) and the bottom isn't parallel to the floor (a similar angle can also be measured). Besides these indicators of imbalance and misfit, the back is located too close to the buttocks. For the imbalance in Fig. 9.1A, the cause is the inadequacy between the width of the body and the clothes at the waist and hip levels due to the special male morphology. For the imbalance in Fig. 9.1B, the front length is shorter than the corresponding body measurement due to the sloping posture.

All defects of the pattern block can be divided into two groups in accordance with the causes of their appearance. The defects of the *first group* take place when the pattern block parameters are bigger or smaller than the corresponding body measurements, usually in the horizontal direction. If the pattern dimension is much smaller than the body measurement, stress folds will arise in the clothes. If the pattern dimension is much

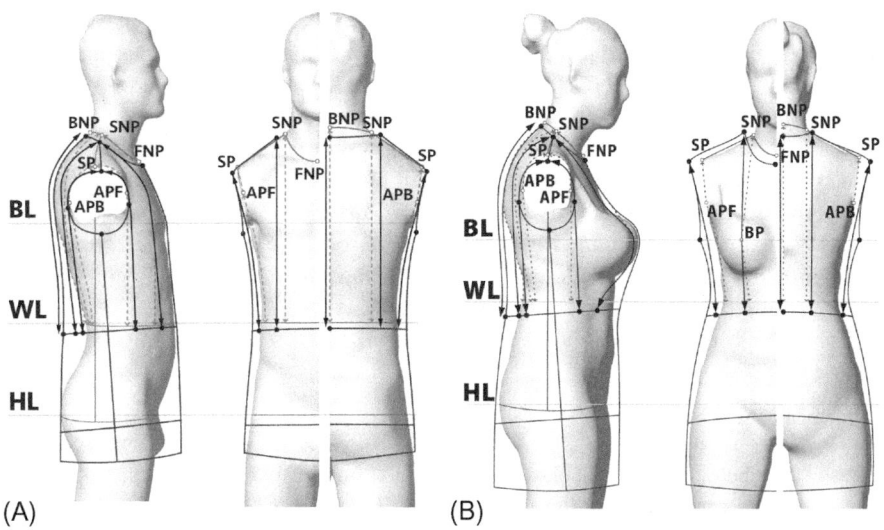

Fig. 9.1 Examples of imbalance when the width of body and clothes on waist and hip levels is not adequate for male (A) and the front length is shorter than corresponding body measurement for female (B). *BL, WL, HL* are, respectively, bust, waist, and hip levels. *BNP, SNP, FNP* are, respectively, back, shoulder, and front neck points. *SP* is shoulder point. *BP* is bust point. *APB, APF* are, respectively, armpit back and front points.

larger than the body measurement, soft folds (vertical or horizontal) will arise. Both kinds of folds destroy the harmonic appearance of the clothes. The defects of the *second group* take place due to imbalance. If the dart value (breast, shoulder) is not sufficient, creases that look like a handheld fan will arise around the bust point *BP* or the blade point *BLP*; the cause is a wrongly calculated bearing balance. If the vertical dimensions of the pattern block are not equal to the body measurements, sloping arch soft folds and stress folds can arise due to vertical or horizontal imbalance.

9.4   Ease allowance

The amount of ease allowance indicates the positioning of the clothing around the body in space. Traditionally, in the pattern block-making process, the ease allowance is the positive or negative difference between two values: body measurements and the corresponding dimension of the pattern block or RTW clothes. The acceptable intervals of ease allowances will lead to well-fitting clothing of the desired type and style made using certain textile materials.

9.4.1   Kinds of ease

A variety of sizes, styles, and silhouettes can be achieved by means of an unlimited combination of ease allowances and configurations of pattern block lines. To calculate an ease, measurements should be taken on the anthropometrical level of the human body and the corresponding construction level of the pattern block or clothes. Because of fashion trends, these two levels sometimes are located in the same place, and sometimes are not. For example, construction of the waist level may be higher or lower than the natural waist line; to describe this difference in the vertical direction, a special ease E_{BL} called "ease to back length" is used. Ease allowance is marked as E_i, where E is the ease symbol and i is a symbol (abbreviation) of body measurement. For example, E_{BL} shows the difference between two heights of waist levels: first, of the body at the narrowest natural anthropometrical level; second, of the clothes at the construction level formed by the waist darts. Fig. 9.1B shows the situation where the construction level is lower than the natural anthropometrical level, so $E_{BL} > 0$.

For upper body clothes, such as suits, jackets, coats, shirts, and so on, there are other eases, as follows:

for bodice (torso):
  – an ease to bust girth E_{BG} and its distribution between back (ease to back width E_{BW}), armhole (ease to armhole width E_{AHW}) and front (ease to front width E_{FW}). These eases can be used as full measurements (E_{BG}, E_{BW}, E_{FW}) or as half ones ($E_{0.5BG}$, $E_{0.5BW}$, $E_{0.5FW}$);
  – an ease to waist girth E_{WG} ($E_{0.5WG}$);
  – an ease to hip girth E_{HG} ($E_{0.5HG}$).

The eases E_{BG}, E_{WG}, E_{HG} are responsible for silhouette or style, but E_{BW}, E_{AHW}, E_{FW} are influenced by the profile contour of clothes;

for sleeve:
  – an ease to armhole depth E_{AHD};
  – an ease to arm girth E_{AG}.
  – an ease to arm length E_{AL}.

In practice, these mentioned eases are used in the following situations:

(1) for new pattern block making when the ease amounts are known and they can be added directly to body measurements to calculate the dimensions of front, back, etc. (straight task);
(2) for analyzing pattern blocks that were drawn earlier when an ease amount was unknown (opposite task). To find the ease values, it's necessary to know the body measurements used for the pattern block drawing, methods of pattern block shaping, number of textile materials and their properties, methods of production (sewing), etc. For example, to analyze the historical pattern blocks, the shrinkage after heat-moisture treatment should be taken into account before an ease calculation (Kuzmichev et al., 2017).

Each ease should be designed as a complex value including the following components:

(1) physiological component, which should provide easy breathing and limited pressure (compression) on lymph nodes and blood vessels;
(2) ergonomic component, which should allow movements in accordance with the functions of the clothes. This ease allowance is equal to the difference between maximal and minimal perimeters of the wearer's body, which is obtained from the human body in standing and moving postures (Chen et al., 2008);

(3) comfort component between the body and the inner layer of clothes, which is important for heat balance and vapor movement to the outside;

(4) friction component, which should allow the clothes to move on the human body, provide sufficient space for body shape and its movements, and take into account the influence of mechanical properties of fabrics (Chen et al., 2008);

(5) design component, for achieving the desirable outline shape in accordance with the fashion trends and human body sizes. The design component very strongly depends on several factors, such as style of clothes, body sizes (fullness or drop), textile material properties (thickness, rigidness, draping, etc.), total aesthetic view of the system "body-garment" in which the morphological features of the body should be in harmony with the outline shape of the garment and should express fashionable or other trends. Due to different criteria of design, this ease validates the applicability of combinations of clothing styles, volumes, and body measurements (Surikova et al., 2017). Fig. 9.2 shows the variants of acceptable combinations of factors that might be taken into account before choosing the design ease to bust girth E_{BG} for women's clothing: X, H, and A silhouettes in tightly fitted, semifitted, slightly fitted, full, extra full styles; kinds of clothing; and bust girth.

As Fig. 9.2 shows, by variation of E_{BG}, the applicable types and styles of clothes for different body sizes are addressed. For example, for women whose bust girth is larger than 100 cm, dresses with design ease allowance more than 2 cm aren't recommended,

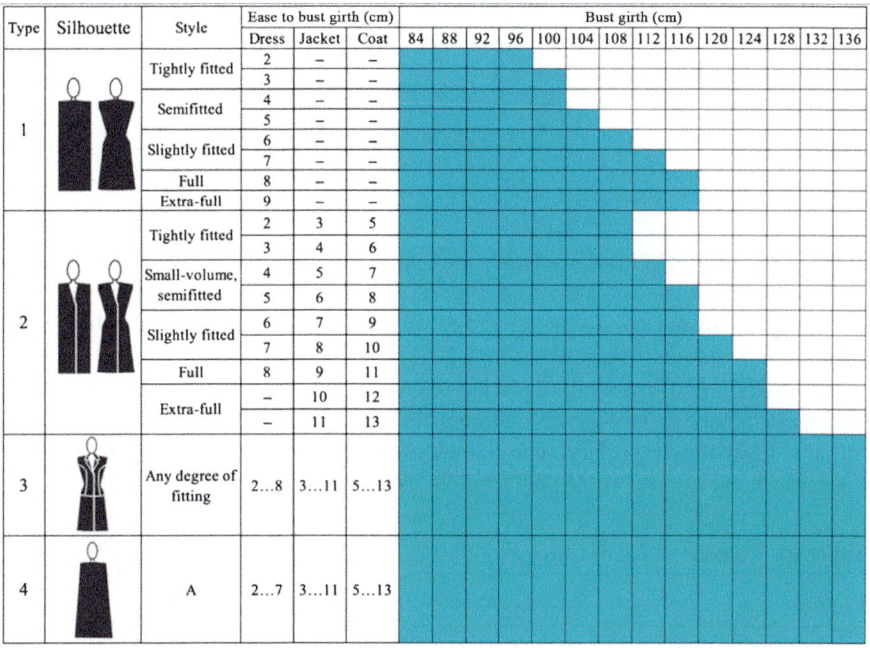

Fig. 9.2 Variants of acceptable design ease E_{BG} in accordance with clothes style and body size (the acceptable variants of combination of ease allowances and bust girth are highlighted) (Surikova et al., 2017).

because these dresses will have many wrinkles and folds. Fig. 9.2 shows the limitations of some styles applied.

All components named can be taken into account and combined during the calculations, in the following order:

(1) minimal physiological ease $E_{min.ph} \cap (E_{stat}, E_{dyn}, E_{heat}, E_{air})$ includes four components that provide ergonomic concordance between the body and clothes in static and dynamic postures and wearer comfort. By means of a fuzzy model, a personalized ease allowance could be generated in accordance with the key body positions and the wearer's movements, permitting further improvements in the wearer's fitting perception (Chen et al., 2008);

(2) minimal ease $E_{min} \cap (E_{min.ph}, E_{tm})$ includes the minimal physiological ease $E_{min.ph}$ and the sum of all textile material thicknesses used in the clothing;

(3) design ease $E_D = E_i - E_{min}$ provides the final outline shape of the clothing. E_D shows the increasing of clothes dimensions under the smallest anthropometrical shell that can cover the human body. The value of E_D is the aesthetic indicator in each period of costume history and presents fashion trends in the pattern-making process, with colors, materials, and decoration.

The values of minimal ease E_{min} and design ease E_D depend on clothes functionality. The common equation for calculating E_i is

$$E_i = E_{min} + E_D \tag{9.1}$$

Let's study the process of choosing each component and calculating E_i (Kuzmichev et al., 2018).

E_{min} can be calculated in two ways:

$$E_{min} = E_{dyn} + E_{tm} \tag{9.2}$$

$$E_{min} = E_{air} + E_{tm} \tag{9.3}$$

To design traditional classical clothes such as coats, suits, jackets, and dresses, it is necessary to calculate only the ease E_{tm}; others eases can be taken from published tables or other resources. Table 9.1 shows the values of E_{tm} and air gaps that should be taken into account to calculate $E_{0.5BG}$.

Before designing new kinds of clothes or clothes with special functions—for example, an astronaut's overalls—all components should be calculated or found experimentally from the beginning. The values of E_{min} for basic width (back, armhole, front), girth (waist, hip, arm), and armhole depth are shown in Table 9.2.

Table 9.1 Components of ease to bust girth $E_{0.5BG}$ including E_{tm} and air gaps (cm)

Clothes	Thickness of material	Air gap	E_{tm}
Women dress	0.05–0.1	0.1–0.3	0.4–1
Men suit, women jacket	0.25–0.3	0.6–0.8	2.4–3.5
Coat	0.35–0.45	0.9–1.1	2.8–3.5

Table 9.2 Minimal eases E_{min} for clothes

Ease allowance	E_{min} (cm)		
	Women dress	Men suit, women jacket	Coat
$E_{0.5BW}$	0.5–0.7	1.0–1.5	1.5–2
E_{AHW}	1.7–2	3.0–3.5	3.5–4
$E_{0.5FW}$	0–0.5	1.0–1.5	2–2.5
E_{AHD}	1.7–2	2.5–3	3.5–4.5
$E_{0.5WG}$	1.5–2	2–2.5	3.5–4.5
$E_{0.5HG}$	1.5–2	2–2.5	3.5–4.5
E_{AG}	3–3.5	4.5–5	6–6.5

E_D must not be smaller than E_{min}, to guarantee comfort:

$$E_D \geq E_{min} \tag{9.4}$$

The combination of different eases while designing in vertical and horizontal directions creates the final outline shape, silhouette, dimensions, and proportions of clothes, which finally influence the fit and balance. The values of the eases should be correlated one to another to achieve an aesthetic, well-fitted, and balanced system of "body-clothes." This combination depends on the fashion.

The ease allowance relating to textile materials varies because the properties are many and lead to modifications of the pattern block to maintain a good fit. Thus, a patternmaker should estimate how the clothing styles are affected by the characteristics of textile materials and also be able to manipulate basic patterns with different mechanical characteristics in order to get a particular garment fit, shape, and style (Lage and Ancutiene, 2017). The length and proportion of a close-fitting dress can be predicted by the tensile strength in the warp direction (Yan et al., 2017). By means of the Kawabata Evaluation System (KES), the index of crease coefficients is calculated by mechanical tests of textile materials, which allows an estimation of the increasing of the fabric surface length compared to the body surface on different longitudinal levels (Zvereva et al., 2012). For producing a suit with satisfactory silhouette, the pattern designers should be able to predict the formability of textile materials from the mechanical properties (Xue et al., 2016). To apply different fabrics to suits from the prototype, these prediction models were created for adapting the pattern block to maintain a good fit.

So, the ease values should be considered in the pattern-making process according to textile material properties to achieve clothes with a good fit.

9.4.2 Ease in shoulder clothes

The ease to bust girth E_{BG} ($E_{0.5BG}$) helps to create the clothes volume at the bust level. During the pattern block drawing, it should be divided into three components between back, armhole, and front. To calculate E_{BG}, the algorithm shown in Table 9.3 can be used.

Table 9.3 Ease to bust girth distribution along bust width of pattern block

No.	Dimension of pattern block	Equation	Body measurements
1	Bust width	$0.5BG$ $+ E_{0.5BG}$	
2	Back width	$0.5BW$ $+ E_{0.5BW}$	BW is the back width
3	Armhole width	$d_{arm} + E_{AHW}$	d_{arm} is the diameter of arm (the distance between two points—armpit back APB and armpit front APF)
4	Front width	$0.5FW$ $+ E_{0.5FW}$ $+$ part of breast dart	FW is the front width

E_{BG} depends on the kind of clothes and the volume and is modified in accordance with fashion trends. E_{BG} is presented as $[E_{BG.min}–E_{BG.max}]$, where $E_{BG.min}$ and $E_{BG.max}$ are E_{BG} corresponding to the smallest and largest thickness of textile materials used for producing this type of clothes, respectively. Usually, E_{BG} is published in manuals for basic sleeve-in style. For raglan and dolman styles, the ease E_{BG} should be increased by 20%–30%. The eases $E_{0.5BW}$, E_{AHW}, $E_{0.5FW}$ can have different units, such as centimeters or percent, as part of $E_{0.5BG}$.

An ease to waist girth E_{WG} ($E_{0.5WG}$) and an ease to hip girth E_{WG} ($E_{0.5HG}$) help in obtaining different styles such as X, H, A, O due to the fit in the waist and hip area. Ease recommendations classified by style of clothes are diverse but the rule of general variation is similar. The recommended ease values gradually increase from close-fitted clothes to very loose-fitted ones. Moreover, clothing styles are more sensitive to E_{BG} and E_{HG}, and the acceptable ease intervals for loose-fitted and very loose-fitted garments are larger than for other kinds. Table 9.4 shows the ease ranges for women's clothes.

An ease to armhole depth E_{AHD} is variable over a huge range under the influence of several factors. The minimum value of E_{AHD} depends on the morphological changing of the joint area located between torso and arm under ergonomic movements; the minimal compression on the armpit; the seam allowance of cup sleeve sewing into the armhole; the air gap; and the thickness of the textile materials. For a woman's dress with sleeve-in, the minimum value of E_{AHD} is near 1.5–2 cm; for a dress without sleeves it is 0.5–1 cm; and for a coat it is 4–4.5 cm.

An ease to arm girth E_{AG} depends on sleeve width, thickness of materials and dynamic postures. For example, the minimum value of E_{AG} for a woman's dress is close to 9%–10% AG and must not be smaller than 3 cm.

All considered eases are very important for clothes fit.

Stable relationships exist between several eases for clothes in the same style. Fig. 9.3 shows the relationship existing between some eases designing in the

Table 9.4 Ease in women's clothes (Gill, 2015; Erwin et al., 1979; Liechty et al., 2010; Holman, 1997; Baezley and Bond, 2003; Palmer and Pletsch, 1995; Creative Publishing International, 2005; Myers-McDevitt, 2009; Keiser and Garner, 2008; Gioello and Berke, 1979; Cock, 1981; Cloake, 2000; Betzina, 2003)

Ease allowance	Clothes style				
	Close-fitted	Fitted	Semifitted	Loose-fitted	Very loose-fitted
To bust girth E_{BG}	0–7	5–11	7–18	12–23	18–35
To waist girth E_{WG}	1–4	2–7.5	7.5–10	10–15	–
To hip girth E_{HG}	0–5	0–10	5–11.5	10–15.5	12–35

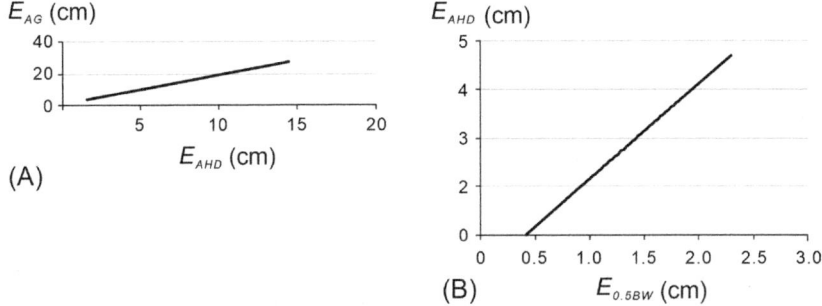

(A)

(B)

Fig. 9.3 Basic diagram for E_{AHD} and E_{AG} for women's dresses from 1950 to 2005 (A) and E_{AHD} and $E_{0.5BW}$ for women's blouses from 2000 to 2005 (B).

horizontal (E_{AG}, $E_{0.5BW}$) and vertical (E_{AHD}) directions in women's clothes. Knowing these mathematical relations, the possibility of correctly choosing the ease combinations and creating of CAD are becoming a reality.

Besides main body measurements, the pattern block-making process needs additional measurements, which also are adapted to clothing style by means of eases. This additional database includes the following eases:

–   an ease to neck width E_{NW} (the minimum value is 1 cm);
–   an ease to neck height E_{NH};
–   an ease to shoulder width E_{SW};
–   an ease to back length E_{BL};
–   an ease to BP height E_{BPH};
–   an ease to the distance between BP E_{BP};
–   an ease to arm length E_{AL};
–   an ease to wrist girth E_{WG};

The values of eases may be as positive (when the dimension of pattern block bigger than corresponding body measurement) as negative (when the dimension of pattern block smaller than corresponding body measurement).

9.4.3 Ease in skirts and trousers

For skirts and trousers, the main eases are:

- an ease to waist girth E_{WG} ($E_{0.5WG}$);
- an ease to hip girth E_{HG} ($E_{0.5HG}$);
- an ease to thigh girth E_{TG};
- an ease to distance measured when the participant is sitting, from sitting surface to waist level E_{SS-WL};
- an ease to knee girth E_{KG} when the leg is bending.

The ease to waist girth E_{WG} ($E_{0.5WG}$) is equal or near to half of the interval between waist girth for mass production. The ease to thigh girth E_{TG} depends on morphological features, dynamic effect. and textile material properties. The ease to distance $E_{Дc}$ depends on the gender: for women the minimum value is 0–0.5 cm, and for men 1–1.5 cm. The recommended minimum values of main eases are shown in Table 9.5.

9.4.4 Ease allowance in the second part of the 20th century to the beginning of the 21st century

Table 9.6 shows the ease allowances used in the second part of the 20th century to the beginning of the 21st century in pattern blocks of women's dresses, jackets, and men's suits. These results were obtained after parameterization of more than 1000 pattern blocks collected from enterprises, fashion magazines, and other resources. Each ease is presented by the average value (mean) and the range from minimum to maximum.

Therefore, the criteria of eases for well-fitting clothing are varied in history with clothing type, style, and body measurements. For examples, the percentage distribution of E_{BG} between the back, the armhole, and the front in men's suits has varied widely by percentage: the back 17–42 (average is 26.5%), the armhole 14–60 (average is 41.5%), and the front 8–67 (average is 32%). By means of combination of these eases, the fit and comfort can be changed in accordance with consumer demands.

Table 9.5 Minimal eases in skirts and pants (trousers)

	Minimum values E_{min} (cm)		
		Pants	
Ease	Skirt	Women	Men
$E_{0.5WG}$	1–1.5	1–1.5	2–2.5
$E_{0.5HG}$	1.5–2	1.5–2	2–2.5
E_{TG}	–	5–5.5	5–5.5

Table 9.6 Ease allowances used in pattern block manufacturing in the second part of the 20th century to the beginning of the 21st century (Kuzmichev et al., 2018)

Ease allowance	Mean (numerator)/range (denominator)					
	1950s	1960s	1970s	1980s	1990s	2000s
Women's dress for typical body 164-96-104						
$E_{0.5BG}$	4.7 / 4–5	4.2 / 4–5	4.6 / 4–5	7.9 / 3.5–13	4.6 / 0.5–8	4.2 / 0–7.5
$E_{0.5FW}$	2.3 / 2–2.5	0.8 / 0–2.5	0.2 / 0–0.5	2.9 / 0–7.8	1.5 / 0–3.5	0.8 / 0–1.5
$E_{0.5BW}$	0.5 / 0–1	1 / 0.8–1	0.9 / 0.5–1.2	2.4 / 0.5–6	1.2 / 0–2	1.3 / 0.5–3
E_{AHW}	1.3 / 1–1.5	2.6 / 0.7–3.5	3.9 / 3.5–4	2.3 / 0–6.5	1.5 / 0–3.5	2.8 / 1.5–4
$E_{0.5WG}$	2 / 1.5–2.3	3.3 / 2–6	7.7 / 3–12	11.1 / 1.5–20.5	5.2 / 1.5–14	6 / 0–12
$E_{0.5HG}$	3 / 0–4	2.7 / 1.5–4	2.8 / 2–4.5	3.6 / 0.5–10	2.7 / 0–8	2.9 / 0–5
E_{AHD}	2.5 / 2–3	3.5 / 3–4	3 / 2.5–3.5	2.9 / 1–5	3.3 / 1.5–6	2.3 / 0–3
E_{SW}	0.6 / 0.5–0.7	0.4 / 0–0.7	0.7 / 0–1.5	-0.2 / -2 to 0.5	0.4 / 0–2.5	0.1 / 0–0.5
Women's jacket for typical body 164-96-104						
$E_{0.5BG}$	4.9 / 3.3–6.2	7.9 / 7.8–8	8 / 5.5–9.2	7.4 / 1.0–12.0	8.3 / 6.8–11.8	8.5 / 7.5–10
$E_{0.5FW}$	2.4 / 1.6–2.9	2.3 / 1.9–3.1	2.8 / 1.1–4.1	2.9 / 0.6–4.5	3.3 / 2.7–4.4	3.9 / 3.5–4.2
$E_{0.5BW}$	0.6 / 0.1–1	1.4 / 1.3–1.6	1.8 / 1.1–2.4	2.6 / 0.8–5.7	2.6 / 2.3–3.2	2.9 / 2.6–3.4
E_{AHW}	1.3 / 1.2–2.2	4.3 / 3.3–4.7	3.3 / 2.3–4.3	3.7 / 1.4–7.3	2.4 / 1.4–5.0	1.3 / 1.0–1.7
$E_{0.5WG}$	7 / 4.5–10.5	10.1 / 7.7–13.1	13.2 / 6.5–19.2	16.4 / 10.8–19.6	10.6 / 6.7–18.0	11.1 / 8.5–14.9

Continued

Table 9.6 Continued

Ease allowance	Mean (numerator)/range (denominator)					
	1950s	1960s	1970s	1980s	1990s	2000s
$E_{0.5HG}$	$\frac{4}{2.4\text{-}6.4}$	$\frac{3.5}{2.2\text{-}4.8}$	$\frac{7.3}{5.2\text{-}8.7}$	$\frac{3.5}{0\text{-}3.7}$	$\frac{3.9}{2.8\text{-}5.1}$	$\frac{4.7}{4.7}$
E_{AHD}	$\frac{4.1}{3.2\text{-}4.8}$	$\frac{6.3}{5.7\text{-}6.8}$	$\frac{6}{4.0\text{-}8.9}$	$\frac{7.0}{3.8\text{-}9.2}$	$\frac{7.4}{6.6\text{-}8.7}$	$\frac{6}{5.4\text{-}6.3}$
E_{SW}	$\frac{0.6}{0.3\text{-}2.3}$	$\frac{0.6}{0\text{-}1.2}$	$\frac{0.1}{0.7\text{-}0.3}$	$\frac{1}{1.6\text{-}4.6}$	$\frac{0.8}{0\text{-}1.6}$	$\frac{1.2}{0.8\text{-}1.5}$

Men's suits for typical body 176-100-82

Ease allowance	1950s	1960s	1970s	1980s	1990s	2000s
$E_{0.5BG}$	$\frac{7.6}{6\text{-}10.3}$	$\frac{7.3}{5.5\text{-}10}$	$\frac{8.7}{5.5\text{-}12}$	$\frac{8}{5.5\text{-}11.8}$	$\frac{8.8}{6\text{-}12}$	$\frac{8.2}{7.2\text{-}9}$
$E_{0.5FW}$	$\frac{2.9}{0.5\text{-}4.5}$	$\frac{2.4}{1.6\text{-}4.3}$	$\frac{1}{0.5\text{-}4.1}$	$\frac{2.8}{1.6\text{-}4.6}$	$\frac{3.3}{0.9\text{-}4.7}$	$\frac{1.7}{1.3\text{-}2.1}$
$E_{0.5BW}$	$\frac{2.3}{1.6\text{-}3}$	$\frac{1.8}{1.2\text{-}3}$	$\frac{2.9}{0.7\text{-}4.9}$	$\frac{1.9}{1\text{-}2.7}$	$\frac{2.7}{2\text{-}3.3}$	$\frac{2.2}{1.5\text{-}3}$
E_{AHW}	$\frac{2.6}{1\text{-}3.8}$	$\frac{3.1}{2\text{-}5}$	$\frac{2.4}{0.8\text{-}5.3}$	$\frac{3.5}{2.6\text{-}4.8}$	$\frac{2.5}{2.1\text{-}3.1}$	$\frac{3.6}{2.5\text{-}4.1}$
$E_{0.5WG}$	$\frac{10}{7\text{-}13}$	$\frac{5.5}{3.5\text{-}7.7}$	$\frac{7.5}{6.5\text{-}8.5}$	$\frac{7.1}{4\text{-}11}$	$\frac{7.9}{4.2\text{-}9.2}$	$\frac{7.5}{5.4\text{-}8.8}$
$E_{0.5HG}$	$\frac{6.7}{2.7\text{-}10.3}$	$\frac{5.1}{3.6\text{-}8.3}$	$\frac{5.8}{3.9\text{-}8}$	$\frac{3.7}{2\text{-}5.2}$	$\frac{4.4}{3\text{-}6}$	$\frac{4.8}{3.5\text{-}6.2}$
$E_{0.5FW}$ as part of $E_{0.5BG}$ (%)	$\frac{36}{8\text{-}54}$	$\frac{30}{19\text{-}42}$	$\frac{34}{11\text{-}67}$	$\frac{32}{27\text{-}38}$	$\frac{38}{18\text{-}47}$	$\frac{22}{19\text{-}26}$
$E_{0.5BW}$ as part of $E_{0.5BG}$ (%)	$\frac{32}{17\text{-}46}$	$\frac{24}{17\text{-}33}$	$\frac{19}{11\text{-}23}$	$\frac{23}{18\text{-}29}$	$\frac{32}{27\text{-}42}$	$\frac{29}{21\text{-}41}$
E_{AHW} as part of $E_{0.5BG}$ (%)	$\frac{33}{14\text{-}46}$	$\frac{46}{29\text{-}64}$	$\frac{47}{22\text{-}66}$	$\frac{45}{40\text{-}55}$	$\frac{30}{26\text{-}43}$	$\frac{49}{38\text{-}60}$
E_{AHD}	$\frac{4.4}{2.5\text{-}7.7}$	$\frac{3.9}{2.5\text{-}5.6}$	$\frac{5.5}{3.1\text{-}7.8}$	$\frac{4.8}{3.3\text{-}6.2}$	$\frac{5.0}{3.1\text{-}8.3}$	$\frac{3.5}{3\text{-}4.8}$
E_{SW}	$\frac{2.1}{1.2\text{-}3.2}$	$\frac{1.8}{0.6\text{-}2.8}$	$\frac{2}{1.3\text{-}3}$	$\frac{1.3}{1\text{-}1.7}$	$\frac{2.9}{1.7\text{-}5}$	$\frac{1.5}{1.3\text{-}1.8}$

9.4.5 Ease calculation from pattern block

From horizontal and vertical cross-sections taken from the scanned system "body-clothes" and consisting of two cross-sections of body and clothes, two types of ease can be calculated: first, the ease as the difference between the lengths of two perimeters; second, the ease as the air gap located between both perimeters.

From the pattern block, the calculation of ease allowances isn't a simple task. They vary drastically with type and style of clothing, from knitted tight-fitting garments with negative ease values to oversize garments with enormous ease values. For example, if the length of shoulder line is presented as the sum of shoulder length (width) taken from the body and the ease amount, this approach can be used only when the shoulder line starts from *SNP* and finishes at *SP* exactly (Gill, 2015). But this situation takes place only for a basic pattern block and this approach couldn't be applied for a shaped pattern block or for a historical pattern block.

For flat pattern blocks and RTW clothes, the eases E_i for girth, width, length can be found by this equation:

$$E_i = D - BM. \tag{9.5}$$

where D is the measured dimension of the pattern block or clothes in centimeters; BM is the body measurement in centimeters.

For analyzing and parameterization of a pattern block by means of ease amounts, the initial database for starting this process should include:

(1) drawing, photo, or sketch of garment;
(2) schedule of body measurements used for pattern block making;
(3) properties of textile materials (shell, lining, interlining) such as thickness, shrinkage after heat-moisture treatment, elongation, and so on;
(4) thickness of underwear clothing.

To present the specific morphological features of the human body and key body measurements, they can be transformed into a flat anthropometrical net. The anthropometrical net is the fragment of the body surface that is oriented on the flat surface and consists of key anthropometrical points in accordance with the body measurements. This net is the morphological passport of the human body with known sizes. Figs. 9.4 and 9.5 show the anthropometrical net and its application for calculating the ease amount from the pattern block. The algorithm of net building is based on the triangle method.

An anthropometric net includes many anthropometric points: back neck point *BNP*, shoulder neck point *SNP*, blade point *BLP*, shoulder point *SP*, bust point *BP*, front neck point *FNP*, armpit back point *APB*, armpit front point *APF*. The anthropometric levels are waist level *WL* and hip level *HL*. To draw the net, the body measurements shown in Table 9.7 are needed.

All body measurements can be measured, calculated, or found out manually or by means of a bodyscanner. Fig. 9.5 shows the location of the main eases obtained by overlapping the anthropometric net and the pattern block. To find out the ease amount,

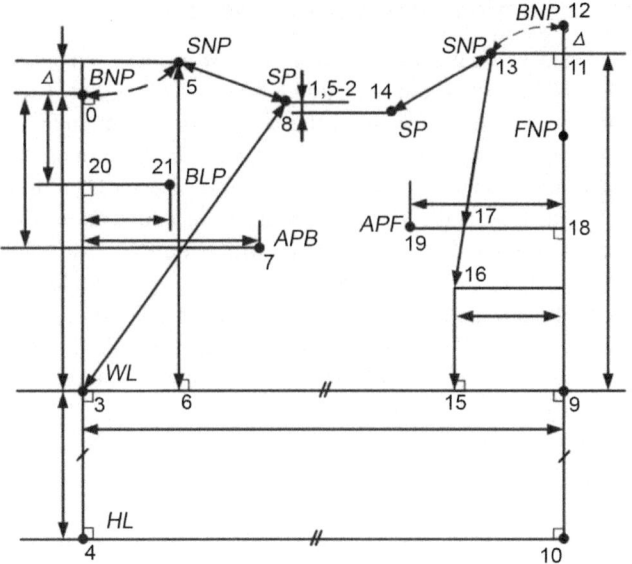

Fig. 9.4 Structure of anthropometric net of upper torso. *BLP* is blade point.

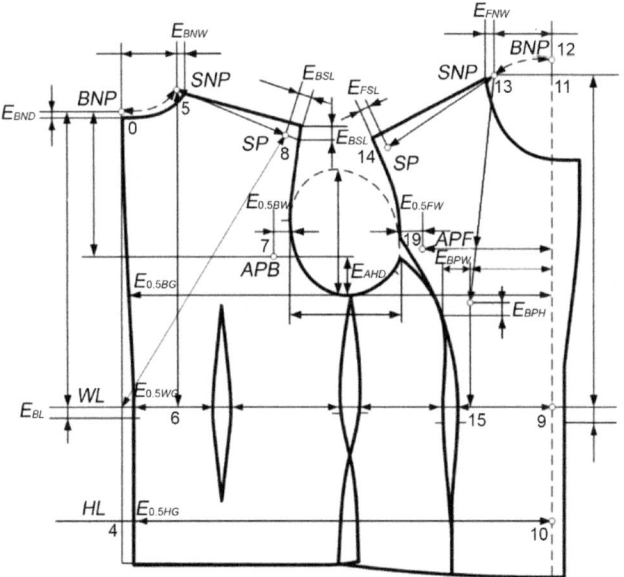

Fig. 9.5 Scheme of ease amount calculation by means of anthropometric net.

Table 9.7 Body measurements for anthropometric net drawing

Part of net	No.	Body measurements	Symbol (Fig. 9.7)
Back	1	Back width	/2-7/
	2	Distance between the blades	/20-21/
	3	Neck diameter	/3-6/
	2	Shoulder width	/5-8/
	3	Length between *BNP* and *SNP* around neck line	/0-5/
	4	Length between left and right *SP* across *BNP*	/0-8/
	5	Length between *ABP* and *APF* across *SP*	/7-8/ + /14-19/
	6	Back length from *BNP* to *WL*	/0-3/
	7	Back length from *SNP* to *WL*	/5-6/
	8	Sloping back length (length between *SP* and cross point of saggital plane and natural waist level)	/3-8/
	9	*SP* height	/11-9/
	10	Natural waist height	
	11	*APB* height	
Front	12	Length between *BNP* and *BP* across *SNP*	/12-13-16/
	13	*FNP* height	
	14	Front length *FL*	/11-9/
	15	Length between *BNP* and *WL* across *BP* and *SNP*	/12-13-16-15/
	16	Length between both *BP*	/17-18/
	17	Front width	/18-19/
	19	Neck girth	/0-5/ + /13-12/
	20	*BP* height	
	21	*APF* height	
Front + back	22	Bust girth	/3-9/

the distances between anthropometric points and lines of the pattern block might be measured. In parallel, other parameters that are also responsible for fit and balance can be measured. Table 9.8 shows the full list of parameters calculated and measured after analyzing the pattern blocks of women's clothes.

9.5 Indicators of balance

A schedule of balance indicators can be created for each kind of clothes. The balance of upper body clothing is the complex of solitary indicators for describing the positions and configurations of clothes parts (front, back, etc.) in the vertical and horizontal directions measured relative to the chosen surfaces of the human body, such as upper bearing area, touching area, or waist level.

Table 9.8 Amounts of ease allowance used in women's clothes in the 1990s to the 2000s

	Mean/range (cm)				
Ease allowance	**Blouse**	**Dress**	**Jacket**	**Coat**	**Down jacket**
Back					
Ease to back neck width E_{BNW}	1.1 0.2–2.6	1.2 0.2–2.8	1.7 0.4–3.5	1.2 1–1.7	1.5 1–2.6
	E_{BNW} depends on configuration of neck line, collar, material thickness, and underwear clothes and accessories				
Ease to back neck depth E_{BND} (or highest point of back seam)	0.4 −0.8 to 1.6	0.2 −1.5 to 1.9	0.3 0.2–0.4	0.4 0.3–0.5	0.6 0.5–0.7
	E_{ND} may be positive and negative, equal 0. E_{ND} depends on collar, configuration of neck line, thickness of underwear clothes, and accessories				
Ease to back shoulder width E_{BSW}	−0.5 −1.8 to 0.8	−0.5 −1.6 to 0.6	0.6 −0.4 to 1.6	0.8 0.1–1.5	2.8 0.1–5.5
	E_{SL} depends on shoulder shape: for natural shape [−0.5 to 1.5 cm]; for dropping shoulder [2–6 cm and more]; for narrow shoulder E_{SL} is negative				
Ease to back width E_{BW}	1.5 0.7–2.3	0.9 −0.5 to 2.3	1.3 0.6–2	2 1.5–2.5	3.4 1–5.8
	E_{BW} depends on armhole seam location				
Ease to back length E_{BL}	0 −1.7 to 1.7	0.3 −1.2 to 1.8	0.2 −1.7 to 2.1	1.1 −0.2 to 2.4	1.7 0–3.4
	E_{BL} depends on waist level position in clothes relatively body natural waist level and material thickness				
Ease to sloping back length E_{BSL}	1 0.4–1.6	0.8 0.1–1.5	1.2 0.7–1.7	1.8 1.4–2.2	1.3 1.1–1.5
	E_{BSL} depends on material shoulder pad thickness. E_{SW} and E_{BL}				
Front					
Ease to front neck width E_{FNW}	0.9 0.2–2.5	1.2 0.2–2.7	1.3 1–2.9	1.2 1–2	1.3 0.7–1.9
	E_{FNW} may be bigger than E_{BNW} due to the difference between the thickness of front and back and the location of shoulder seam				
Ease to front shoulder width E_{FSW}	E_{FSL} is equal to E_{BSL} or smaller due to shrinkage in 0.5–0.7 cm				
Ease to front width E_{FW}	1.7 0.5–2.9	1 0.3–1.7	1.6 0.9–2.3	2.1 1.3–2.9	3.6 0.9–6.3
	E_{FW} depends on location of armhole seam and methods of front shaping				

Table 9.8 Continued

Ease allowance	Mean/range (cm)				
	Blouse	Dress	Jacket	Coat	Down jacket
Ease to BP height E_{BPH}	2.5 0.7–4.3	1.6 0.6–2.6	1.6 0–3.2	1.1 −1.3 to 3.5	0.7
Ease to BP width E_{BPW}	E_{BPH} may be positive or negative and depends on breast dart length				
	2.2 1–3.4	1.8 0.8–2.8	2.3 0.5–4.1	3.1 1.7–4.5	5.2
	E_{BPW} always positive and depends on methods of front shaping				
Front + back					
Ease to armhole depth E_{AHD}	3.7 2.5–4.9	2.6 1.4–3.8	3.7 2.4–5	4.4 2.6–6.4	7.1 2.9–11.3
	E_{AHD} depends on sleeve shape and volume, thickness of underwear clothes, armhole design				
Ease to half bust girth $E_{0.5BG}$	6.3 2.3–8.5	4.3 0–5.5	6.0 3.5–10.0	8.0 5.0–13.0	7.0 5.0–15.0
	$E_{0.5BG}$ depends on fashion				
Ease to half waist girth $E_{0.5WG}$	8.4 9.0–15.0	5.3 5.0–12.0	11.0 4.5–15.5	14.0 8.5–20.0	8.0 6.0–13.5
	$E_{0.5WG}$ depends on fashion				
Ease to half hip girth $E_{0.5HG}$	4.0 5.0–9.5	3.0 0–9.5	4.0 3.1–4.7	6.0 2–13.0	7.0 6.0–11.0
	$E_{0.5HG}$ depends on fashion				

The balance of belt clothing is the complex of solitary indicators for describing the positions and configurations of the clothing parts (front, back), in the vertical and horizontal directions and measured relative to the chosen surfaces of the human body, such as lower bearing and touching areas. The indicators of balance exist not only for clothes in total but also for their part combinations (bodice and sleeve), parts, and lines (armhole, neck).

Fig. 9.6 shows the location of the bearing and touching areas of the human body, which are responsible for the balance of clothes. The bearing area (in *black*, Fig. 9.6) has the closest contact with the clothes covering, which can be obtained by correct calculation of darts and appropriate line configurations located in this area. The touching area *(dotted)* gives the directions to clothes and its distribution between the front and the back depends on body morphology.

The indicators of balance are calculated between anthropometrical points and they include:

(1) *bearing balance,* showing the adequacy of the bearing surface of the body and covering parts of the front and back. Bearing balance includes the amounts of two darts, the breast in front and the blade in back;

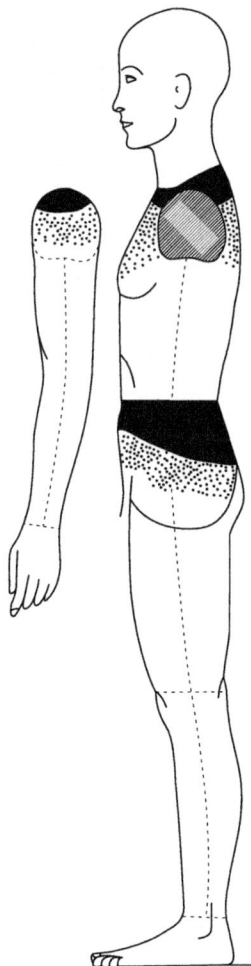

Fig. 9.6 Location of bearing area *(black)* and touching area *(dotted)* on upper torso for upper body clothing and below the natural waist for skirts and trousers, and the arm for sleeve.

(2) *longitudinal balance*, consisting of linear and arc indexes and promoting the horizontal positions of structural level clothes, respectively, at anthropometric levels of the body. The linear longitudinal balance includes three indicators: upper, down, and initial. Arc linear longitudinal balance includes two indicators: front-back and side;

(3) *latitudinal balance* promotes the concordance between the width of the front and back on upper and lower bearing areas and the vertical directions of the back seam and front edges. The transverse balance includes the upper indicator related to the width of the front neck and back neck and the down indicator related to width of front and back on the hip and bust levels;

(4) *angle balance* promotes the difference of sloping similar lines, such as shoulder and side.

The classification of balances is shown in Table 9.9.

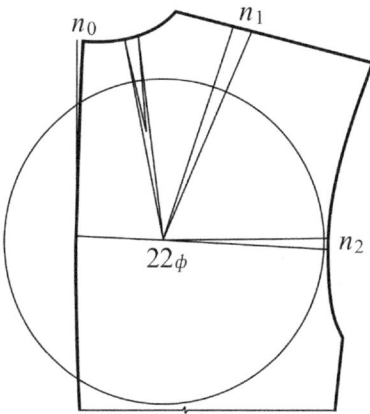

Fig. 9.7 Scheme of shaping to adapt the clothing back to the bearing surface of the body back.

The algorithms for calculation of balance indicators are shown in the following section. These algorithms can be used in two ways: during pattern block making and during its checking before clothing manufacturing.

9.5.1 Indicators of bearing balance

To calculate the indicators of bearing balance, it is necessary to compare designed parameters that should be measured in the pattern block with the real parameters of the body or mannequin. The designed parameters include the dart values; the real parameters of the body can be found after flattening the shell into the surface. Fig. 9.7 shows the scheme of the back shaping to get the adequacy with the human body shape. There are three values that shape the back together with the shoulder area: n_1 is the dart from neck line, n_2 and n_3 are the shrinkage along the shoulder and armhole lines, respectively.

The equation to calculate the back bearing indicator of balance is

$$6_{bb} = \beta_{pattern} - \left(\beta_{body} + K_1 + K_2 + K_3 \right) \tag{9.6}$$

Table 9.9 Classification of bodice balances

Indicators of balances							
Balance							
Bearing (on shoulder area)		Longitudinal (along system "body-garment")					Transverse (latitudinal)
		Linear		Arc			
Indicator							
Back	Front	Upper	Down	Initial	Front-back	Side	Down
6_{bb}	6_{bf}	6_{lu}	6_{ld}	6_{li}	6_{lfb}	6_{ls}	6_t

where δ_{bb} is the back bearing indicator, degree; $\beta_{pattern}$ and β_{body} are the values of blade darts in the pattern block and calculated from the body, respectively, degrees; K_1 is the coefficient showing the flattering effect of the back under the influence of the shoulder pad, degree/cm, $K_1 = 1.7t_{sp}$ where t_{sp} is the shoulder pad thickness, cm; K_2 is the coefficient showing the flattering effect of the back by means of ease allowance to the back width $E_{0.5BW}$, degree/cm, $K_2 = 0.9E_{0.5BW}$; K_3 is the coefficient showing the flattering effect of the back by means of ease allowance to shoulder length E_{SL}, degrees/cm, $K_{yc3} = 0.1–0.3E_{SL}$.

The equation to calculate the front bearing indicator of balance is

$$\delta_{bf} = \beta_{pattern} - (0.5BG - 0.5BG_1 - a) - K_4. \tag{9.7}$$

where δ_{bf} is the front bearing indicator, degrees; BG and BG_1 are the bust girth measured across BP and upper breast, respectively, cm; a is the coefficient depending on gender: for women $a = [0.65 \pm 0.05]$, for men $a = [0.5 \pm 0.05]$; K_4 is the coefficient showing the flattering effect of the front which depends on the method of pattern block making and shaping, $K_4 = [-1.0–0.1]$ cm. A negative amount for K_4 is used for the pattern block when the ease allowance to front width E_{FW} is increased. If the front shape is moving for the shape of the real breast, K_4 is increasing.

The differences between $(\delta_{bb} - \beta_{pattern})$ and $(\delta_{bf} - \beta_{pattern})$ can explain the modifications that were made during shaping of the basic block.

9.5.2 Indicators of longitudinal balance

To calculate the longitudinal indicators, it is necessary to collect the body measurements and to measure some dimensions of the pattern block. The principal scheme of the measuring of the body and the pattern block is shown in Fig. 9.8. Fig. 9.8A shows the profile of the female torso, the scheme of body measurements *(dotted lines)*, and the pattern block *(solid line)* coordinated under the waistline. Fig. 9.8B shows the pattern block with numeric symbols: the symbols with first number *1* are related to the shoulder area, with first number *3* to the bust area, with first number *4* to the waist area, with first number *9* to the bottom, under the instruction of the Universal Method of pattern block making (formerly the USSR—Poland—Germany—Hungary—Czech—Slovak—Romania—Bulgaria, 1982).

Longitudinal linear balance has three indicators—upper, lower, and initial—which are connected to the points of neck, shoulder, and waistline of the pattern block with the anthropometrical waist level.

Longitudinal upper balance δ_{lu} is used to identify the distance between the highest points of the neckline and waist level of the front and the back. The equation for calculation of this balance is:

$$\delta_{lu} = (/16 - 461/ - /121 - 42/) - (LF_{SNP} - LB_{SNP}) = 2M_{SNP}. \tag{9.8}$$

where δ_{lu} is the longitudinal upper balance, cm; $/13\text{-}15/$ is the length of the pattern block front between point *16* and point *461*, cm; $/121\text{-}42/$ is the length of the pattern

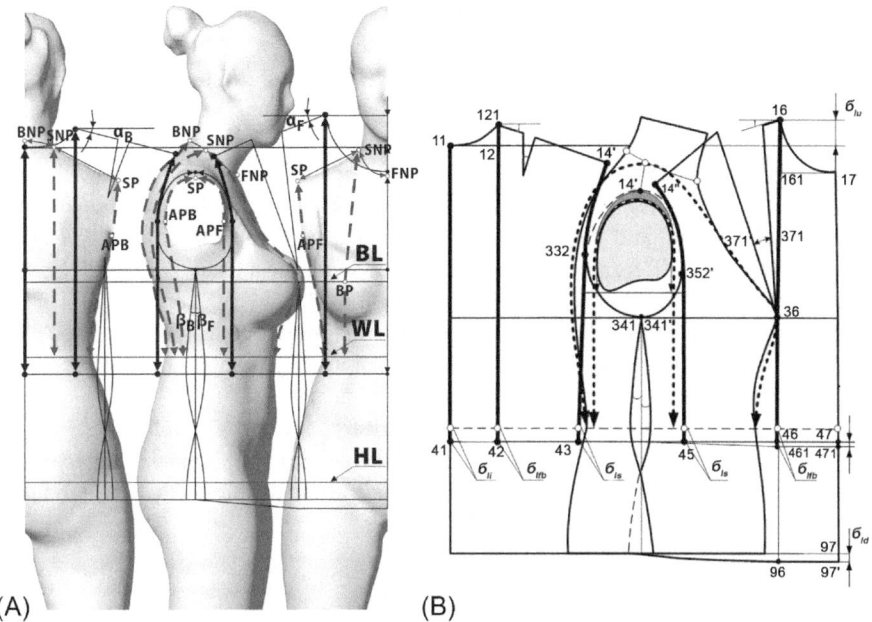

(A)   (B)

Fig. 9.8 Scheme of pattern block and body measuring to calculate the indicators of longitudinal balance. *BL,Wl,* and *HL* are, respectively, bust, waist, and hip levels. *BNP* (11), *SNP* (12 or 16), and *FNP* (17) are, respectively, back, shoulder, and front neck points. *SP* (14) is shoulder point. *BP* (36) is bust point. *APB* (332) and *APF* (352) are, respectively, armpit back and front points. α_B and α_F are, respectively, the slopinigs of back (12–14') and front (16–14'') shoulder lines. $\angle(371\text{-}36\text{-}371')$ is breast dart. β_B and β_F are, respectively, the angles between vertical line from 341 (341') and back and front side lines.

block back between point *5* and point *6*, cm; LF_{SNP} is the body measurement of front from *SNP* to *WFC* on the natural waist line (waist front center), cm; LB_{SNP} is the body measurement of back from *BNP* to *WBC* on the natural waist line (waist back center), cm; M_{SNP} is the distance between *SNP* and point *121*, cm.

Longitudinal lower balance δ_{ld} is used to identify the lowering of the waist structural line of the pattern block relative to the natural waistline of the body (designed as /97-97'/ on Fig. 9.8B).

Longitudinal initial balance δ_{li} is one of the most important indicators because its amount influences other indicators: δ_{li} indicates the difference between the natural waist level *WL* as a more narrow place of the body (anthropometrical level) and the clothing waistline /41-471/ (structural level) as a more narrow place; δ_{li} is measured along the back line; δ_{li} is the same as the ease allowance to back length E_{BL} (Fig. 9.8B). From the pattern block δ_{li} can be found by the equation:

$$\delta_{li} = /11 - 41/ - BL = E_{BL}. \tag{9.9}$$

$$E_{BL} = T_{tm.b1} \pm M_{BNP} \pm M_{WL}. \tag{9.10}$$

where /11-41/ is the length of the back line from highest point 11 to point 41 located on the structural waist line (Fig. 9.8B), cm; BL is the body measurement "back length," cm; E_{BL} is ease allowance to back length that consists of three components, cm: $T_{tm.b1}$ is the ease allowance that equals to the textile material thickness, cm; M_{BNP} is the vertical distance between BNP and point 11, cm; M_{WL} is the vertical distance between the natural waistline WL and the narrowest structural level of the clothes /41-471/, cm.

M_{BNP} can have different values depending on style: for a basic pattern block with closed neckline $M_{BNP} \approx 0$; for clothes with a collar $M_{BNP} = \pm 0.3$ cm; for raglan and dolman styles $M_{BNP} = 0.5$–0.8 cm. M_{WL} depends on the silhouette and may be positive (as 1920s style) and negative (as an empire style of the beginning of the 18th century).

So, the total amount of δ_{lu} may be positive or negative.

Longitudinal arc balance includes front-back δ_{lfb} and side δ_{ls} indicators.

Longitudinal front-back balance δ_{lfb} shows the position of points 42 and 46 on the structural waist level /41-471/; δ_{lfb} is equal to half of the difference between the similar dimensions belonging to body and clothes:

$$\delta_{lfb} = 0.5\left[(/121\text{-}42/ + /16\text{-}36\text{-}46/) - (FL_{SNP} - BL_{SNP}))\right]. \tag{9.11}$$

The equation for calculation of δ_{lfb} is:

$$\delta_{lfb} = 0.5\left(T_{tM.f2} + T_{tM.b2} + T_{tm.sp} + E_{bending}\right) \pm M_{WL}. \tag{9.12}$$

where $T_{tM.f2}$ and $T_{tM.b2}$ are the ease allowance, which equals the textile material thickness of front and back, respectively, cm; $T_{tm.sp}$ is the ease allowance that equals the thickness of the shoulder seams, cm; $E_{bending}$ is the ease allowance appearing as an additional value due to the bending of clothes under the shoulder area, cm.

Eq. (9.12) can be simplified by using the average thickness $T_{tm.av}$ of textile materials instead of the two values $T_{tm.f2}$ and $T_{tm.b2}$:

$$T_{tm.av} = 0.5\left(T_{tm.f2} + T_{tm.b2}\right) \tag{9.13}$$

$$\delta_{lfb} = (T_{tm.av} \pm M_{WL}) + 0.5\left(E_{bending} + T_{tm.sp}\right). \tag{9.14}$$

1st balance concordance between δ_{li} and δ_{lfb} is

$$\Delta_1 = \delta_{lfb}\text{-}\delta_{li} = (1 - 1.1)\, T_{tm.av}. \tag{9.15}$$

For example, for one-layer clothes made of thin fabrics, when $T_{tm.f2} = T_{tm.b2}$ and $M_{BNP} = 0$, 1st balance concordance will be:

$$\delta_{lfb} = \delta_{li}. \tag{9.16}$$

Longitudinal side balance $б_{ls}$ shows the position of points *43* and *45* on the structural waist level */41-471/*; $б_{ls}$ is equal to half of the difference between the similar dimensions belonging to the body and the clothes:

$$б_{ls} = 0.5[(/14'\text{-}332\text{-}43/ + /14''\text{-}352\text{-}45/) - (AL + 2(BL - BL_{APB})]. \tag{9.17}$$

where */14'-332-43/* is the length of back armhole, cm; */14''-352-45/* is the length of front armhole, cm; *AL* is the armpit length measured between *APB* and *APF* across *SP*, cm; *BL* is the back length, cm; BL_{APB} is the part of *BL* measured from *BNP* to the horizontal level across *APB*, cm.

The equation for calculation of $б_{ls}$ is:

$$б_{ls} = 0.5\left(T_{tm3} + T_{tm.sp} + E_{bending} \pm 2M_{WL} + 1.6T_{pad} + E_{AHS}\right). \tag{9.18}$$

where T_{tm3} is the ease allowance that equals to the thickness of the textile material located along the side balance line */11-41/*, cm; $T_{tm.sp}$ is the ease allowance that equals to the thickness of the shoulder seams, cm; $E_{bending}$ is the ease allowance appearing as an additional value due to the bending of clothes around the shoulder area, cm; M_{WL} is the distance between the natural waistline *WL* and the narrowest structural level of clothes */41-471/*, cm; T_{pad} is the thickness of the shoulder pad, cm; E_{AHS} is the ease allowance to the armhole shaping, which is equal to the shrinkage along the back armhole, cm (see Fig. 9.7, n_2).

2nd balance concordance between $б_{ls}$ and $б_{lfb}$ is

$$\Delta_2 = б_{ls}\text{-}б_{lfb} = 0.8\,T_{pad} + 0.5\,E_{AHS}. \tag{9.19}$$

If a shoulder pad is absent and $E_{AHS} = 0$, Eq. (9.19) becomes simpler:

$$б_{ls} = б_{lfb}. \tag{9.20}$$

For a balanced pattern block of one-layer clothes such as a man's shirt, woman's dress, and blouses (without shoulder pad), the proportion between the three indicators is

$$б_{li} = б_{ls} = б_{lfb}. \tag{9.21}$$

Fig. 9.9 shows the misfit problems arising from wrong calculations of longitudinal balance.

Because textile materials of a shirt are usually very thin, $T_{tm.b1} = T_{tm.f2} = T_{tm.b2} = T_{tm3}$. $E_{bending} = 0$. $T_{tm.sp} = 0$. Fig. 9.9 presents three possible situations with different proportions between the indicators of longitudinal balance:

(a) $б_{li} = б_{lfb} = б_{ls}$; $\Delta_1 = \Delta_2 = 0$. As Fig. 9.9A and D shows, the surface of the shirt is smooth, the bottom is in a well-designed position, and the side seam is vertical. So, the pattern block is balanced and the shirt is well-fitted;

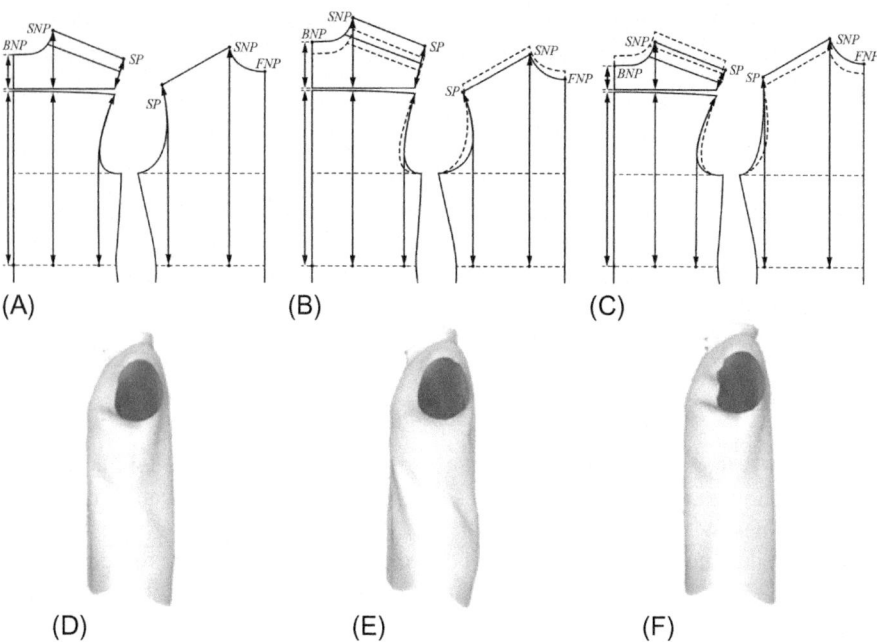

Fig. 9.9 Examples of men's shirt misfit due to wrongly designed indicators of longitudinal balance. *BNP*, *SNP*, and *FNP* are, respectively, back, shoulder, and front neck points. *SP* is shoulder point. (A, D) Balanced pattern block and well-fitted shirt, (B, E) nonbalanced pattern block (*dotted line*) and badly fitted shirt when the back is too long, (C, F) nonbalanced pattern block (*dotted line*) and badly fitted shirt when the back is too short. Improved pattern blocks after correction (B, C) shown by *solid line*.

(b) $\delta_{li} > \delta_{lfb} = \delta_{ls}$; $\Delta_1 \neq \Delta_2$. Fig. 9.9B and E shows three defects of fit: the stress folds and creases located on the front and the back and starting from *BP*; the bottom is deformed and looks bad; the side seam is sloping to the front. This pattern block is not balanced and the shirt is very badly fitted;

(c) $\delta_{li} < \delta_{lfb} = \delta_{ls}$. $\Delta_1 \neq \Delta_2$. Fig. 9.9C and F shows only one defect of fit—the stress folds and creases located near the front armhole—because the front is bigger than the back. This pattern block isn't balanced and the shirt is badly fitted.

If the 1st and 2nd balance concordances are larger than 0.5 cm, the clothes will have fit problems.

9.5.3 Transverse (latitudinal) balance

Transverse (latitudinal) balance δ_t provides the concordance between the width of the pattern block and the width of the body:

(1) if the waist girth is smaller than the hip girth, δ_t will provide the concordance between the structural levels of hip and bust;

(2) if the waist girth is bigger than the hip girth, δ_t will provide the concordance between the structural levels of waist and bust.

By means of this balance, the vertical direction of back seam, front edges, etc., can be achieved. To get the vertical directions, it is necessary to find agreement between the minimal ease allowances on both compared levels. Fig. 9.10 shows the defect when the side seam is not vertical and moves forward in parallel with the structural waist level, and the bottom *(solid lines)* slopes from initial and desirable horizontal directions *(dotted lines)*. These defects look like the defects of wrongly calculated indicators of longitudinal balance, but the cause is different.

The more obvious way to improve the misfit (Aldrich, 2008) is to compare the corresponding width of pattern block and body width calculations after dividing the girth between the front, the side (located under armpit or armhole), and the back, and to establish the proportionality between them. The proportionality, or transverse (latitudinal) balance, reveals the allocations of different parts of clothing on a body. From the latitudinal aspect, clothes and body are usually distributed into front, side, and back sections. If the ratio between certain sections of clothes and body are too big or too small, the vertical direction of the side seam will be changed and stress folds and creases will appear, caused by the shearing force due to the unbalanced proportionality.

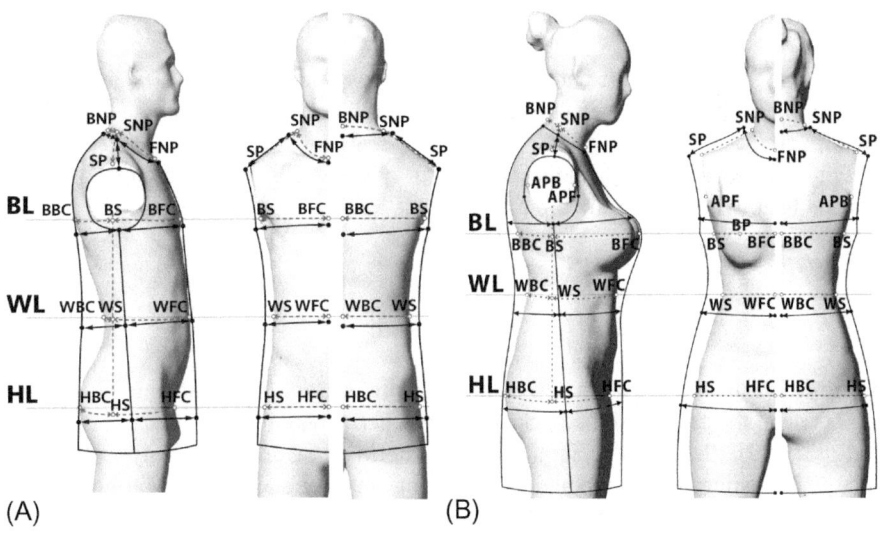

Fig. 9.10 Examples of destroying of transverse (latitudinal) balance δ_t. *BS*, *WS*, *HS* are the points of side seam on bust, waist, and hip levels. *BFC*, *WFC*, *HFC* are the front points of the sagittal plane on bust, waist, and hip levels. *BBC*, *WBC*, *HBC* are the back points of sagittal plane on bust, waist, and hip levels. *BL*, *WL*, and *HL* are, respectively, bust, waist, and hip levels. *BNP*, *SNP*, and *FNP* are, respectively, back, shoulder, and front neck points. *SP* is shoulder point. *BP* is bust point. *APB* and *APF* are, respectively, armpit back and front points.

Fig. 9.11 shows the real situations that take place during the girth measuring. Fig. 9.11A shows three torsos of female bodies overlapped with the similar bust girth BG, waist girth WG, and hip girth HG but which have different projections on front and profile views and positions of girth levels. Fig. 9.11B shows overlapped three torsos of male bodies with the same amounts of bust girth. waist girth, and hip girth.

As Fig. 9.11 shows, the proportions between the front and the back parts can be changed widely. To find the correct proportionality between the front and back, the following steps should be carried out to calculate $б_t$:

(1) to draw the vertical line from the chosen point of the armpit as the position of the designed side seam. Fig. 9.11B shows the possible beginnings of the side seam at the deepest point APD of the armpit. The side seam connects three points BS, WS, HS. To describe the position of the side seam around APD, the amount Δ_{ss} is used;
(2) to measure the bust, waist, and hip girth and to divide both measurements into two widths. The width of the front part of the bust girth should be measured from BFC to BS; the width of the front waist part is from WFC to WS, and the width of the front part of the hip is from HFC to HS. The same scheme should be applied to find the back width on three anthropometrical levels;
(3) to measure the corresponding width of the pattern block, as Fig. 9.12 shows.

The common approach to transverse (latitudinal) balance $б_t$ is the site on the pattern block will be sufficient to cover the corresponding body site if its size is equal to or less than the body site by permissible lack of width Δ:

$$P_i\text{-}HG_i \geq \pm\Delta. \tag{9.22}$$

where P_i is the width of the pattern block site across the hipline, cm; HG_i is the body measurement across hip level, cm; Δ is the permissible lack of the clothes width across the hip, cm.

For example, for the typical female body the proposed criteria to calculate the front and the back width along the hip girth are:

$$HGF = 0.235...0.245\,HG. \tag{9.23}$$

$$HGB = 0.255...0.265\,HG. \tag{9.24}$$

where HG is the hip girth, cm; HGF and HGB are the segments of the hip girth dividing between the front and the back width, respectively, cm.

Eqs. (9.23), (9.24) can be applied as the criteria of the proportionality and the balance $б_t$ for pattern block making and checking only for standard typical bodies.

Table 9.10 shows the results of both object measurements and how they can be used to calculate transverse (latitudinal) balance $б_t$.

Fig. 9.13 illustrates the influence of transverse (latitudinal) indicators of balance on the view of a virtual system "male torso—shirt."

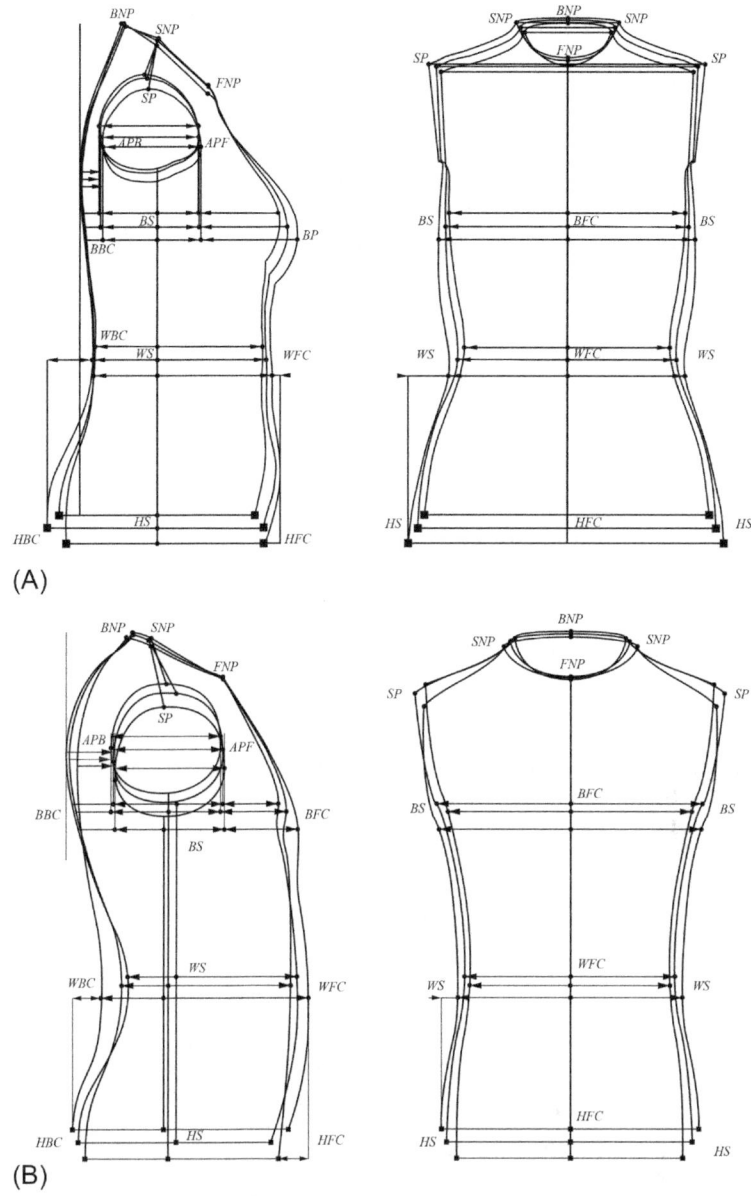

Fig. 9.11 Profile views of female (A) and male (B) bodies with the same or similar perimeters but with different distributions between the front and the back.

Fig. 9.13A, and D shows a balanced pattern block and well-fitted men's shirt. Fig. 9.13B, C, E, and F shows different defects. After pattern block analysis, we can find the following proportions between the indicators:

(a) $\Delta_{bf} - \Delta_{hf} = \Delta_{bb} - \Delta_{hb}$. As Fig. 9.13A shows, the shirt is well-fitted because the pattern block is balanced in the latitudinal direction;

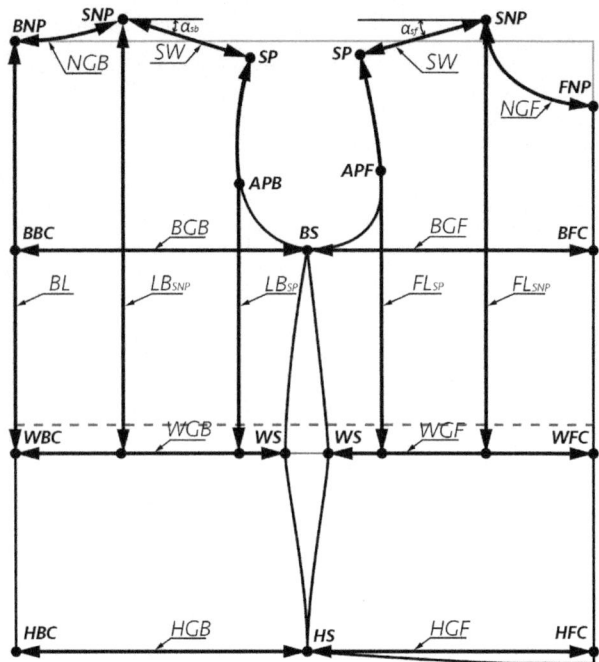

Fig. 9.12 The scheme of pattern block measuring to calculate transverse (latitudinal) balance. *BGF, BGB* are, respectively, the segments of the bust girth dividing between the front and the back. *WGF, WGB* are, respectively, the segments of the waist girth dividing between the front and the back. *HGF, HGB* are, respectively, the segments of the hip girth dividing between the front and the back.

(b) $\Delta_{bf} - \Delta_{hf} > \Delta_{bb} - \Delta_{hb}$. Fig. 9.13B shows that the upper part of the side seam is moved to the back because the back width of the pattern block is smaller than the body measurements;

(c) $\Delta_{bf} - \Delta_{hf} < \Delta_{bb} - \Delta_{hb}$. Fig. 9.13C shows an increased number of folds and creases due to shearing deformation in textile materials under wrong proportions at the hip level.

So the shirts with the mentioned defects were not adapted to the morphological features of the male body used as avatar for a virtual try-on.

9.5.4 Angle balance

The balance between two side lines belonging to the front and back means that the angles of its sloping should be equal. Fig. 9.8 shows both balanced side lines when the angles are equal:

$$\beta_F = \beta_B. \tag{9.25}$$

where β_F is the angle between the front side line and the vertical line, degrees; β_B is the angle between the back side line and the vertical line, degrees.

Table 9.10 Indicators of transverse (latitudinal) balance

Place of measurements	Results of measuring (Fig. 9.13)		Indicator of transverse (latitudinal) balance δ_t	Indicators of balance in transverse (latitudinal) direction
	Body	**Pattern block**		
Bust level	/BFC-BS/	BGF	$\Delta_{bf}=$/BFC-BS/ $-BGF.$	$\Delta_{bf}-\Delta_{hf}=\Delta_{bb}-\Delta_{hb}=\Delta_{ss}$, where is the distance
.	/BS-BBC/	BGB	$\Delta_{bb}=$/BS-BBC/ $-BGB$	between side seam and vertical line which is
Waist level	/WFC-WS/	WGF	$\Delta_{wf}=$/WFC-WS/ $-WGF.$	dividing two girths (bust and hip) between the front
	/WS-WBC/	WGB	$\Delta_{wb}=$/WS-WBC/$-WGB$	and the back (Fig. 9.13B and C shows the initial
Hip level	/HFC-HS/	HGF	$\Delta_{hf}=$/HFC-HS/ $-HGF.$	(before correction of pattern block) and final
	/HS-HBC/	HFB	$\Delta_{hb}=$/HS-HBC/$-HFB$	(after correction of pattern block) BGB and BGF, HGB and HGF)

The angle β depends on the silhouette and clothes style (X, A, H, O). If the proportion $\beta_F \neq \beta_B$ exists, it will cause creases appearing along the side seam and side seam sloping.

The balance between the two shoulder lines belonging to front and back means that the difference between the two angles should reflect the shoulder morphology and clothes construction. Fig. 9.18A shows the scheme of both angle measurements. For typical bodies, the difference is

$$\alpha_F\text{-}\alpha_B = -5 - +5. \tag{9.26}$$

where α_F is the angle between the front shoulder line and a horizontal line drawn from the highest point of the front neck line, degrees; α_B is the angle between the back shoulder line and a horizontal line drawn from the highest point of the back neck line, degrees.

If the difference is bigger or smaller than -5 to $5\,cm$, the indicators of longitudinal balance δ_{lfb} and δ_{ls} will be destroyed. For atypical bodies, the difference depends on body features such as posture (straight, normal, sloping), shoulder sloping, distribution of fat and muscles, and clothes construction (thickness of shoulder pad, position of shoulder seam along SNP-SP line). For balanced pattern blocks in the second part of the 20th century to the beginning of the 21st century (Kuzmichev et al., 2018) the angles were (degrees):

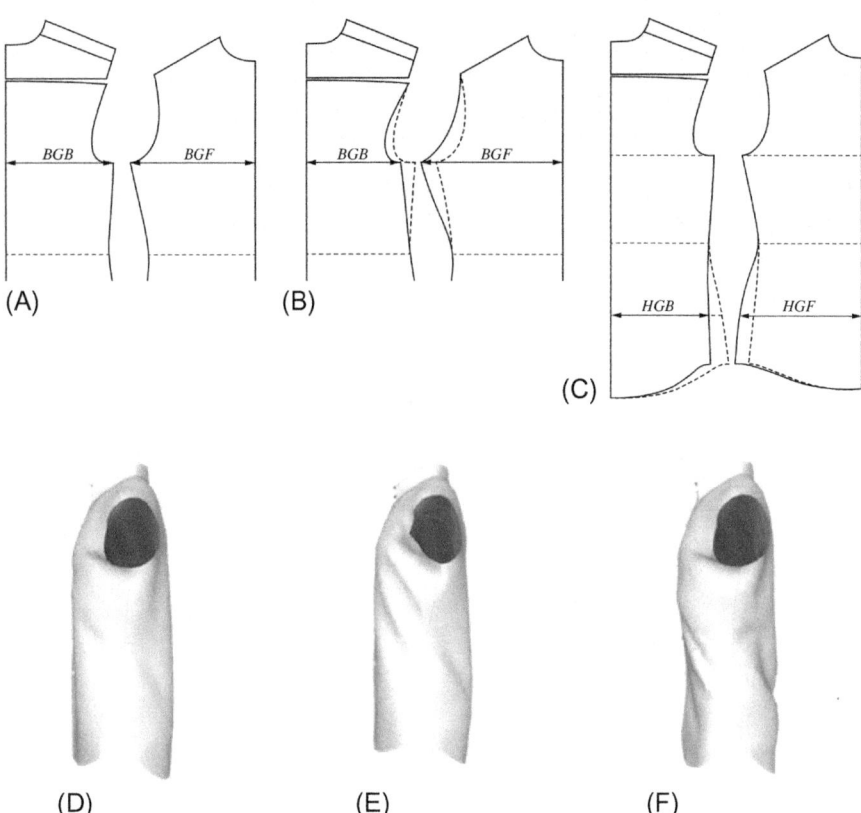

Fig. 9.13 Examples of men's shirt with misfit due to wrongly designed indicators of transverse (latitudinal) balance. (A, D) Balanced pattern block and well-fitted shirt, (B, E) nonbalanced pattern block (*dotted line*) and badly fitted shirt when *BGB* and *BGF* do not reflect the body morphology at bust level, (C, F) nonbalanced pattern block (*dotted line*) and badly fitted shirt when *HGB* and *HGF* do not reflect the body morphology at hip level. Improved pattern blocks after correction (B, C) shown by *solid line*.

(1) for women's dresses $\alpha_F = 15–30$ (mean is 17). $\alpha_B = 12–25$ (mean is 16); the average difference is 1;

(2) for women's jacket $\alpha_F = 5–33$ (mean is 18.7). $\alpha_B = 0–25$ (mean is 17.7); the average difference is 1;

(3) for women's coat $\alpha_F = 8–35$ (mean is 19.1). $\alpha_B = 7–33$ (mean is 17.3); the average difference is 1.8;

(4) for men's coat $\alpha_F = 4–32$ (mean is 21.3). $\alpha_B = 10–30$ (mean is 19.7); the average difference is 1.7;

(5) for men's suit $\alpha_F = 10–27$ (mean is 18.5). $\alpha_B = 12–30$ (mean is 20.5); the average difference is 2.

9.6 Future trends

Fit and balance, as two corresponding indicators of well-designed and well-fitted clothing, are important areas of practical and scientific exploration. The practical application of the results obtained is obvious, and the scientific interest is related to many areas. There are several important factors that are pushing professional knowledge of fit and balance: to develop areas that contribute to fit and balance, including anthropometry, pattern block making, construction of new textile materials, clothing comfort, and on-line selling.

(1) Traditional anthropometry used a limited number of body measurements for producing RTW clothes, which are enough to draw the basic pattern blocks for typical bodies. Due to the acceleration process and popularization of special physical exercises (fitness, bodybuilding, etc.), body measurements are changing and several groups of consumers are not satisfied with existing combinations of body dimensions used in production and labeling. To improve the situation, new body measurements to present these morphological features should be investigated. New body measurements could be measured by means of new non-contact technologies, including contemporary devices. Other aspects of anthropometry belong to dynamic changes of body measurements, which should be supported by correct construction of textile fabrics and ease allowance choices. The combination of traditional and new measurements can describe the features of human bodies of different consumers more accurately and increase the anthropometrical database, as first key sources for well-fitted clothing production.

(2) Development and improving of pattern block methods are following the anthropometrical database. New body dimensions should help to draw the pattern blocks in accordance with customization trends and should reflect features of human bodies not only in total dimensions, but also in detailed items such as the distribution of waist dart, sloping of shoulder lines, distribution of ease to bust girth between back, armhole, and front, and so on.

(3) New textile materials have properties different from the traditional fabrics for which the manuals of pattern block making and the supporting databases were created. The weight of contemporary fabrics becomes lighter and lighter, and its ability to preserve the clothing shape and silhouette becomes weaker. The elasticity of contemporary fabrics produced from new fibers and threads such as Lycra can influence the properties over a wide range. To design clothes from new materials, the patternmakers need information about ease allowances, which should be used during pattern block drawing and shaping and reflect the properties of contemporary textile materials.

(4) Comfortability has become the leading trend and to achieve clothing with such an important property, new information based on sensory analysis should be included in the design process. Comfort of clothes reflects the quality and well-established database of all steps of the design.

(5) On-line selling has become more popular due to the possibilities of CAD. Virtual mirrors, virtual ateliers, and virtual try-ons are several new concepts being used in online purchasing, replacing traditional means of purchase. To move so many steps of the design process into virtual reality, digitalization and formalizations of professional knowledge, especially concerning fit and balance, should be carried out. This trend is based on numeric values and a complex approach to analysis of the system "body-clothes."

So, fit and balance are complex characteristics of system "body-clothes" and the progress of clothes design will involve new explorations concerning both indicators.

Acknowledgments

Thanks to my good students Jiaqi Yan and Xia Peng for their contributions in picture drawing.

References

Aldrich, W., 2008. Metric Pattern Cutting for Women, fifth ed. Wiley-Blackwell. 218 p.

Baezley, A., Bond, T., 2003. Computer-Aided Pattern Design and Product Development. Blackwell Science, Oxford.

Beazley, A., 1997. Size and fit: procedures in undertaking a survey of body measurements. J. Fash. Mark. Manage. 2 (1), 55–85.

Betzina, S., 2003. Fast Fit: Easy Pattern Alterations for Every Figure. The Tauton Press, Newtown, CT.

Chen, Y., Zeng, X., Happiette, M., Bruniaux, P., et al., 2008. A new method of ease allowance generation for personalization of garment design. Int. J. Cloth. Sci. Technol. 20 (3), 161–173.

Cloake, D., 2000. Lingerie Design on the Stand: Designs for Underwear and Nightwear. B.T. Batsford, London.

Cock, V., 1981. Dressmaking Simplified, third ed. Granada Publishing, Forgmore.

Creative Publishing International, 2005. The Perfect Fit: A Practical Guide to Adjusting Patterns for a Professional Finish. Apple Press, London.

Erwin, M.D., Kinchen, L.A., Peters, K.A., 1979. Clothing for Moderns. Macmillan, New York.

Fujii, C., Takatera, M., Kim, K., 2017. Effects of combinations of patternmaking methods and dress forms on garment appearance. Autex Res. J. 7 (3), 1–10.

Gill, S., 2015. A review of research and innovation in garment sizing. prototyping and fitting. Text. Prog. 47 (1), 1–85.

Gioello, D.A., Berke, B., 1979. Figure Types and Size Ranges. Fairchild Publications, New York.

Gu, B., Liu, G., Xu, B., 2016. Individualizing women's suit patterns using body measurements from two-dimensional images. Text. Res. J. 87 (6), 669–681.

Holman, G., 1997. Pattern Cutting Made Easy. B.T. Batsford, London.

Hu, P., Li, D., Wu, G., Komura, T., Zhang, D., Zhong, Y., 2018. Personalized 3D mannequin reconstruction based on 3D scanning. Int. J. Cloth. Sci. Technol. 30 (2), 159–174.

Keiser, S.J., Garner, M.B., 2008. Beyond Design: The Synergy of Apparel Product Development. Fairchild Publications, New York.

Kim, K., Innami, N., Takatera, M., Narita, T., Kanazawa, M., Kitazawa, Y., 2017. Individualized male dress shirt adjustments using a novel method for measuring shoulder shape. Int. J. Cloth. Sci. Technol. 29 (2), 215–225.

Kuzmichev, V., Moskvin, A., Surzhenko, E., et al., 2017. Computer reconstruction of 19th century trousers. Int. J. Cloth. Sci. Technol. 29 (4), 594–606.

Kuzmichev, V.E., Ahmedulova, N.I., Udina, L.P., 2018. Garment Design: System Approach. Uright Publishing House, Moscow, p. 392.

LaBat, K.L., DeLong, M.R., 1990. Body cathexis and satisfaction with fit of apparel. Cloth. Text. Res. J. 8 (2), 43–48.

Lage, A., Ancutiene, K., 2017. Virtual try-on technologies in the clothing industry. Part 1: investigation of distance ease between body and garment. J. Text. Inst. 108(10). https://doi.org/10.1080/00405000.2017.1286701.

Li, Y., 2001. The science of clothing comfort. Text. Prog. 31 (1–2), 1–135.

Liechty, E., Rasband, J., Pottberg-Steinecksert, D., 2010. Fitting and Pattern Alteration, second ed. Fairchild Books, New York.

Lin, Y.L., Wang, M.J.J., 2016. The development of a clothing fit evaluation system under virtual environment. Multimed. Tools Appl. 75 (13), 7575–7587.

Monobe, A., Kim, K., Takatera, M., 2017. Effect of the difference between body dimensions and jacket measurements on the appearance of a ready-made tailored jacket. Int. J. Cloth. Sci. Technol. 29 (5), 627–645.

Myers-McDevitt, P.J., 2009. Complete Guide to Size Specification and Technical Design, second ed. Fairchild Publications, New York.

Naglic, M., Petrak, S., Stjepanovič, Z., 2016. Analysis of 3D construction of tight fit clothing based on parametric and scanned body models. In: 7th International Conference on 3D Body Scanning Technologies.

Palmer, P., Pletsch, S., 1995. Easy Easier Easiest Tailoring. Vancouver, WA, Palmer/Pletsch Publishing.

Pei, J., Par, H., Ashdown, S.P., 2018. Female breast shape categorization based on analysis of CAESAR 3D body scan data. Text. Res. J. 4(1).

Surikova, O., Kuzmichev, V., Surikova, G., 2017. Improvement of clothes fit for different female bodies. Autex Res. J. 17 (2), 71–79.

Turner, J.P., 1994. Development of a commercial made-to-measure garment pattern system. Int. J. Cloth. Sci. Technol. 6 (4), 28–33.

Wang, Z., Zhong, Y.Q., Chen, K.J., et al., 2014. 3D human body data acquisition and fit evaluation of clothing. In: Advanced Materials Research. vol. 989. Trans Tech Publications, pp. 4161–4164.

Xue, Z., Zeng, X., Koehl, L., 2016. An intelligent method for the evaluation and prediction of fabric formability for men's suits. Text. Res. J. 88 (4), 438–452.

Yan, J.Q., Zhang, S.C., Kuzmichev, V., Adolphe, D.C., 2017. New database for improving virtual system "body-dress." In: IOP Conference Series: Materials Science and Engineering. vol. 254(17). pp. 172027–172029.

Zvereva, J.S., Kuzmichev, V.E., Adolphe, D.C., et al., 2012. Identification of textile materials properties in "body-clothes" scanned systems. In: Proceeding of 3rd International Conference on 3D Body Scanning Technology, Lugano, Switzerland, pp. 335–342.

Part Four

Designing for specific functions

Sizing and fit for swimsuits and diving suits

10

Slavenka Petrak[a], Maja Mahnić Naglić[b], Jelka Geršak[c]
[a]Department of Clothing Technology, Faculty of Textile Technology, University of Zagreb, Zagreb, Croatia, [b]Faculty of Textile Technology, University of Zagreb, Zagreb, Croatia, [c]Faculty of Mechanical Engineering, Institute for Textile and Garment Manufacture Processes, University of Maribor, Maribor, Slovenia

10.1 Introduction

Clothing intended for wearing in the water is specific from the various aspects, designs, and materials used for the production and the aspect of high demands on functionality and manufacturing technologies for such clothing. Whether it is used for recreation or during particular water sport activities, such clothing needs to satisfy list of criteria to ensure appropriate body protection and perfect fit according to body shape and dimensions during dynamic wearing conditions. In most cases, clothing intended for wearing in water must completely lay against the body. That puts in front of the designers and constructors a special demand to develop the attractive model design with a perfect dimensional and shape fit according to targeted body shape, together with shape stability and comfort in the water wearing conditions. During the development of such clothing, the very important issue is a right selection of primary and secondary materials and the manufacturing technology. To ensure total body fit of different kinds of swimwear, the complete development process is best conducted digitally, with the application of specialized CAD systems that enable complete 3D design and construction process on the 3D body model of a particular size.

For the number of years, computer-based 3D design, using the specialized CAD systems and automated manufacturing technology, is present in different industries in which final products are composed of rigid parts. Intensive and continuous development of computer graphics systems, in the last decade, enabled a significant improvement of 3D design process and prototype development in clothing, footwear, automotive, and upholstery industries. That mainly refers to the design of prototypes for which it is necessary to construct coverings from textile or some other materials that need to completely fit the shape of the product, like upholstery furniture or automobile seats. That level of fit is difficult to achieve with the use of conventional 2D construction method because of complexity of products' 3D shapes and material characteristics from the aspect of physical and mechanical properties. The same problem is present in the swimwear design and construction process. Regardless the different clothing size systems according to which the clothing is developed in a huge range of sizes for different body figures, those are still sizes defined based on average body

Anthropometry, Apparel Sizing and Design. https://doi.org/10.1016/B978-0-08-102604-5.00010-X

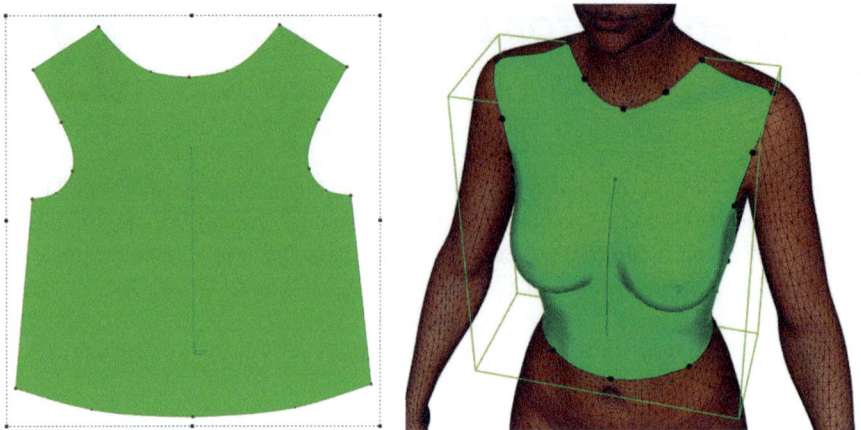

Fig. 10.1 3D flattening of human body model.

measurements of a targeted population. Given that clothing for water activities needs to fit the body perfectly and support particular body parts, it is very difficult to satisfy those criteria and to achieve a high fit of clothing models for different sizes and body types using only clothing size systems and conventional 2D construction method.

Furthermore the characteristics and high variability of physical and mechanical properties of materials used for manufacturing of such clothing all affect the complexity of 3D computer design application and prototype testing. In that sense, this chapter will present possibilities of computer 3D systems and technology application in the process of 3D design and development of swimwear, with the special attention on diving suits. Diving suit pattern and cutting parts need to be constructed in a precise way so that the final product perfectly fits the size and shape of the human body. The computer 3D prototype needs to be further transformed into 2D cutting parts in order to enable cutting of the pattern from the particular material. The method for computer transformation of irregular 3D surfaces into 2D cutting parts is called 3D flattening (Yih-Chuan et al., 2006; Petrak, 2007; Fang et al., 2008; Wen-liang et al., 2011). The method enables results that completely respond to 3D shape of the used body model, Fig. 10.1.

10.2   Swimsuits for different purposes

Clothing intended for wearing in the water can be classified in several different categories, depending on purpose and activity for which it is used, Fig. 10.2:

- fashion swimwear (monokini, bikini, and swimming pants)
- sports swimwear
- sports surfwear
- diving suits

Fig. 10.2 Clothing intended for wearing in the water.

Each of the mentioned categories has its specific characteristics, while demand for functionality during use and particular activities is defining specific suit design for different activities. However, they all have the same demands for dimensional stability and tight fit against the body, without unwanted stretch and movements against the body in water. Depending on the activity for which the suits are used, the demand on functionality is different and specially increased for the suits intended for professional diving and water sports.

10.2.1 Materials and fabrics for swimsuits

To ensure the functionality of swimwear, it is necessary to satisfy particular criteria while selecting materials for swimwear production. Since most of swimwear models fit tightly to the body and helps shaping it in the right figure, the elastic properties of used material are highly important. This is especially pronounced in professional sport activities where swimwear needs to ensure high fit with no restriction on body movement of sportsman in water. Materials used for swimwear production, especially for sport activities, have to match characteristics defined by the particular standard. In that sense, swimsuit materials are laboratory tested and evaluated in terms of quality of use. That includes testing of elastic properties of knitted materials, puncture resistance, dimensional stability, coloration stability, friction resistance and a tendency for piling, chlorine resistance, and sometimes UV resistance properties (Tomljenović and Sinel, 2015) (https://www.kiefer.com/swimwear).

Swimsuits are usually made from knitted materials. The raw material composition is mostly a mixture of polyamide and elastane or polyamide, polyester, and elastin fibers. Polyamide (PA, PA6, and PA6.6) is a chemical synthetic fiber with a characteristic of good strength and endurance and incapability of wrinkling. Materials are maintained by washing at low temperatures, which gives them characteristic of smoothness and glow. Elastin fiber, frequently called by commercial name Spandex or Lycra, is also a chemical synthetic fiber with high elasticity. It is often used in a mixture with other fiber for production of tight-fit clothing. Elastin mixture materials also have the characteristic of easy relaxation after stretching, meaning it will easily return in primary state after strain. That is the property that ensures dimensional stability of garment model and high body fit. Polyester fiber (PES) is a highly strong synthetic fiber that can endure high elastic strains. Polyester materials are very endurable; they do not wrinkle or shrink while washing and are dried quickly. Thus maintenance of such materials is easy, which is very important for swimwear. Properties of polyester fibers can be modified in a way that materials have touch similar to wool, silk, or cotton. Also, polyester materials can offer high UV resistance, which is important for all outdoor clothing, especially swimwear (https:/www.kiefer.com/swimwear).

Modern materials used for swimwear today are results of technological innovations in the field of development of new chemical fibers with enhanced properties and targeted modifications of fibers, yarns, and materials, enabling to achieve wanted material functionality and higher comfort in use. Additionally, in terms of design, construction, and production technology of swimsuits, such materials provide great possibilities for production of suits for different age group and body figures.

Quality and expensive swimsuits do not change their shape during use or maintenance. Additionally, implemented materials and suspensors enable body shape stability, which mostly refers to production of women's bras. Equal extension of knitted material in longitudinal and transversal direction provides dimensional stability during wear, meaning that swimsuits will not be stretched in only one direction during wear, but will tightly fit against the body. Breathability and quick dry of materials are also characteristics that are significantly affecting quality and use of the final product.

Considering today's growing need for safety from unwanted UV radiation and body thermal protection, many products and swimwear models are designed to cover most of the body surface. Such swimwear models are made of contemporary, functional materials, proving wearing comfort and thermal protection at low temperatures of water and high protection of UV radiation from the sun. The brand Kiefer can be mentioned as one of the industrial producers of such clothing (https:/www.kiefer.com/swimwear).

10.3   Diving suits

Diving suit presents the type of clothing for special purposes with high demands and criteria for comfort, fit, and functionality (Gupta, 2011). There are three types of diving activities: scuba diving (diving with oxygen tanks), spearfishing (diving with underwater rifle), and free diving (diving on breath in length or depth). Each of the three mentioned activities has their own particular characteristics that define demands and necessary properties of diving suits (Diving Unlimited International Inc, 2013).

10.3.1   Scuba diving suits—Diving with oxygen tanks

Scuba diving is mostly recreational activity. For intensive diving, divers often use so-called dry suits that are completely waterproof, and it is possible to wear usual day clothing underneath. That kind of suit is loose fitted enabling air from the tanks to enter the space between body and the suit, regulating the diver's upthrust that way. It is important for seams and patents to be waterproof (Diving Unlimited International Inc, 2013). Three-layered laminate materials called Cordura are used for production of such suits (http:/www.subcraft-store.com/, 09.11.2016; https:/www.ursuit.com/fab rics, 10.10.2016). Form and shape of the suits are not defined. Most commonly used suits for scuba diving are wet diving suits made out of specific material with thermal isolation properties polypropylene, commercially called neoprene. For wet suits, it is important to fit perfectly and tightly to the body, that way preventing penetration and circulation of water between the body and the suit. Materials for wet suits are usually laminated with a knitted layer from both sides. That provides additional material strength, especially at the seams that are glued and then sewed, and makes it easier for the diver to put the suit on or to take it off. Considering that scuba diving activities usually last for maximum 50 min, it is not such a great problem if the suit does not fit perfectly to the diver's body. That is why scuba diving suits are usually made in standard clothing sizes.

10.3.2 Spearfishing suits—Diving with underwater rifle

This type of diving considers diving on the breath. That is why it is important for the diver to feel comfort without a body temperature decrease during diving. Diving activities usually last for 5–6 h, which is a great effort for the body. In those conditions, it is important that the suit fits well against the diver's body to disable even smallest circulation of water that can cool the body. Suits are usually made from lined/open-cell material, which is layered with knitted material from the outside for additional strength and smooth from the inside to create vacuum and completely lie on the skin (http:/www.cordura.com/en/fabric/classic-fabric.html, 10.10.2016). Knitting material in this sense provides friction durability but needs more time to dry off. This can be very important since the diver can change diving locations using a boat; if the suit dries off faster, more heat will remain in the body. For this type of diving, suits are made according to diver's individual measurements. Method for pattern construction differs from one manufacturer to another, and those are usually conventional construction methods developed by particular diving suit manufacturers. The firm Subcraft from Zagreb, Croatia, is a successful diving suit manufacturer, with the brand present on the market all over the world. Subcraft supported the realization of diving suits developed and presented in this chapter (http:/www.subcraft-store.com/, 09.11.2016). Construction of diving suits is developed based on the 24 anthropometric body measurements and knowledge on necessary functionality characteristics that suit needs to provide for diver. Suits are made from neoprene material of different thickness, from 1.5 to 9 mm, depending on user needs. Considering great elasticity of neoprene, which differs according to type and thickness, it is necessary to modify and scale particular pattern measurements according to elastic properties. Scaling ensures tight fit of the suit on the diver's body, without unwanted pressure. Great attention is oriented to pattern details like special ways of pattern modeling on elbow and knee areas with a purpose of preventing unwanted model wrinkling. Furthermore, modeling of headcap is very important since the diver can experience a great and unpleasant pain if the shape compresses ears on a high level. Particular suit areas are additionally supported by double neoprene layer, while soft endings on trousers and sleeves enable normal circulation to the extremities.

Neoprene density is also one of the important parameters that affects diving suit functionality (The Guide to Spearfishing in New South Wales, 2008). Divers who dive intensively at depths higher than 20 m will certainly choose suit of neoprene with higher density, considering it provides a lower compression during a dive in greater depth and does not allow big changes in lift force. If the suit compresses during increased pressure, lift force will be lower, and there is a chance for lead that divers carry around their belts to enter the zone of negative flow; in other words, the diver will begin to sink. As an additional quality, the characteristic of this kind of suits is a design of knitted material that is laminated outside of neoprene. Spearfishers believe that suit with camouflage colors and pattern is less visible to the fish. However, the professionals in the field have a strong and opposite opinion about this issue. That's why there is no standardized coloring that will be best for protection and reduction of visibility of a diver in the sea depth. Because of the reduction of sunlight under the sea surface, visibility of color quickly disappears. For example, red colors become invisible at approximately 5 m under the sea. Selection of diving suit design is most dependable on the diver's choice of colors and patterns.

10.3.3 Free diving suits—Diving on breath in depth or distance

This type of diving suits is most complicated for realization, considering the high demands on pattern construction and production technology. This type of activity requires with a perfect fit on human body. The suits must provide a feeling of soft compression on the body, without unwanted wrinkling in the joint areas and with a low friction coefficient and a whole other list of demands referring to lots of details in order to become high-quality diving suits. Divers in this discipline are exhausted to the limits, and the quality of diving suit, even in small detailed modifications, can have a big role on a diver sport score. Professional suits are usually very thin, made out of 1.5-mm or 3-mm neoprene (Naebe et al., 2013). Rarely, neoprene of 5-mm thickness is used. Dive lasts for a few minutes, and it is important that suit ensures minimal lift force, which is in this case an advantage considering preservation of body heat. In that sense, outer side of the suit should have minimal friction coefficient, which can be achieved with an overlay of titanium foil. Since the position of a diver in motion is with arms spread above the head, construction of diving suit is a little bit different comparing with suits for spearfishing, which is why this type of suit is selected and developed in the research project presented in this chapter.

10.4 Modern anthropometry approach

The application of the 3D body scanners has an increasing implementation in the field of body measurement for garment construction in the last two decades (Fan et al., 2004; D'Apuzzo, 2009; Chun and Oh, 2004). In addition, to determine the linear body measurements that are most commonly used data in the clothing industry and on which conventional clothing construction is based, 3D scanning is used to obtain data on body shape, anthropometric relationships of individual body parts, deviations from the normal proportions, and body posture (Simmons and Istook, 2003). In this manner, all relevant data necessary for computer-aided design and modification of garment patterns according to the individual body anthropometric characteristics can be determined.

The application of the 3D body scanning began for the needs of the military industry. As a consequence of further development, these systems began to be used in Western developed countries for the needs of realizing the project of the systematic anthropometric population measurement for the purposes of the clothing industry. The existing 3D body scanners designed by different manufacturers differ from each other in the number of the cameras used for scanning, the scanning range, and the light source as well as in the sophistication of the accompanying computer program used for the visualization of the scanned body and the determination of measurements, Table 10.1 (D'Apuzzo, 2009; Bragança et al., 2016).

In view of the mentioned differences, 3D scanning systems available on the market can be divided into five basic groups:

1. laser scanning systems using lasers as a light source;
2. light scanning systems, which project a sample of structured, mainly white light;
3. LED scanning systems using infrared detectors;
4. systems providing body shadows on the opposite side of the camera and scanning 2D body contours in different body postures;
5. systems using radio waves for scanning the body surface through the clothing.

Depending on the scanner type, body scanning is performed in a very short period of time of about 10 s. By using electronic circuits and microprocessor, data are scanned, processed, stored as a file, and visualized as a three-dimensional group of points, which outline a body shape on the monitor screen. The representation shows a full, two-dimensional precise replica of the scanned object or body, which can be viewed from different views, rotated, enlarged, and reduced, serving as support data for automatic computer-aided determination of body measurements. International standard ISO 20685 has been developed to ensure the comparability of body measurements determined by ISO 7250 (Basic Human Body Measurements for Technological Design) and ISO 8559 (Garment Construction and Anthropometric Surveys—Body Dimensions) using various 3D body scanners, Fig. 10.3 (ISO 20685:2010, 2010).

10.4.1   3D scanning and body model processing—Vitus smart 3D laser scanner

The principle of the operation of laser scanners is based on the measurement of polar coordinates and horizontal and vertical angle, respectively, to an individual point of the space. The instrument transmits a sequence of laser impulses according to a beforehand specified distance. Registering the total shift of the system in relation to its initial position and the measured length, space coordinates of each point are

Table 10.1 Technical properties of different 3D body scanners

System/ scanner	Scanning technology	Captured body segments	Resolution (mm)	
			Horizontal	**Vertical**
Breuckmann Body scan	Structured light	Whole body	2.0	2.0
Cyberware WBX, WB4	Laser Class 1	Whole body	5.0/0.019	2.0
Hamamatsu Body line scanner	Laser/ distance detection	Whole body	2.0/2.0	App. 2.0
Hamano VOXELAN	Laser Class 1	Trunk	3.4/3.4	3.4
INTELLIFIT Intellifit	Radio waves	Whole body	Approx. 4.0	Approx. 4.0
InSpeck 3D Mega Capturor II	Structured light	Whole body	1.9	1.0
[TC]² NX12	Structured light	Whole body without hands and head	2.5/1.0	2.5
Telmat SYMCAD 3D	Structured light	Trunk/whole body parts only as silhouette	1.4/0.8	1.4
VITRONIC Vitus Smart Vitus Pro	Laser Class 1	Whole body	2.0/2.0 1.2/1.2	2.0 app. 1.0
Wicks and Wilson TriForm	Structured light	Whole body	App. 1.5	App. 1.5

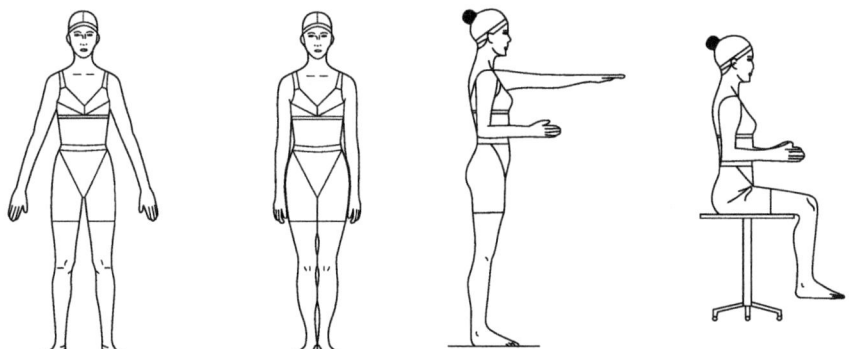

Fig. 10.3 Body positions according to ISO 20685 (ISO 20685:2010, 2010).

calculated. A 3D group of points is arranged in parallel, horizontally placed planes with a resolution of points amounting to about 1 mm in the horizontal plane and to about 2 mm in the vertical plane. Each camera records a body segment and processes the data of the recorded segment in the computer. The individual segments are combined into a group of 3D points, which outlines the body shape. Laser scanning opens up the possibility of collecting a huge amount of 3D data on the object being recorded. A group of points in the 3D coordinate system is called point cloud. For several applications, such as determination of body measurements, it is enough to use the data in their original form, with minimum subsequent processing or without it. The measured point cloud enables an almost instant measurement using the computer without physical access to the real object and the body, respectively. To achieve simpler visualizations and presentations, it is sufficient to generate the topology of the measured objects (irregular surfaces) by automatic algorithms on the basis of the collected geometric data, whereby their very true models are obtained. All laser scanner manufacturers also deliver the computer program, which enables the visualization and determination of body measurements and the execution of triangulation of 3D points, whereby certain points are interconnected, resulting in a surface grid, which is used for object visualization and which can be further processed in a 3D program. The triangulation process is performed automatically, whereby the program uses complex mathematical algorithms due to a very high density of point that are to be connected; a correct selection of points presents the most complex part of the triangulation process. Each point has its own coordinates (x,y,z), which determine the position of the point in the three-dimensional Cartesian coordinate system. In this system, x, y, and z are the coordinates orientated in such a way that x- and y-coordinates determine the plane on which the object stands, while z-coordinate is directed vertically at the plane (x,y) in the direction of the object height. The Vitus Smart 3D body scanner allows to make body scans in a range from 1200×800 mm and 2100 mm in height. The scanning is performed using a system of eight cameras and takes 10 s, whereby between 500,000 and 600,000, space coordinates of the scanned body are extracted, Fig. 10.4.

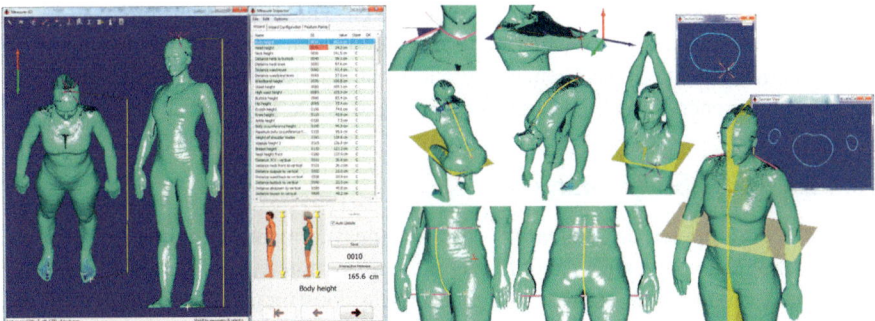

Fig. 10.4 3D group of points as a result of scanning the person using the Vitus Smart 3D scanner with a computer-controlled determination of the positions of individual measurements on the body.

It takes 40 s to process data, and after that, the software package ScanWorx or Anthroscan extracts the measurements of the human body necessary for the implementation into the computer program to alter the cut of a garment according to the found measurements (VITUS, 2018). The software package provides the possibility to correct the obtained body measurements and to add new ones. Using the video cameras the position of the laser beam on the object is detected outside the fixed angle. Concerning the triangulation angle and the formation of an optically static object, the position of a part of the object can be calculated in the direction of the x- and y-coordinates. To obtain the third object dimension, the triangulation sensor has been shifted in exactly defined steps in the direction of the z-axis. Using the information about the distance between steps, the object or body shape is scanned piecemeal in layers. The Vitus Smart 3D body scanner uses triangulation sensors directed at the body from different directions (a total of 360 degrees) to scan the human body in one passage (http://www.human-solutions.com/fashion/front_con tent.php?idcat=139&lang=7, 12.06.2018).

In order to ensure the determination of accurate body measurements, it is necessary to check and to calibrate the scanner periodically. The person to be scanned should stand in suitable erect body posture while being scanned and wear undergarment in bright color tightly fitted. The hair, if longer, should be tied back. To obtain the best measurement results, it is recommendable to wear a swim cap, Fig. 10.4. It is also necessary to take off jewelry.

The automatic computer-aided determination of body measurements within a 3D group of points is used to find the values for a total body measurement; automatic computer-aided measurements determine the distances between the computer-defined positions of the measuring points and by measuring the circumference of individual body parts, Fig. 10.4. The results of automatic computer-aided determination of measurements are generated in a chart upon completion of measurements whereby a graphical representation of the position of body measurements is given for each taken measurement. Of the total number of the determined measurements, several measurements were taken separately for the left and right body side, such as arm and shoulder length; then left and right shoulder length, values of angles that determine the slope of the individual shoulders; and leg length in the crotch. The values of the circumferences of individual body parts were separately measured, such as left and right upper arm circumference, wrist circumference, left and right leg circumference at thigh height, knee circumference, lower leg circumference, and ankle joint circumference.

10.4.2 Static body analysis

Despite the computer determination of body measurements, scanned body model analysis provides information about body posture, symmetry, and shape of particular body parts, with a great significance in clothing construction (Mahnic Naglic and Petrak, 2017). With body model rotation, it is possible to visualize all details and possible insufficiencies on the body, Fig. 10.5. Body symmetry can be analyzed according to the central median body axis and with transversal cross sections in different body areas.

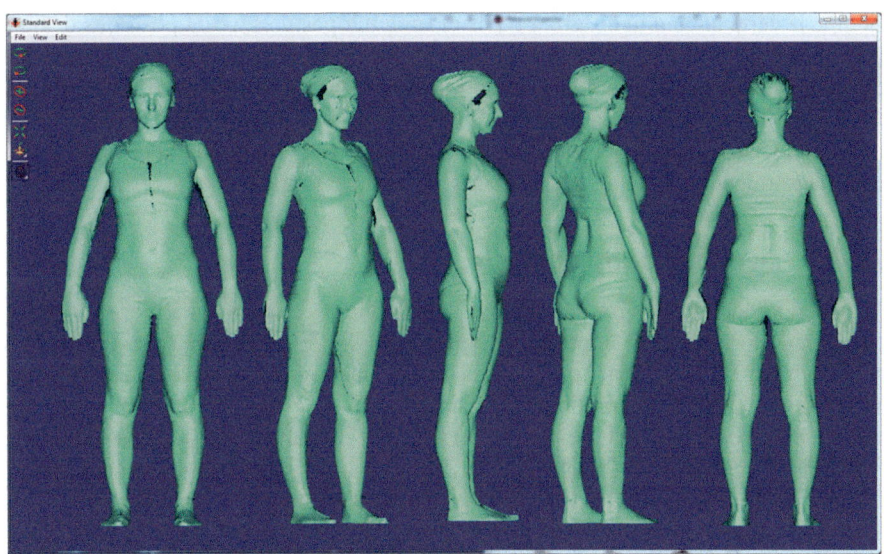

Fig. 10.5 Rotation of scanned body model.

Proper body posture gravity line passes through the middle of all vertically aligned joints. Gravity line is presented by a drawn vertical line that passes through body gravity center located in second sacral vertebra, which connects the middle point between the eyes, the center of the chin, the tip of the sternum, the center of the pubic area, and the middle point between the ankles, Fig. 10.6. The line that connects the middle point on the upper part of the ear, the middle of the shoulder, the hips, the knee and the ankle joint was analyzed in the sagittal plane (Palmer and Epler, 1998). With proper posture

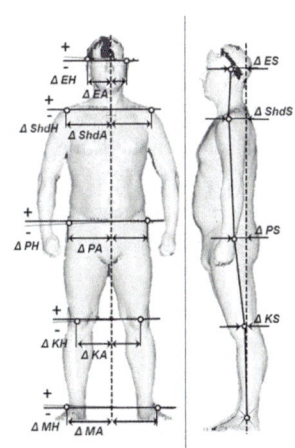

ΔEH – vertical deviation of upper ear edge on right in front view

$\Delta ShdH$ – vertical deviation of acromion on right in front view

ΔPH – vertical deviation of spine iliace anterior superior on right in front view

ΔKH – vertical deviation of epicondylusa medialisa on right in front view

ΔMH – vertical deviation of on right in front view

ΔEA – horizontal deviation of upper ear edge on right in front view

$\Delta ShdA$ – horizontal deviation of acromion on right in front view

ΔPA – horizontal deviation of spine iliace anterior superior on right in front view

ΔKA – horizontal deviation of epicondylusa medialisa on right in front view

ΔMA – horizontal deviation of on right in front view

Fig. 10.6 Body balance analysis in frontal plane and body posture assessment in sagittal plane.

the line is vertical, and in different types of irregular postures, point positions deviate from the gravity line. The method is used for detection and analysis of upper back curvature, curvature of shoulders, lifted knees, and head tilted forward. The application of 3D body scanner and innovative computer-based method for 3D body model analysis not only enables the assessment of body posture but also enables a very precise measurement of parameters for body posture assessment (Petrak et al., 2015).

10.4.3 Dynamic body analysis

3D scanning technology can also be used for obtaining data on dynamic anthropometry, which is especially important when developing clothing for special purposes with high demands on functionality and fit (Geršak and Marčić, 2013). Depending on the body position and motion, body surface areas are deforming, and body measurements are changing, which demands additional ease allowances on corresponding segments of the garment (Gill and Heyes, 2012; Lee et al., 2013). Dimensional change of body in dynamic positions is a complex issue that cannot be considered only from the aspect of basic anthropometric measurement, but the body morphology parameters must be taken into consideration.

Standard measurement protocols used by 3D body scanning technologies usually enable automatic measurement only in standard upstanding position. If it is necessary to determine body measurements in any of the dynamic positions, it is performed interactively, with precision measurement being largely dependent on the user/measurer. During the process, the measurer manually and interactively positions characteristic points on the surface of scanned model after which the linear or curved dimensions are taken in between. Such a measurement method is not very appropriate for serial measurements of the test subjects, since it is impossible for measurer to repeat the position of marker points in the same way for every test subject, which is necessary to ensure the comparability of the measurement results of a sample. For a better precision and easier detection of characteristic points, it is possible to put markers on a test subject before scanning or to create an automatic measurement protocol for every particular position. Creation of an automatic protocol is a complex and time-consuming process, but it provides a greater precision in defining anthropometric points compared with manual processing of individual models. It reduces measurer mistake and is repeatable since the points are always positioned in the same way, according to the measurement file defined within the protocol, Fig. 10.7. This enables comparison of obtained measurement results since the measurements are always taken in the same manner, providing possibilities for complex studies of body dimensional changes in dynamic conditions (Petrak and Mahnic Naglic, 2017).

Dimensional body changes must be taken into consideration while designing and constructing tight-fit garments with high criteria for functionality such as diving suits. For a construction of diving suits, it is important to select and analyze specific body positions that occur during diving. For a free diving discipline, there are five selected dynamic postures, the first refers to the mobility of the upper extremities and presents the terminal positions of the arms at characteristic body movements when diving; the fourth posture refers to the length of elongation of the body, especially in the area of

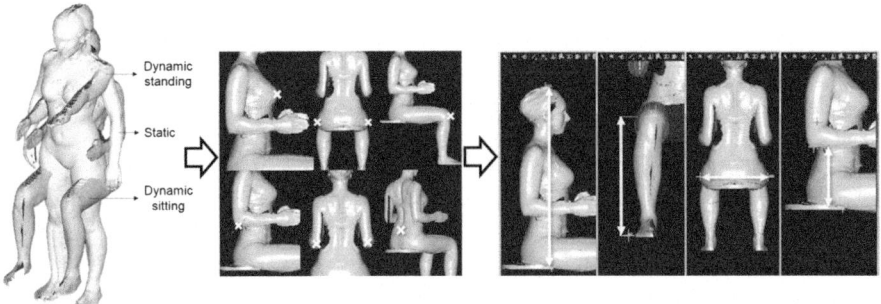

Fig. 10.7 Automatic measurement protocol of sitting position.

the rear part of coccyx, at maximum flexion of the torso; and the fifth posture refers to the mobility of the lower extremities, respectively, the flexion of 90 degrees in the hip and knee, wherein the elongation length on the back line of the leg and the change in length in the area of the knee joint are analyzed, Fig. 10.8 (Mahnic Naglic et al., 2017).

Measurements on the back body, such as shoulder width (SH), defined by the length between the acromion points and width of the back at the armpit level area (BW), have the most prominent issues related to the changes of body dimensions. When moving the hands forward, a shift of shoulders forward and stretching of the

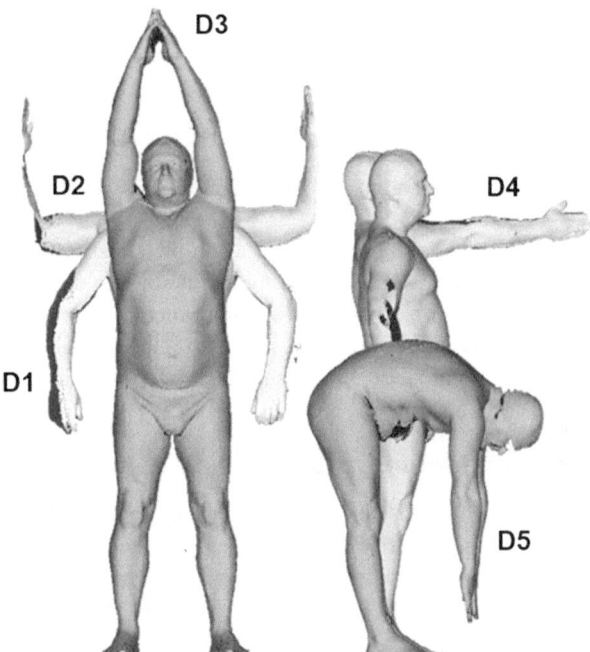

Fig. 10.8 Characteristic dynamic body positions for diving.

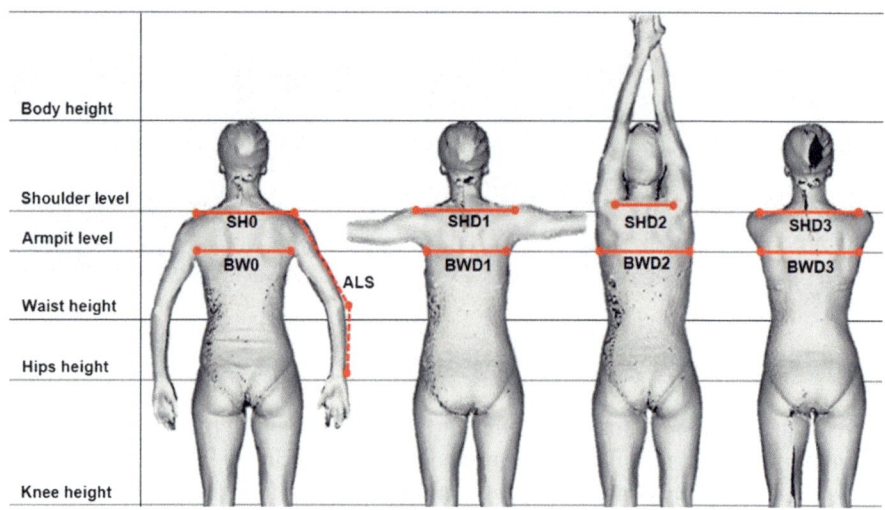

Fig. 10.9 Changes of shoulder and back measurements in different body positions.

width of the back occur, Fig. 10.9. Therefore also the garment moves with the body, causing the pressure on the body and hindering hand dexterity. When lifting the arms above the head, which is a characteristic posture in diving, the acromion points approach, which decreases the shoulder width, Fig. 10.9. Due to compression the garment is separated from the body, and wrinkles and buckling appear. This can be stated as a critical area because of the possibility of penetration of water in the space between the body and the garment (Mahnic Naglic et al., 2016).

10.5 Computer-aided 3D swimsuits and diving suit design

Computer-aided 3D design of clothing presents a complete and complex development process, from application of 3D body scanning technology to obtain information on body size and shape, to 3D construction of clothing model directly on body surface and flattening of 3D surfaces into 2D cutting parts, to simulation of clothing and strain testing of virtual prototype in static and dynamic conditions of use, Fig. 10.10.

10.5.1 Virtual mannequins as design and construction templates

The scanned body models have to be processed using appropriate software. Processing includes closing the surface, creating a single-layered polygonal model, and converting into a suitable format for import into a CAD system, Fig. 10.11. Closing of scanned surface can be performed using interactive, manual, or automatic process. Closing is performed by implementing additional polygons on shaded areas, which are connecting present points using the principle of closest distance. The main insufficiency in the process is the way of closing the areas with flat surfaces, where curved

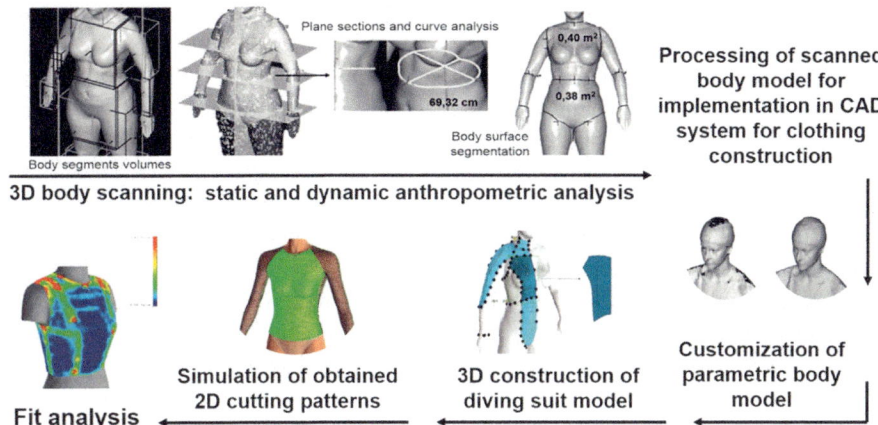

Fig. 10.10 Computer-aided 3D development of clothing prototypes.

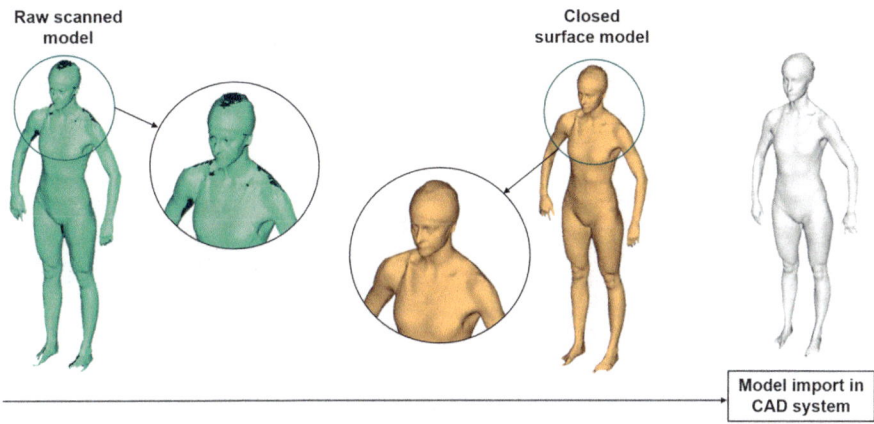

Fig. 10.11 Processing of scanned body model for implementation in CAD system for 3D construction.

ones would be a better option. That causes deviations of scanned model from real shape on particular body parts. Deviation on such parts also depends on the size of shaded area that needs to be closed. At small distances, curves can be completely approximated with the flat surface, without great effect on the shape, while on bigger shaded areas, deviations of surface shape will be significant, which needs to be taken into account in further process and application of such models for 3D clothing construction.

Parametric body models enable adjustment of a greater number of body measurements and some characteristics of body shape and posture. Body measurements are usually organized according to the type of measurements and grouped as basic body measurements; body circumferences; body lengths; and additional advanced measurements that are defining shape and posture like body depth and width, upper and lower back curvature, position of the neck and shoulders, and breast and buttock shape. Circumference, length, and height measurements are defined numerically, while measurements of shape and posture are usually defined with a measure of expression in a particular range of values where every characteristic is adjusted visually. This is why the adjustment of parametric body model is never complete, and deviation in shape and posture can be seen compared with the scanned body model. During the adjustment, it is also necessary to control the positions on which particular measurements are taken and placed, because of possibility that different systems use different measurement standards (Petrak et al., 2012).

Adjustment of the parametric body model can be performed interactively by entering a numerical value of every particular measure manually or automatically using the .ord measurement file. Measurement file contains a list of measurements that needs to be set in a right order to provide accurate adjustment of body model. First, it is necessary to adjust measurement that defines body depth and width. After defining characteristic circumferences and body lengths, body model obtains its final shape and dimension. Particular body characteristics like upper and lower back curvature, position of shoulders, and breast and buttock shape that cannot be defined numerically, but can be interactively adjusted between a defined range of expression, should also be set before implementation of .ord measurement file (Petrak et al., 2012). Fig. 10.12 shows

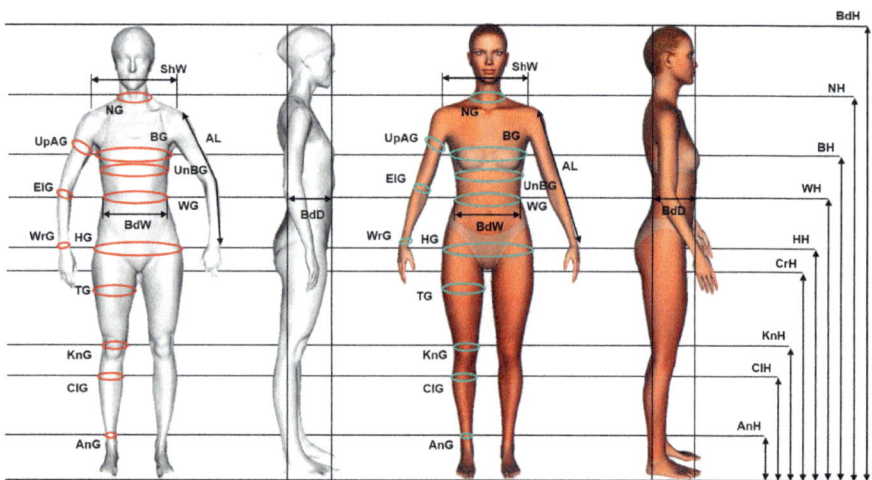

Fig. 10.12 Comparison of the scanned and customized parametric body model with the presentation of body measurement positions. BdH, body height; NH, neck height; BdW, body width; BdD, body depth; UnBG, under-breast girth; BG, breast girth; WG, waist girth; HG, hips girth; NG, neck girth; UpAG, upperarm girth; ElG, elbow girth; WrG, wrist girth; TG, thigh girth; KnG, knee girth; ClG, calf girth; AnG, ankle girth; BH, breast height; WH, waist height; HH, hips height; ShW, cross-shoulder width; CrH, crotch height; KnH, knee height; ClH, calf height; AnH, ankle height; AL, arm length.

the comparison of the scanned and customized parametric body model with the presentation of 26 body measurement positions and adjustments, selected as the most important for the construction of female diving suit (Mahnic Naglic et al., 2017).

10.5.2   3D construction of diving suits and 2D pattern flattening

The 3D garment construction considers the construction of a garment directly on computer body model by drawing and creating pattern lines directly on body surface and application of 3D flattening method for separation of discrete 3D surfaces and transformation into the 2D cutting parts (Zhang et al., 2016). The methods find its application mostly for the construction of tight-fit garments, garments for special purposes, and garments that have high fit demands.

Application of 3D construction method for the design of swimwear and diving suits presents an innovative approach in a clothing creation process. The method integrates clothing design and construction process and enables designer to create and modify clothing model segments between precisely positioned points and to directly visualize its idea on body model surface during the construction process, Fig. 10.13. During the process, the designer needs to think about all possible cutting segments that can be necessary out of esthetic or industrial reasons. The method makes the construction of complex cuts easier. The use of tools for contour segment modification fastens further modeling of cutting parts and development of model variations for the collection, Fig. 10.14.

Diving suits are a type of clothing for special purposes, with high demands on functionality and fit. Body models obtained by 3D scanning or parametric body models adjusted according to measurements and shape of targeted individual are used for the construction of that type of clothing in order to achieve a high fit of suit to individual body characteristics. Body activity and motion need to be considered during the

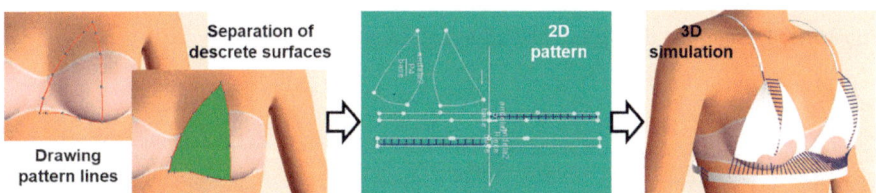

Fig. 10.13 3D construction of swimwear model.

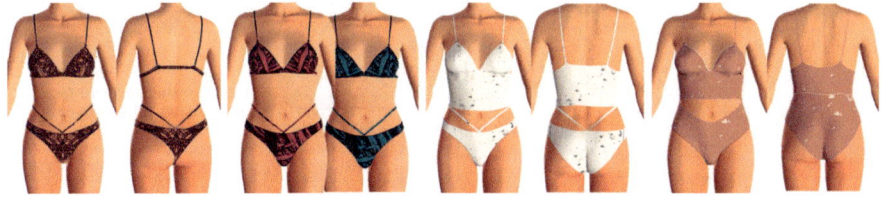

Fig. 10.14 Collection of swimwear developed using 3D construction method.

design and construction process in order to provide high functionality of diving suit in static and dynamic conditions. Since the 3D body models enable clear visualization and precise detection of anthropometric and kinematic body points, during the design of pattern segments, it is necessary to ensure mobility of particular kinematic body areas and the additional support on critical stress zones. Shoulders and acromion point can be pointed out as one of the most important kinematic areas with great mobility. In order to provide required undisturbed mobility of upper extremities, transverse cutting over the shoulders and acromion point can be avoided with the construction of diving suit models with raglan sleeve type, Figs. 10.15 and 10.16. On the sides of the model, they usually inserted a triangular ending in the armpit area for ensuring the required width and preventing tightening of the sleeve when a person is in motion. In general the complete suit should be designed with a minimum number of distribution lines, Figs. 10.15 and 10.16. After creating all the necessary points and pattern lines on the body model, it is necessary to define separate detachable surfaces and to determine

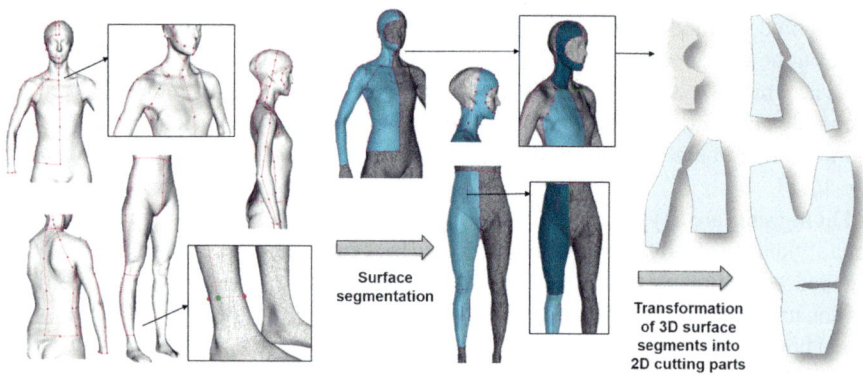

Fig. 10.15 Design of a diving suit pattern using scanned body model.

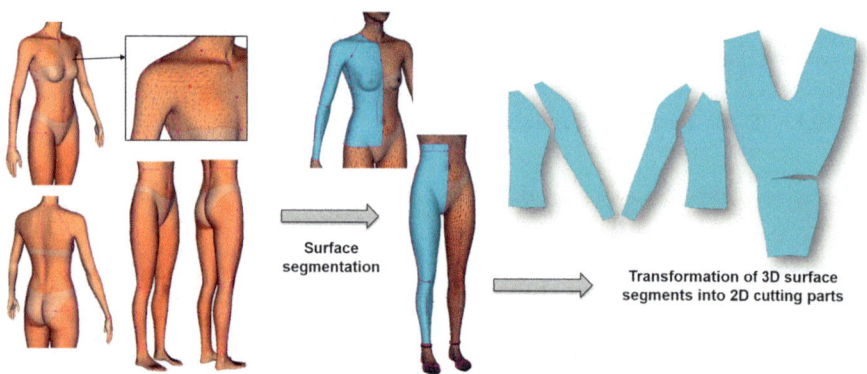

Fig. 10.16 Design of a diving suit pattern using adjusted parametric body model.

the direction of the baseline in the 3D space after which the extraction of all separated 3D surfaces and transformation to 2D cutting parts is derived using the 3D flattening method. Extracted 2D cutting parts have to be elaborated in terms of smoothing the distribution segments, due to irregularities in the surface of the body caused by the 3D flattening, which influences the shape and smoothness of the curves (Mahnic Naglic et al., 2016).

The use of scanned body models for the 3D construction provides better possibilities for precise adjustment of model pattern according to anthropometric characteristics, but the construction process itself is a more complex task, Fig. 10.15. Beside mentioned shaded areas and surface closing error, some of the biggest issues in implementation of scanned models for 3D flattening method are caused by irregular body posture and body asymmetry. The presence of body deformations complicates process of designing pattern segments on body model surface. Because of the irregular triangulated surface, it is difficult to create regular cutting segments, especially straight segments like orientational median line, which divides left and right body side. Median line is created on symmetrical patterns when only one-half of the pattern is constructed. After transformation of 3D surfaces into 2D cutting parts, the final pattern is obtained after additional modification of pattern segments and symmetry over median line.

Main insufficiency of parametric body models is their inability for complete adjustment according to individual measurements and shape of a scanned body, which causes minor deviations in final pattern dimensions and shape in comparison with a pattern constructed using the scanned body model. However, parametric body models have a regular surface topology and are perfectly symmetrical, which makes the process of design and construction on body model easier and faster. An additional advantage of parametric models is the regular upstanding position in relation to basic anatomy planes, while on scanned body model, position depends on the individual in the capturing moment. Regular body position provides more regular segments and surfaces of model cutting parts, Fig. 10.16.

Considering the differences in position and adjustment of the parametric body model compared with scanned body model, cutting parts obtained by 3D flattening method can be analyzed in terms of calculation and comparison of surfaces and segment dimensions. Minor differences in dimensions and greater in contour shape of cutting parts can be seen by folding the contour line of patterns obtained using scanned and adjusted parametric body model. However, comparison of determined surfaces of particular cutting parts and calculations of total suit coverage of body surface did not show significant differences between patterns separated from the scanned and the parametric body model, Fig. 10.17.

10.5.3 3D simulation of 2D patterns

Cutting patterns of diving suits obtained using the 3D flattening method can be verified with 3D simulations. Simulations can be performed on adjusted parametric or scanned body models. Simulation process includes predefining set of parameters that are referring to positioning of particular cutting parts against the body and layers of

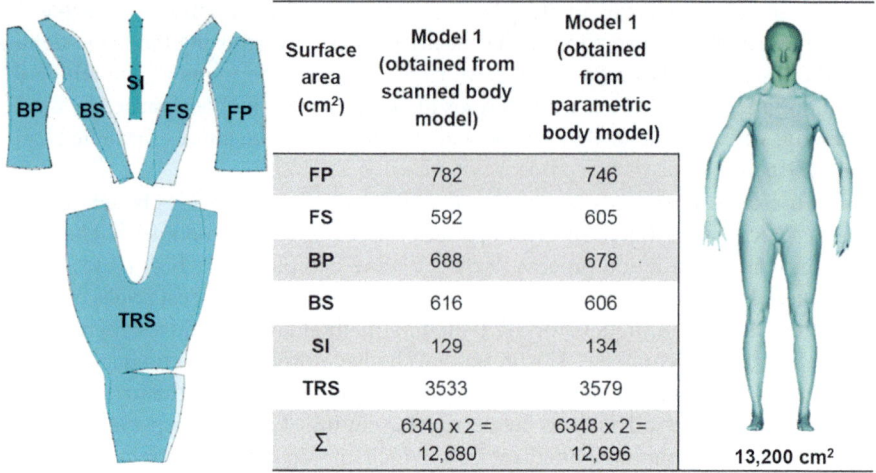

Surface area (cm²)	Model 1 (obtained from scanned body model)	Model 1 (obtained from parametric body model)
FP	782	746
FS	592	605
BP	688	678
BS	616	606
SI	129	134
TRS	3533	3579
Σ	6340 x 2 = 12,680	6348 x 2 = 12,696

13,200 cm²

Fig. 10.17 Calculations and comparisons of suit model surface areas and body area.

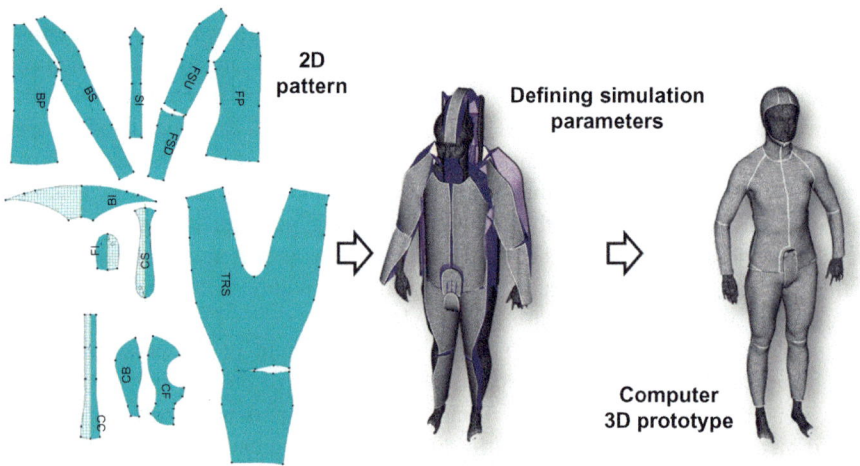

Fig. 10.18 3D simulation of diving suit on scanned body model.

clothing model. In order to facilitate calculation and improve simulation result, it is possible to define initial deformation of cutting parts in a form of cylindrical curvature, which also depends on cutting part position against the body, Fig. 10.18. It is also necessary to define joining segments and to apply additional joining properties like specific width or folding if necessary.

In the simulation process, cutting parts are transforming from flat 2D form to a 3D shape of a body model on which it is simulated. The way on which cutting parts will be deformed into three-dimensional garment model depends on the possibilities of simulation (Volino and Magnenat-Thalmann, 2000; Magnenat-Thalmann and Volino, 2005). CAD systems used in the clothing construction process for visualization and prediction of clothing model behavior before the production of a real prototype enable 3D simulation with application of physical and mechanical properties of particular textile material, intended for the production of a real garment model. Physical and mechanical properties define deformation of cutting polygonal surface parts according to particular real material, obtaining realistic visualization of computer 3D prototype (Igarashi and Hughes, 2002). Systems for objective evaluation of textiles and clothing that are evaluating properties at low loads, such as Kawabata Evaluation System (KES) or Fabric Assurance by Simple Testing (FAST), are usually used for determination of physical and mechanical property parameters (Geršak, 2001). Important simulation parameters are elongation at maximum load in warp and weft direction (load depends on the system), bending rigidity in warp and weft direction, shear rigidity, thickness, and weight.

Physical and mechanical properties of neoprene material were applied to the simulation of diving suits. With regard to the specific elasticity of the neoprene, simulations of patterns obtained directly from the body showed too high ease allowance. Suit ease allowance can be tested using the transversal cross sections of the simulated garment on body model where dimensions of both suit and body cross-section girth can be obtained. Ease allowance is calculated as a difference between obtained girths. Because of the neoprene high elasticity in the next step, it is necessary to perform the scaling of patterns in the transverse direction, in accordance with the extensibility of the material, Fig. 10.19. Usually a value of 5%–15% shrinkage is applied depending on the position of segment in total pattern. The results of visualization of simulated prototypes of both models after scaling confirmed appropriate fit of garment patterns. Beside analysis of ease allowance, visualization of simulated prototype is also analyzed in terms of checking the positions of pattern cutting segments on the body and particular length measurements, Figs. 10.20 and 10.21.

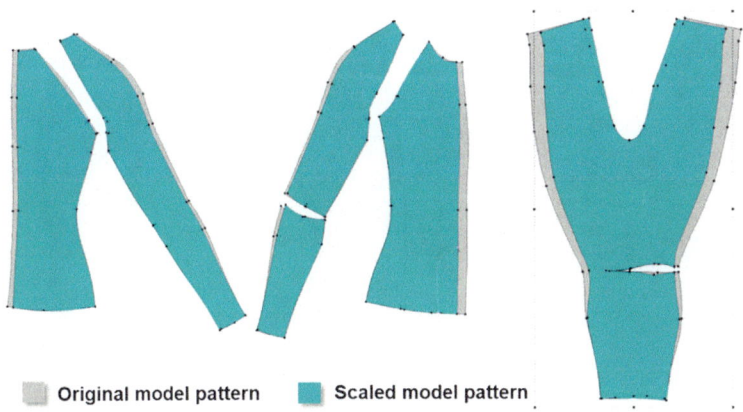

■ Original model pattern   ■ Scaled model pattern

Fig. 10.19 Comparison of original and scaled diving suit pattern.

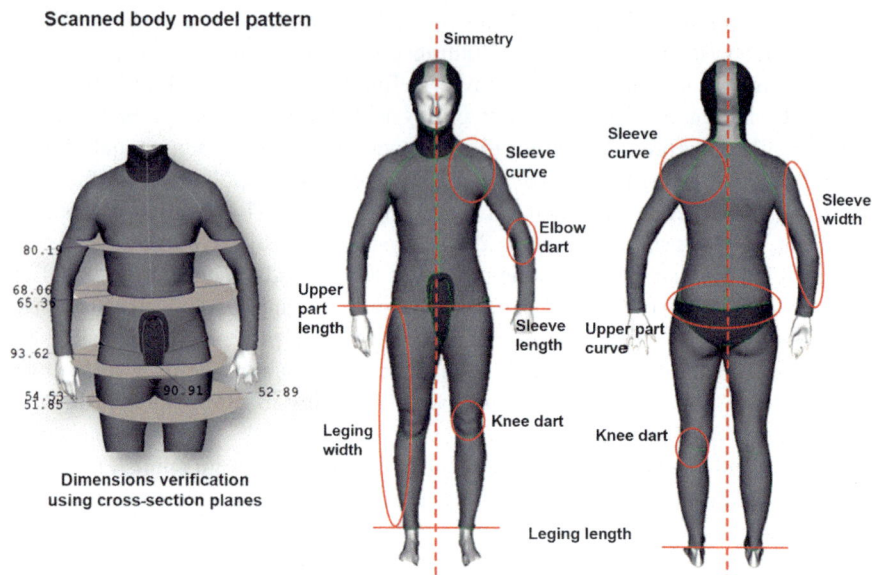

Fig. 10.20 Analysis of ease allowance and cutting part segments on pattern obtained from the scanned body model.

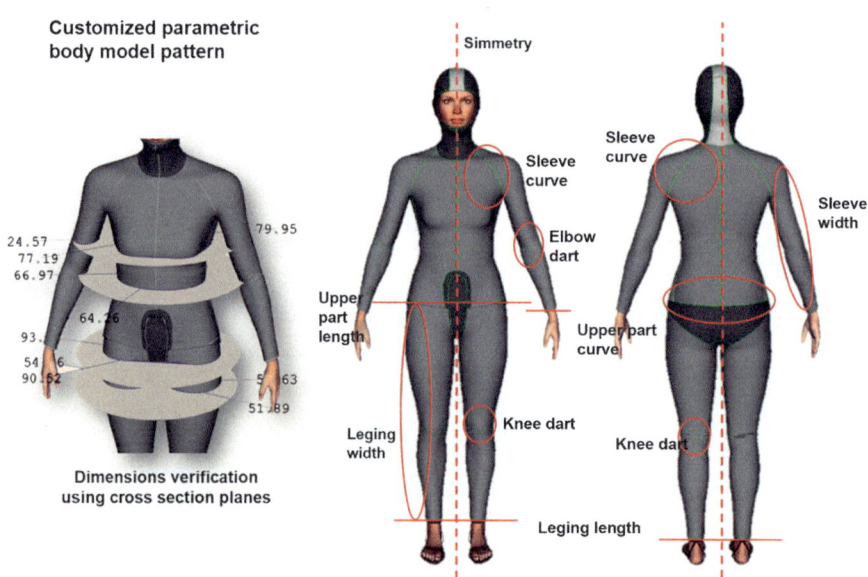

Fig. 10.21 Analysis of ease allowance and cutting part segments on pattern obtained from the adjusted parametric body model.

In terms of process optimization and applicability of the method in real manufacturing condition, additional criteria, such as the number of required additional interactive corrections on 2D patterns, position and correctness of pattern connection segments, and symmetry of pattern parts, can be evaluated, showing better optimization and applicability for the flattening method on a custom parametric body model.

10.6 Fit analysis and testing of diving suits

10.6.1 Virtual prototype testing

Beside visualization of computer prototype and analysis of pattern dimensions, innovative CAD systems for clothing construction enable additional assessments of pattern deformability, like testing the garment stretch on the body, in longitudinal and transversal direction or testing the pressure of garment against the body, depending on applied physical and mechanical properties. Diving suits have a characteristic of compressibility, which is caused by specific elastic properties of neoprene material. In that sense, computer prototypes can be further tested in terms of material stretching in transverse direction on different body areas in static upstanding position. In order to assume appropriate fit in wearing conditions, determined values should not increase the limits of values obtained by evaluation of real materials, Fig. 10.22.

Pressure analysis considers normal collision pressure of garment against the body, Fig. 10.23. It is necessary to predefine pressure points on which the suit model will be tested. Points for pressure analysis are defined based on the characteristics and demands of the suit being tested. According to professional consultations with experienced spearfisher and professional awarded free diver, specific points are selected for

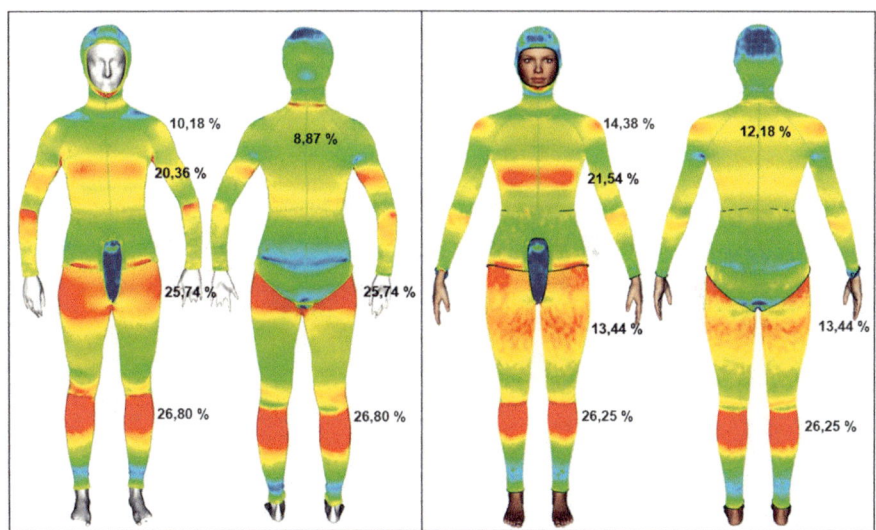

Fig. 10.22 Computational analysis of the material stretching on 3D prototypes in static position.

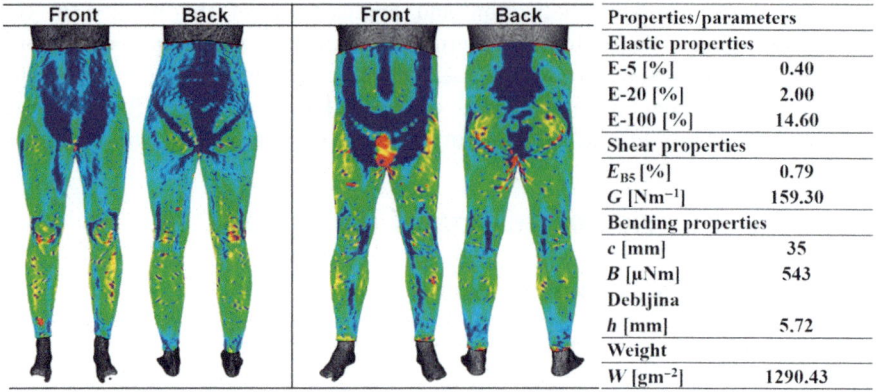

				Properties/parameters	
Front	Back	Front	Back	Elastic properties	
				E-5 [%]	0.40
				E-20 [%]	2.00
				E-100 [%]	14.60
				Shear properties	
				E_{B5} [%]	0.79
				G [Nm^{-1}]	159.30
				Bending properties	
				c [mm]	35
				B [µNm]	543
				Debljina	
				h [mm]	5.72
				Weight	
				W [gm^{-2}]	1290.43

Fig. 10.23 Normal collision pressure in static position obtained using the presented parameters of physical and mechanical properties: (A) female prototype and (B) male prototype.

the pressure analysis: forehead middle (FH), ears (ER), top of chin (CH), and front neck middle (NH), for assessment of cap pressure on head area; front breast points (CF), front side points on under-breast circumference (UB), and side back points on breast circumference (CF), for assessment of pressure in chest area, which is also one of the most important areas considering its effect on breathing; front side points on waist circumference (WF), frontmost prominent points on maximum belly circumference (MB), front side points on hips circumference (HS), and points on gluteus (GF), for assessment of pressure on the belly and pelvic area; and points on wrists (WR) and inner ankles (IA), for assessment of pressure on limbs, which are also very important critical areas where it is possible for water to enter the suit, Fig. 10.24.

Results of pressure measurement of computer garment prototypes on female and male static body models in defined measuring points are shown in Table 10.2.

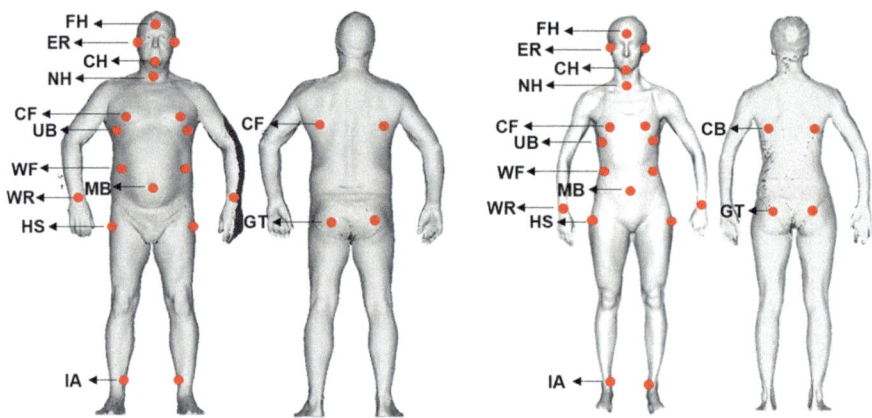

Fig. 10.24 Pressure measurement points.

Table 10.2 Garment pressure values in static conditions measured on virtual prototypes simulated on parametric and scanned body models

Norm. coll. press. (N/m²)	Male prototype		Female prototype		Norm. coll. press. (N/m²)	Male prototype		Female prototype	
	Parametric body	Scanned body	Parametric body	Scanned body		Parametric body	Scanned body	Parametric body	Scanned body
UB	415.84	428.41	443.37	424.71	FH	**712.82**	**897.82**	**202.59**	**837.43**
CF	1000.45	1107.79	**2092.09**	**1583.73**	ER	214.37	258.94	**152.95**	**672.44**
CB	318.67	311.08	767.99	722.82	WR	**1348.01**	**2184.43**	277.03	**1966.00**
MB	774.15	808.25	80.27	118.78	MB	712.70	884.54	**343.66**	**131.77**
HS	562.73	497.13	400.96	449.45	HS	739.63	672.84	650.10	742.98
NF	226.50	344.51	1020.29	1014.73	GT	1382.63	1314.26	1048.41	1243.57
CH	**1264.63**	**2213.27**	2123.79	2208.17	IA	1272.05	540.79	718.25	514.04

Determined pressure values showed differences depending if simulation was performed on scanned or parametric body model. Significant differences are visible on measurement points and body areas where the parametric body model does not allow precise adjustment according to individual shape. This refers to measurement points on the head, wrist, and ankles (CH, FH, ER, WR i IA). Pressure analysis of the female suit also showed significant differences on measurement point CF between scanned and customized body model, which can also be explained by the limitations of parametric body model.

Result analysis showed better correlation between garment pressure values measured on scanned body models and real test subjects, which is mostly visible from body areas where the parametric body model does not allow precise shape adjustment. This refers to measurement points on the head, wrist, and ankles (CH, FH, ER, WR i IA). Pressure analysis of the female suit also showed significant differences on measurement point CF between scanned and customized body model, which can also be explained by the limitations of parametric body model adjustment in the breast area.

Due to the limited possibilities of CAD system and inability for a fast and accurate body animation of the parametric models, where there are only few predefined dynamic postures accessible, the 3D simulation and computational analysis of stretching and pressure in dynamic postures are performed only in a dynamic posture with arms spread across. That dynamic posture reveals critical zone on the shoulder area, showing the unwanted wrinkles that appear during the rising arm movement. Unwanted wrinkles can be removed with additional corrections and modeling of curved segments in the shoulder area. After corrections, stretch analysis of diving suit model on back, in a dynamic position of arm spread across, showed satisfactory fit, Fig. 10.25.

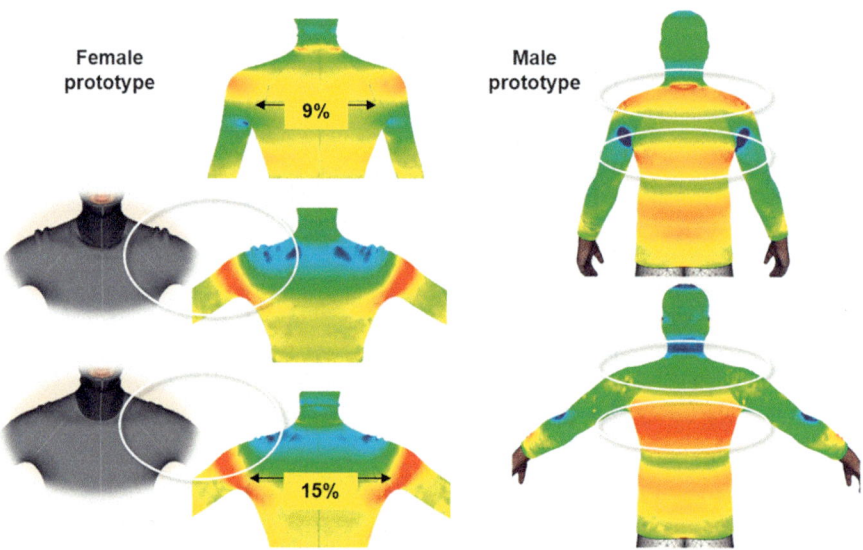

Fig. 10.25 Prototype simulation in dynamic position and computational analysis of the material stretching.

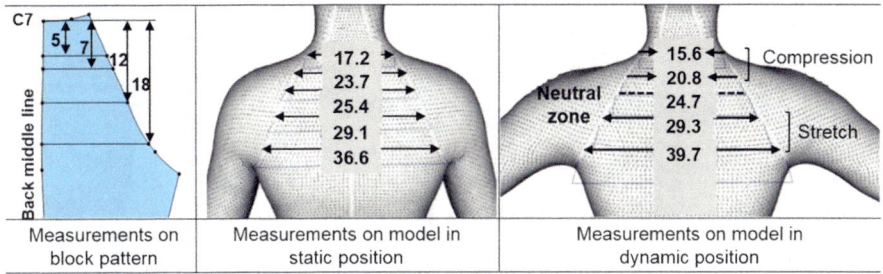

Fig. 10.26 Linear stretch analysis in dynamic position—arms spread across.

Additionally, it is possible to perform analysis of linear deformations of the suit on the body. First, it is necessary to define linear measurements on suit pattern on which the measurements will be taken. Measurements are first taken on suit unstressed, flat 2D pattern, after which the same position measurements are taken on the suit prototype simulated on a body model in static and dynamic conditions. From the differences in linear dimensions, it is possible to calculate the percentage of material stretch and compare the results with the elongation values obtained by real material measurements in order to evaluate fit and functionality of computer prototype in dynamic conditions.

Linear stretch analysis of simulated diving suit prototype on linear dimensions on upper back body area, in position D2, is presented in Fig. 10.26. The figure shows the differences between values measured in static and dynamic position, caused by the arm movement. Diving suit requires a skin-tight fit, and elastic properties of neoprene materials enable that fit but in a combination with precise construction and right scaling of garment pattern to achieve the right pressure of the garment in static and functionality in dynamic conditions. A stretch of neoprene material can be seen from the differences between linear segment measurement on garment pattern and values measured on garment on the body in a static condition, Fig. 10.26.

10.6.2 Physical prototype testing

Verification of the whole diving suit computer-based design and development process is, as usual, performed with the real physical prototype production, Fig. 10.27.

Fig. 10.27 Production of real diving suit prototype.

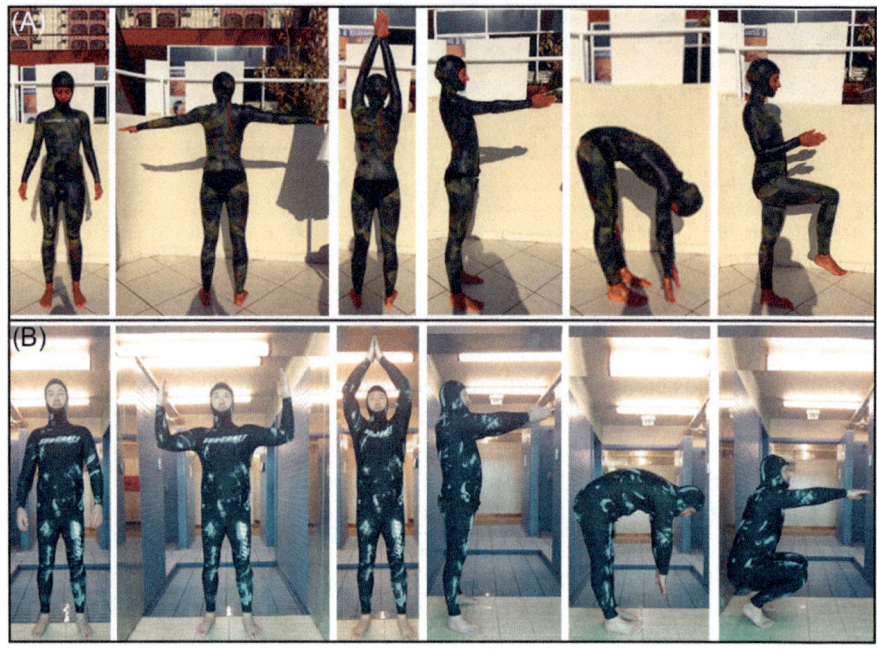

Fig. 10.28 Diving suit evaluation in five dynamics positions: (A) female prototype for free diving sport discipline and (B) male prototype for spearfishing.

The presented female and male diving suit prototypes were tested by the female professional diver and experienced male spearfisher in real conditions of use and in various dynamic positions, Fig. 10.28. According to the results of evaluation of the real prototype in wearing conditions, it can be estimated that the diving suit assures an appropriate fit. Therefore the methods can be verified as applicable for virtual prototyping and production of tight-fit garments with high demands on fit and functionality.

Real prototype normal collision pressure analysis is performed using the Picopress M-1200 measuring equipment of Microlab Electronics and compared with the analysis results of a computer prototype in the same measurement points. Table 10.3 presents the determined pressure values of a real prototype and prototype simulated on a scanned body model. Prototype simulated on the scanned body model showed much better correlation with pressure values determined on real prototypes in comparison with simulation on a parametric body model. Small differences in the determined values can be explained by a difference between computer body model, which is a rigid model and cannot be compressed with clothes like real body due to the underlying fat tissue that is depending on body area more or less compressible. Bigger differences are visible in female prototype measurements considering the female body constitution has a greater tendency of accumulation of fat tissue in comparison with a male muscular constitution.

Table 10.3 Garment pressure values in static conditions measured on real and virtual prototype simulated on scanned body models

Norm. coll. press. (N/m²)	Male prototype		Female prototype		Norm. coll. press. (N/m²)	Male prototype		Female prototype	
	Computer prototype on scanned body	Real prototype	Computer prototype on scanned body	Real prototype		Computer prototype on scanned body	Real prototype	Computer prototype on scanned body	Real prototype
UB	428.41	400.00	424.71	533.33	FH	897.82	933.33	837.43	1333.33
CF	1107.79	1200.00	1583.73	1600.00	ER	258.94	266.67	672.44	533.33
CB	311.08	266.67	722.82	666.67	WR	2184.43	2133.33	1966.00	2000.00
MB	808.25	800.00	118.78	133.33	MB	884.54	933.33	131.77	133.33
HS	497.13	533.33	449.45	400.00	HS	672.84	666.67	742.98	666.67
NF	344.51	266.67	1014.73	1066.67	GT	1314.26	1200.00	1243.57	1200.00
CH	2213.27	2533.33	2208.17	2000.00	IA	540.79	533.33	514.04	666.67

Table 10.4 Pressure measurements on characteristic points for dynamic postures D2, D5, and D6 measured on virtual and real prototypes

Pressure (N/m²)	CF		CB		GT		
	Static	D2	Static	D2	Static	D5	D6
Virtual prototype	1107.8	1536.9	311.1	371.9	1314.3	–	–
Real prototype	1200.0	1466.7	266.7	400.0	1200.0	1333.3	1600.0

Real prototype pressure analysis in dynamic conditions is performed in positions of arms spread across (D2), total front body flexion (D5), and squat (D6) on access points: front chest (CF), back chest (CB), and gluteus (GT) for testing the lower extremities, Table 10.4. Position D2 enables comparison with a computer prototype, and the other two positions were tested only in real conditions. Measured garment pressure values in the chest area showed satisfying correlation between simulated and real prototype.

Linear stretch analysis of simulated and real male diving suit prototype, in a dynamic position D2, is presented in Table 10.5. Data analysis of garment stretch and compression values caused by body movement, Fig. 10.26, confirmed tight fit with the same behavior that can be observed on naked body dimensions in dynamic postures.

10.7   Future trends

The presented method of 3D construction of diving suits is a complex process of computer garment prototype development. Parametric body models are appropriate, and they provide good clothing fit when used for flattening method and construction of tight-fit clothing. 3D flattening method and transformation of body surface segments into the clothing cutting parts provide a very good fit not only to particular body measurements but also to shape and posture of an individual. The method can be used for design and development of swimwear and other types of sport clothing that has to closely fit the body. In that sense, further research and development will be focused on testing of computer diving suit prototypes on body models of different sizes, shapes, and postures. Differences in construction methods for different body types and correlation between characteristic pattern measurements depending on different types of neoprene, the influence of physical and mechanical properties of different materials on development process of swimwear, are also important issues that need to be further investigated. Beside the segment of computer design and construction of tight-fit clothing prototypes according to different body types, future research will be oriented on development of clothing prototypes with integrated electronic components as a part of complete development of intelligent clothing model.

Table 10.5 Linear segment dimensions measured on block pattern and on the virtual prototype in static and dynamic conditions

Linear segment dimension (cm)		Block pattern measure	Static						Dynamic					
			Virtual (P)	%	Real	%	Virtual (P)	%	Δ	%	Real	%	Δ	
L1	(0)	14.7	17.2	17	17.2	17	15.6	6	10 ↓	6	14.6	0	18 ↓	
L2	(↓ 5)	20.7	23.7	14	24.2	17	20.8	1	14 ↓	1	21.1	2	15 ↓	
L3	(↓ 7)	22.8	25.4	11	26.2	15	24.7	8	3 ↓	8	25.8	13	2 ↓	
L4	(↓ 12)	27.6	29.1	5	28.4	3	29.3	6	1 ↓	6	31.4	14	11 ↓	
L5	(↓ 18)	34.6	36.6	6	39.2	13	39.7	15	9 ↓	15	44.0	27	14 ↓	

The presented concept can greatly contribute to accelerating the process of developing and testing new model prototypes, increasing the quality of finished products, and reducing the cost of real prototype production processes. Scientific research in this area is the subject of research conducted at the Department of Clothing Technology at the Faculty of Textile Technology of the University of Zagreb. This work was supported by the Croatian Science Foundation under the project number 3011: Application of mathematical modelling and intelligent algorithms in clothing construction.

References

Bragança, S., et al., 2016. Current state of the art and enduring issues in anthropometric data collection. Dyna Dyna Rev. Fac. Nac. Minas 83(197). https://doi.org/10.15446/dyna. v83n197.57586.

Chun, J., Oh, S., 2004. 3D body scanning posture to collect anthropometric data for garment making. J. Asian Reg. Assoc. Home Econ. 11 (4), 301–307.

D'Apuzzo, N., 2009. Recent advances in 3D full body scanning with applications to fashion and apparel. In: 9th Conference on Optical 3D Measurement Techniques, Vienna, Austria.

D'Apuzzo, N., 2009. Recent advances in 3D full body scanning with applications to fashion and apparel. In: 9th Conference on Optical 3-D Measurement Techniques, Vienna, Austria.

Diving Unlimited International Inc, 2013. Dui Drysuit Owner's Manual, EN 14225-2:2005,San Diego, CA 92102-2499, USA. .

Fan, J., Yu, W., Hunter, L., 2004. Clothing Appearance and Fit: Science and Technology. Woodhead Publishing Limited and The Textile Institute, Cambridge.

Fang, J., Ding, Y., Huang, S., 2008. Expert-based customized pattern-making automation. Part II. Dart design. Int. J. Cloth. Sci. Technol. 20 (1), 41–56.

Geršak, J., 2001. Objektivno vrednovanje plošnih tekstilija i odjeće. Sveučilište u Zagrebu, Tekstilno—tehnološki fakultet, Zagreb.

Geršak, J., Marčić, M., 2013. The complex design concept for functional protective clothing. Tekstil 62, 38–44.

Gill, S., Heyes, S., 2012. Lower body functional ease requirements in the garment pattern. Int. J. Fash. Des. Technol. Educ 5, 13–23.

Gupta, D., 2011. Design and engineering of functional clothing. Indian J. Fibre Text. Res. 36, 327–335.

Igarashi, T., Hughes, J.F., 2002. Clothing manipulation. In: 15th Annual Symposium on User Interface Software and Technology, ACM UIST'02, pp. 91–100.

ISO 20685:2010: 3-D Scanning Methodologies for Internationally Compatible Anthropometric Databases (n.d.).

Lee, H., Hong, K., Lee, Y., 2013. Ergonomic mapping of skin deformation in dynamic postures to provide fundamental data for functional design lines of outdoor pants. Fibers Polym. 14, 2197–2201.

Magnenat-Thalmann, N., Volino, P., 2005. From early draping to haute couture models: 20 years of research. Vis. Comput. 21 (8), 506–519.

Mahnic Naglic, M., Petrak, S., 2017. A method for body posture classification of three-dimensional body models in the sagittal plane. Text. Res. J. 1–17. https://doi.org/10.1177/0040517517741155 (Online first).

Mahnic Naglic, M., Petrak, S., Gersak, J., Rolich, T., 2017. Analysis of dynamics and fit of diving suits. IOP Conference Series: Materials Science and Engineering. In: 17th World Textile Conference AUTEX 2017—Shaping the Future of Textiles; 2017 May 29–31; Corfu (Greece).

Mahnic Naglic, M., Petrak, S., Stjepanovič, Z., 2016. Analysis of 3D construction of tight fit clothing based on parametric and scanned body models. In: 7th International Conference on 3D Body Scanning Technologies; 2016 Oct, Lugano, Switzerland, pp. 302–313.

Naebe, M., et al., 2013. Assessment of performance properties of wetsuits. J. Sports Eng. Technol.

Palmer, L.M., Epler, E.M., 1998. Fudamentals of Musculoskeletal Assessment Techniques. Lippincott Williams & Wilkins.

Petrak, S., 2007. Metoda 3D konstrukcije odjeće i modeli transformacija krojnih dijelova [dissertation]. Univ. of Zagreb, Zagreb.

Petrak, S., Mahnic, M., Rogale, D., 2015. The impact of male body posture and shape on design and garment fit. Fibres Text. East. Eur. 6 (114), 150–158.

Petrak, S., Mahnic Naglic, M., 2017. Dynamic anthropometry—defining a protocols for automatic body measurement. Tekstilec 60 (4), 254–262.

Petrak, S., et al., 2012. Research of 3D body models computer adjustment based on anthropometric data determined by laser 3D scanner. In: 3rd International Conference on 3D Body Scanning Technologies; 2012 Oct, Lugano, Switzerland, pp. 115–126.

Simmons, K., Istook, C., 2003. Body measurement techniques: comparing 3D body-scanning and anthropometric methods for apparel applications. J. Fash. Mark. Manag. 7 (3), 306–332.

The Guide to Spearfishing in New South Wales. Australian Government, Department of Agriculture, Fisheries and Forestry, Underwater Skindiver's and Fishermen's Association (USFA), Sydney, Australia.

Tomljenović, A., Sinel, M., 2015. Ženski kupaći kostim—modeli, značajke i odabir materijala. In: International Conference MATRIB 2015, Materials, Wear, Recycling, pp. 328–343.

VITUS, 2018. 3D Body Scanner. http://www.vitus.de/.

Volino, P., Magnenat-Thalmann, N., 2000. Virtual Clothing Theory and Practice. Springer.

Wen-liang, C., et al., 2011. Surface flattening based on linear-elastic finite element method. Int. J. Mech. Aerosp. Ind. Mech. Manuf. Eng. 5(7).

Yih-Chuan, L., et al., 2006. Fast flattening algorithm for non-developable 3D surfaces. In: International Conference on Modeling, Simulation & Visualization Methods, MSV; 2006 June 26–29, Las Vegas, Nevada, USA.

Zhang, Y., et al., 2016. Optimal fitting of strain-controlled flattenable mesh surfaces. Int. J. Adv. Manuf. Technol. 1–15. http://homepage.tudelft.nl/h05k3/pubs/AMTOptimalFitting.pdf.

Sizing and fit for protective clothing

Inga Dāboliņa, Eva Lapkovska
Riga Technical University, Institute of Design Technologies, Riga, Latvia

11.1 Introduction

"Of all things the measure is man." This saying from Protagoras is known as the "man-measure statement": the notion that knowledge is relative to the knower (http://www.iep.utm.edu/protagor/), which can be widened to all environments—micro and macro—can be stated according to the inhabitants of the environment. Clothing can be considered as a type of microenvironment, as its use and functionality depends heavily on the wearers themselves—their physical fitness, metabolism, and other aspects. Clothing features affect the functionality of the wearer; clothing is intended to protect the human body from the surrounding environment, to promote human activities, to provide the ability to perform tasks, and to facilitate physiological processes.

Not unusually, to improve human functionalities and the development of clothing functions, some parts of clothing can be supplemented by protective elements, such as helmets, gloves, guards, armor, masks, etc. Such a system is referred to as personal protective equipment (PPE). PPE is a comprehensive term involving the clothing and equipment of firefighters and rescuers, soldiers, policemen, chemists, factory workers, and other specialized workers, which allows tasks to be safely carried out and protects the wearers from environmental influences. People also tend to use PPE for everyday purposes, such as the specialized clothing of motorcyclists, helmets, and items required for household work, such as gloves, aprons, etc.

The standard definition of protective clothing is clothing that includes protectors to cover or replace personal clothing, and that is designed to provide protection against one or more hazards. A hazard is a situation that can be the cause of harm or damage to the health of the human body, i.e., a potential source of harm. In addition, the term can be qualified in order to define its origin (e.g., mechanical hazard, electrical hazard) or the nature of the potential harm (e.g., electric shock hazard, cutting hazard, toxic hazard, fire hazard) (ISO 13688:2013, 2013; EN 13921:2017, 2017).

The origin of PPE is associated with the desire of people to protect themselves. Even some means used in ancient times are considered to be PPE. People used individual protective equipment in the Middle Ages, when blacksmiths wore protective leather linings for hands and large aprons to avoid burns from molten metal. Iron armor allowed ancient soldiers to conquer fortresses in assaults that would otherwise have failed because of bodily injuries. Although these PPE were not comfortable for

Anthropometry, Apparel Sizing and Design. https://doi.org/10.1016/B978-0-08-102604-5.00011-1

wearing and their weight exceeded human lifting capacity—if he fell off his horse, a trooper could not stand up—their role in protecting against the enemy's spears was important.

As industrialization and safety concerns were growing more advanced, the PPE also developed. The development of protective clothing respected the specificities of employees' routine work, traditions, and uniformity, particularly important for special task performers such as armed forces, firefighters, rescuers, etc.

Although soldiers had always worn special clothing, including the armor already mentioned, it has been only since the end of the 17th century that similarity with modern concepts of a uniform can be observed. Before that, only individual military units such as bodyguards or guards of royal families wore unified garments. In the battlefields, most of the soldiers were dressed in a varicolored variety of plain clothes. Only in the late 17th century, when regular armies emerged, were soldiers supplied with uniform clothing. Perhaps it was more about thrift rather than how to distinguish friends from enemies at the time. In the 19th century, military uniforms became overly ornate compared to men's civil robes, and they became so pompous that military personnel were confident of the need for more functional types of clothing. First, army uniforms had to be appended by so-called noncombatant working clothes, such as overalls, overclothes, and service caps, so that the more expensive sets of formal uniforms could be saved for parades and other ceremonial activities. Since the 19th/20th centuries, almost all armies introduced functionally safer combat outfits that began to be made of more inconspicuous, even-looking, but more durable textile fabrics in fallow or khaki colors or green-gray "protective colors." The experience of both world wars and the demands of thrift put a point on the custom of wearing traditional parade uniforms almost everywhere, except for some special-assignment departments. For example, in the United States the parade uniform with a shako hat, worn since 1814, is preserved in the military academy cadet corps. Even more ancient past remnants are observed in the Vatican Swiss Guard outfit, whose appearance has not changed since the 16th century.

In the aftermath of World War II, the development of military uniforms was increasingly less impressed by the traditions, and the practical considerations and protective capabilities of the outfitting came into the foreground. There were specialized uniforms for hostilities in jungle, desert, and arctic conditions. The customers began to require the clothing to be more functionally thought-out, to be made of light but at the same time durable materials, and to be constructed in such a way as to enable soldiers to fulfill their service obligations without restrictions on movements. Special needs also created special uniforms, such as antiflame uniforms for crews of tankers, tanks, and helicopters, and protective clothing against chemical assault. The colors of garments, for masking or camouflage, were planned to be worn in an environment with a similar coloristic background.

This highlights the different aspects of the use of the PPE: that is, the PPE must ensure both human protection and the masking of visibility, or, on the contrary, ensure highlighting (doctors, rescuers, transport road workers). Also, given that the wearing of PPE is most often linked to a representation of a particular profession, the PPE must ensure the functions of the uniform, i.e., uniformity and recognition.

11.2   Protective clothing size and fit requirements

To ensure external look and functionality requirements of the PPE, it must be of good appearance and fit. Side effects of using PPE can range from discomfort to severe constraint and physical load. The application of ergonomic principles to PPE allows optimization of the balance between protection and usability (EN 13921:2017, 2017). Although the corresponding standard EN 13921:2017 "Personal protective equipment—Ergonomic principles" covers many aspects, it cannot be expected to identify all the possible problem points. It will remain the responsibility of the relevant experts to identify and quantify the hazards in the work and to foresee the potential ergonomic problems, and thus to ensure that the PPE specified and manufactured is fit for the purposes intended in all respects (EN 13921:2017, 2017). Principles in standard for writers of PPE product standards are related to:

* anthropometric characteristics related to PPE;
* the biometrical interaction between PPE and the human body;
* the thermal interaction between PPE and the human body;
* the interaction between PPE and the human senses: vision, hearing, smell and taste; and skin contact (EN 13921:2017, 2017).

Depending on the application, the PPE may vary; for instance, PPE is just a simple smock for laboratory conditions (lab coat), the task of which, first of all, is to protect the individual clothing of a human and the parts of the body from contact; moreover, the more complex/dangerous the working conditions, the higher the level of protection the PPE must ensure.

11.2.1   Types of PPE

To understand the importance of PPE fit and sizing, it is necessary to classify the PPE according to its target. For example, PPE protecting an individual from the effects of dangerous or harmful environmental factors for his or her health must target:

* temperature—heat and cold,
* fire,
* humidity,
* dirt,
* radiation,
* poisons,
* acids,
* alkali,
* biological pollution, etc.

A separate division of PPE is that of protecting against physical damage, such as strokes, mechanical injuries, falling objects, bullets, electrical risks, etc. Groups such as respiratory protective equipment (effects of different substances, dust, bacteria), eye and ear protection products, and limb protectors such as gloves and boots (against all types of hazards) must be observed. Warning or high-visibility protective clothing (fluorescent or reflecting) is also widely used (ISO 13688:2013, 2013).

Individual protective equipment may be divided into categories as follows:

- *Working clothing*—overalls, bib-and-brace overalls, trousers, jackets, rain jackets, coats, suits.
- *Head protection*—protective helmets for skull protection, hats, hoods, nets for hair and head protectors.
- Face and eye protection—glasses, goggles, protective screens, facemasks, welding shields.
- *Hearing protection*—headphones, earplugs, acoustic helmets.
- *Respiratory and digestive system protection*—respiratory-type devices, masks with appropriate dust and/or gas filters, devices with air supply.
- *Arm and hand protection*—protective gloves against mechanical effects, chemical impacts, ionizing radiation and radioactive contamination, against cold and heat, wrist protectors for physically hard work, dielectric gloves.
- *Leg and foot protection*—protective shoes, footwear with additional protection for fingertips, footwear and overboots with heat-resistant soles, vibration-damping footwear, electric-insulation shoes, protective boots for working with chainsaws, knee supporters, etc.
- *For protection of the abdomen and other parts of the body*—protective guards, jackets, aprons against moving parts of machinery, aprons for protection against chemicals, X-ray radiation, life jackets, protective equipment for working with knives, belts, etc.
- *For protection against falls*—safety systems and equipment against falls from altitude, such as devices for holding the body, antidrop systems, safety belts, ropes, coupling hooks.
- *Protective clothing for all-body protection*—different types of clothing for protection against mechanical injuries, chemicals, melted metals, infrared and electromagnetic radiation, dust, gases, heat, and protective clothing for working with chainsaws, fluorescent or reflecting protective clothing or accessories (road repairs, rescue services).
- *For skin protection*—protective creams, protective ointments (Амирова and Сакулина, 1985; Wang and Gao, 2014; Song et al., 2016; Chapman, 2012; Geršak, 2013; McCann and Bryson, 2009a; Song, 2011a; Mattila, 2006).

Parts of protective garments and PPE products are worn or applicable as different sets and as layering, depending on the tasks or objectives of the PPE, as well as the conditions of wearing, i.e., the season and weather conditions. Moreover, if the overclothes and the various protective equipment directly protects the body of the wearer, then a suitable layer of underwear will be relatively crucial for the comfort of the wearer, protecting the human body from cooling or overheating and facilitating the removal of moisture from the surface of the skin.

11.2.2   *Necessity of well-fitted PPE*

Clothing as a constituent of a personal protective equipment (PPE) system must comply with numerous and sometimes even incompatible or difficult-to-combine safety, comfort, performance, and other conditions. Taking into consideration these aspects, key design factors can be outlined: protection, comfort, mobility, connectivity, and ease of use (McCann and Bryson, 2009a). In modern PPE, the look and the overall appearance are also important—protective clothing must be provided not just with PPE quality, but with a fashionable look and aesthetics as well (Podgórski, 2017).

Consequently, the task of designing becomes even more complex, to provide contemporary materials and integrative consumables with the fit of work wear that not only allows productive execution of the task and use of PPE smart devices, but also offers an aesthetic appearance.

Protective wear must comply with safety and performance requirements (not affecting the health and hygiene of the user), size designations and clothing marking (measuring procedures and checking standards for compliance with standards (ISO 8559-1:2017, 2017; ISO 8559-2:2017, 2017; ISO 7250-1:2017, 2017) and wear-out (changes in one or more initial qualities of protective clothing over time). Within all these requirements, users must be provided with a level of comfort that is in line with the level of protection against hazards, environmental conditions, user activity levels, and expected wear time (ISO 13688:2013, 2013).

In the case of unsatisfactory appearance and fit, situations may arise when a person may decide not to use the necessary protective clothing and the PPE, unless it is strictly determined by the rules of the entity or institution (e.g., in the armed forces, police, rescue services, etc.). For example, clothing that is too loose or too tight burdens human movements, dynamics, and also breathing, and sometimes threatens the functioning of internal organs and human well-being in general. Skin contact information, such as irritation, tickle, cold, hot, pressure, and pain, is also important (EN 13921:2017, 2017). The anthropometric relevance of uniforms/working apparel parts is most commonly understood in this respect. Because parts of the garment are exposed to the skin, their properties, the feelings of the wearer, and the compliance with the need must be taken into account first of all.

Anthropometric fit ensures the implementation of protective functions of the PPE. Similar to everyday clothing, a noncompliant size of the PPE results in a downgrade of human capacity, which may lead to injuries, or even fatal outcomes, during the execution of specific tasks. It is therefore essential that the PPE is suitable or adapted to the wearer's body, also is in accord with specific demands, in the proper size and shape. Anthropometric fit (proportionality, anthropometric compliance) is defined as an overall dimension configuration (tightness or looseness) corresponding to the constructive purpose of the garment in relation to the overall measurements of the recipient's (wearer's) figure. As mentioned previously, the PPE consists not only of clothing but also of parts of equipment, gear, and accessories; therefore application of human body measurements to the design of PPE, including variation in dimensions within the user group, is called the anthropometrics of PPE (EN 13921:2017, 2017). Furthermore, biomechanics of PPE is the application of principles and methods from physics and engineering to describe the effect undergone by the human body and various body segments and the forces acting on these body segment, including any physical loading that may be caused by PPE (EN 13921:2017, 2017).

The ergonomic characteristics of PPE affect: the comfort, mobility, and dexterity of users; the rate at which they develop fatigue; the efficiency with which they can work in the PPE; the interaction of the PPE with other PPE; and the effectiveness of the protection provided (EN 13921:2017, 2017).

Factors to be considered in specifying requirements to take into account the anthropometric factors of PPE include:

- the hazards against which the PPE is intended to provide protection;
- the body part(s) it will be in contact with or cover;
- the physical activities expected to be performed during its use;
- the intended PPE user group (EN 13921:2017, 2017).

The hazard against which the PPE is to provide protection will determine whether or not closeness of fit is likely to be important. The body part(s) the PPE will be in contact with will serve to identify those parts where anthropometric data is needed. If the PPE crosses or covers a body joint, then more anthropometric dimensions should be specified based on different joint positions. The physical activity expected to be performed during the use of PPE may alter body dimensions. This should be taken into account in specifying anthropometric dimensions in a PPE standard. Excessively close fitting or otherwise poorly dimensioned PPE may prevent or hinder the performance of necessary activities. One example is the well-known apparent lengthening of the arm when extended and the increased girth of the thigh muscles when squatting or kneeling. Both of these have been shown to impair fit and therefore reduce the comfort or effectiveness of PPE. Loose fitting or bulky PPE may restrict access to working areas or may present a potential safety hazard by snagging on equipment parts or other features of the environment (EN 13921:2017, 2017).

11.3   Anthropometrics for protective clothing

Even at the beginnings of mass production of clothing, it was necessary to think about the number of sizes of a single model series, what would be the anthropometrically different garment sizes that would match the sizes distributed, and how to mark them. In the beginning, the manufacturing and marketing firms of the ready-made garments had to establish their own sizing, grading, and size marking rules independently because otherwise production of an assortment series of various sizes could not be started. The purpose of anthropometric typifications is to introduce the classifications of confection (for mass production) sizes for quantity production garments corresponding to a variety of human figures. The typification requirements are clearly satisfied by the types of two body measurements known in physical anthropology as "total morphological characteristics," namely the height of the body and horizontal chest or bust girth, which replaces the anthropological thorax perimeter measurement. However, in order to comprehensively describe the diversity of human figure dimensions, at least for the degree of precision required for the commercial business of the confection, the anthropometric standardization is not sufficient with two control dimensions only, and therefore the following morphological characteristics are to be introduced: the hip girth for women's clothing and the waist circumference for children and men's garments.

The size designation systems are based on body measurements, not garment measurements. The choice of garment measurements is determined by the designers and manufacturers who make appropriate allowances, style, cut, and elements of the garment. Standard ISO 8559-2:2017 "Size designation of clothes—Part 2: Primary and secondary dimension indicators" specifies primary and secondary dimensions for specified types of garments, which are explained as:

- *Primary dimension*—body dimension, in centimeters, that is used to designate the size of a garment for the customer. For example, chest girth, waist girth, bust girth, hip girth, body height, neck girth.
- *Secondary dimension*—body dimension, in centimeters (or body mass in kilograms), that can additionally be used to designate the size of a garment for the customer. For example, body height, waist girth, back shoulder width, inside leg length, hip girth, chest girth, underarm length, arm length, back neck point to wrist length, neck girth.

These primary and secondary dimensions are listed separately for men's, women's, boy's, and girl's clothing. These body measurements provide the basis for designation of sizes, and this standard is used in combination with ISO 8559-1:2017 "Size designation of clothes—Part 1: Anthropometric definitions for body measurement," which comprises the definition and generation of anthropometric measurements that can be used for the creation of size and shape profiles and their application in the field of clothing (ISO 8559-1:2017, 2017; ISO 8559-2:2017, 2017).

The design of protective clothing shall facilitate its correct positioning on the user and shall ensure that it remains in place for the foreseeable period of use, taking into account ambient factors, together with the movements and postures that the wearer could adopt during the course of work. For this purpose, adequate adjustment systems or adequate size ranges shall be provided so as to enable protective clothing to be adapted to the morphology of the user (ISO 13688:2013, 2013).

The intended user group will need to be defined to ensure that appropriate dimensions are specified to encompass that population. Variations in size may be accommodated by means of adequate adjustment systems or the provision of size ranges as appropriate. Because different body dimensions are not necessarily closely correlated, standards writers should consider the need to specify more than one essential anthropometric dimension in order to ensure satisfactory fit for the intended user group. Where necessary, the body mass of users may be specified. Writers of PPE product standards should address at least the topic related to anthropometric factors detailed in EN ISO 15537:2004 (EN 13921:2017, 2017; ISO 15537:2004, 2004).

For assessment of the anthropometric characteristics of PPE and their impact on the wearer, it is necessary to determine the:

- range of users;
- purpose for which the PPE is intended (e.g., activities and environment);
- adequate fit requirements (e.g., closeness, coverage);
- anthropometric measurements of users it is necessary to define to ensure adequate fit across the range of users;
- means of establishing actual PPE dimensions (including tolerances) for intended size category(ies) within the range of users;
- means of describing size category(ies) to ensure correct selection of PPE by users (EN 13921:2017, 2017).

11.3.1 Traditional anthropometry and databases

Anthropometry is a study and measurement of the physical dimensions and mass of the human body and its constituent (external) parts. Taken from the Greek words *anthropos* (human being or Man) and *metron*, to measure (ISO 15535:2012, 2012).

Standard anthropometry (traditional contact methods) dimensions are measured from one point on the body (or fixed surface such as the floor) to another. In the case of circumferences, the dimension is measured around a part of the body at a specified level. These measurements are taken with specific instruments: anthropometer, measurement tape, caliper.

Sufficient extensive anthropometric studies are required to ensure the fit of produced uniform/working clothing series, with sufficient understanding of the measurements for that part of the population for which the designed PPE is to be produced. For example, when designing firefighters' clothing, it is appropriate to establish only morphological characteristics of firefighting and rescue workers. If work is not associated with a severe physical load, the sizing system of the common population will also be essential. The design of the garment may then be used for existing anthropometric databases, which are collections of:

- individual body measurements (anthropometric data—dimensional measurements, such as heights, lengths, depths, breadths, and circumferences, of the human body and its component parts),
- background information (demographic data—such as sex, dwelling or working place, occupation, education).

Anthropometric databases are used to describe members of the user population (population segment or segments for whom a technological design is intended) and/or population segment (group of people having one or more common background characteristics that influence their anthropometric distribution) recorded on a group of people (the sample) (ISO 7250-1:2017, 2017).

It is desirable that the number of subjects needed for a database be established using a statistical power formula based on the accuracy of results desired by the investigator (ISO 15535:2012). However, in reality, the selection of subjects is often influenced by various factors, such as a population size, number of people who agree to participate, and cost and period of time required for the investigation. Sample size shall be sufficient to estimate the value of a given measurement in a specified group. In most cases, anthropometric data for technological design are of interest at the 5th and 95th percentiles.

The majority of the body measurements and derived measurements will serve to fulfill multiple design and sizing purposes and will be used in some of the following categories: basic body descriptors, key measurements, garments including personal equipment, clothing manikins, load-carrying systems, head and face equipment, footwear, handwear, workspace and body clearance, vehicle accommodation, biomechanical body links, computer manikins, and body templates.

11.3.2   *User population characteristics*

The standard figure of the human body is a spatially simplified form of anthropometric classified bodies, for which a set of the rest body measurement values are proportionally calculated, so that this form is representative of a certain age-sex group, which would be a commercially or otherwise significant number of consumers (for example, representing a separate type at least 0.1% of a morphologically homogeneous population of individuals). Parameters of standard figures necessary for industrial production and trade are reproduced in reference tables of proportionally calculated measurements, in standard figure dummies and in computerized clothing design input databases.

Body measurement [ISO 8559; ISO 7250] is a determination of human body measurements or other measurement values (angle, mass), which measurement, scan, or calculated numeric values are used to achieve anthropometric fit required for the wearers of clothing (ISO 8559-1:2017, 2017; ISO 8559-2:2017, 2017; ISO 7250-1:2017, 2017).

A nonstandard figure (type) is a figure of which the measurements differ from the corresponding size by individual but important secondary measurement values for the fit of the garment. These measurements have such large deviations from standard size values that they can significantly impair the fit of the finished clothing if patterns were constructed by standardized measurements. For example, nonstandard figures are stately, stooped, reclusive, tilted, thinned, plateau, lordotic, delorotic, wide-hip, shrugged shoulders, O leg, X leg, etc. figures.

The regular male body (type by drop value) (normal-waist male standard figure) is a male standard figure in which the difference of the chest and waist circumferences is within 12 ± 3 cm. Usually the waist indication is labeled with the first letter "R" of the English word "Regular."

A corpulent male body (type by drop value) (men figure with developed belly) is a male standard figure in which the waist circumference is larger than the chest circumference exceeding 3 cm. In this case, usually the waist indication is labeled with the first letter "C" of the English word corpulent, following the recommendations of ISO/TR 10652.

The EN ISO 15537:2004 (Principles for selecting and using test persons for testing anthropometric aspects of industrial products and designs) standard establishes methods for determining the composition of groups of persons whose anthropometric characteristics are to be representative of the intended user population of any specific object under test. It is applicable to the testing of anthropometric aspects of industrial products and designs having direct contact with the human body or dependent on human body measurements, e.g., machinery, work equipment, personal protective equipment (PPE), consumer goods, working spaces, architectural details, or transportation equipment.

SLIM BODY TYPE—person for whom at least two width measurements (preferably shoulder width and hip breadth) and two depth measurements (preferably chest depth and abdominal depth) are smaller than the figure representing the 25th percentile or, where this figure is not available, the average value of the 5th and the 50th (mean) percentile for the population in question.

CORPULENT BODY TYPE—person for whom at least two width measurements (preferably shoulder width and hip breadth) and two depth measurements (preferably chest depth and abdominal depth) are bigger than the figure representing the 75th percentile or, where this figure is not available, the average value of the 50th (mean) percentile and the 95th percentile for the population in question.

11.3.3 Dynamic anthropometry

Anthropometric standards (EN 13402, ISO 7250) characterizing types of human body sizes used in the design of clothing, including protective wear, are based on data derived from measurements of male and female bodies in the standard posture. Therefore, in the various methods of design of clothing, the design of the structures is based on the measures obtained in the static position of the human body. However, a person spends relatively little time of his life in such a static position and performs various kinds of commonplace or working specific movements, such as gait or walk, squatting, bending, stretching arms, etc. (Koblyakova et al., 1974). Therefore, results from standard posture may not always fully satisfy requirements of pattern makers who are to deliver clothing patterns for a man in motion, rather than to a tailor's dummy with "fixed posture" (Амирова and Сакулина, 1985).

There are standards that define the movements for the development of special projects (ISO 7250-1:2017, 2017), but the ergonomic standards only describe different modes of motion, e.g., arm range, without any size references, just percentile type. Due to this, correlation between the two measuring systems is missing (Loercher et al., 2017a).

Studies on anthropometric dynamics are based on observations that human body measures depend on body condition or posture, and, as the human moves, the distance between individual body points measured on the surface of the body constantly changes as well. For example, the studies carried out in Soviet times have revealed significant changes (increases/reductions) in the size of body measures (lengths, circumference) at the maximum positions of the head, torso, and extremities, inherent in the working positions (Koblyakova et al., 1974). Different research has been carried out at different times, differing both in terms of the size and gender of groups in question, the methods of measurement used, as well as the selected body postures and their amplitudes. An earlier study carried out in 1968, including CSR (Czechoslovak Socialist Republic) and the GDR (German Democratic Republic) populations, as well as the USSR (Union of Soviet Socialist Republics) participating specialists, resulted in the compilation of various dynamic indicators. Large groups of people (340 individuals, 170 females and 170 males) were studied using traditional anthropometric manual methods; measurements were made in positions such as deep inhalation, back torsion, seated posture, chest torsion, bending, arms upwards (hands together), arms frontwards (wrists together), hands bent in the elbow at 90-degree angle, hands horizontally to the front at 90-degree angle, leg on a chair, and squat. In today's research, using contactless measurement methods, for example, changes in the measurements of 24 men's upper bodies in five movement positions were found, which show both significant decreases (negative, -22% chest width) as well as increases (positive, 27% back width) (Hayes and Venkatraman, 2016). For the purpose of exploring the dynamic suitability of diving suits, test persons were scanned at five positions specific to the diving process and testing of prototypes in real-world conditions was performed (Naglic et al., 2017). By analyzing and classifying work positions to identify the most characteristic forms of motions, as well as to define poses so that they can be reproducible, 10 different body poses were scanned (Loercher et al., 2017b). For the study of the lower body, women's conditional mundane poses such as a standing position, 120-degree knee bending, one-step climbing, and sitting with knees bent at 90-degree angle were selected and the dynamic measurement deviations from the standard position measurements were analyzed (Choi and Ashdown, 2010). Another lower body study examines changes in values within the larger range of movements and with a more detailed set of measurements when examining changes in the leg surface by comparing the results obtained in a standard posture and at five different heights in leg positions (Xiao and Ashdown, 2013).

In various studies, the measures obtained in the static position may be called static, while those obtained in another body or limb position (based on basic anthropometric methods) the dynamic ones. With regard to the design of ergonomic spaces, furniture, instruments, and appliances, dynamic anthropometry is also called functional or ergonomic anthropometry.

Professionals must be familiar with human anatomy and functions to understand the body in its kinetic positions and to be able to design clothing that moves properly along with the body (Ashdown, 2011). The adult body consists of 206 bones, most of which are joined by moving joints, creating many combinations of positions that are employed in everyday movements. In addition, their usability and frequency of use depend on the task to be performed and areas in which the person works. There are also a large number of variations for the potential volume/amplitude of movements between individuals in the population that are associated with the total morphological features as well as the differences between the members of the population: age, race, gender, health status, physical fitness, body proportions, and temporary effects of exhaustion—these can all be significant factors that can affect movement (Ashdown, 2011). For example, sports and active leisure design studies have shown that as a result of the movement, the skin of the human body significantly extends on the knee—35%, and elbow—in 45% of zones. Depending on the movements that can be made (movement segmentation of legs, lower back, thighs, shoulders, belly, hands, etc. in the human body), the wearer stretches in different ways, and the stretching of the skin in different areas ranges from 10% to 50% (Hayes and Venkatraman, 2016).

It should be taken into account that the resulting dynamic anthropometry data will be indicative and their use in the design of clothing structures will depend on the type and function of the particular clothing. In the design of protective clothing, when selecting a motion program for the acquisition of dynamic indicators, it would be necessary to select those that are typical of the daily working movements and are most affected by the distances between the anthropometric points. In determining and researching dynamic indicators, the following tasks must be addressed:

- The selection of anthropometric points,
- The selection of a package of movements,
- Determination of measurement location in dynamics,
- Determination of the type of statistical processing of dynamic indicators,
- The designation of the direction of utilization of the results obtained,
- Selection of dynamic indicators required for the construction of patterns (Koblyakova et al., 1974).

For testing and analysis of dynamic body changes, different postures should be measured and analyzed. For example, Fig. 11.1 displays body reproductions obtained by 3D scanning in standard standing position and in step-up, bending, and squat positions. The maximum possible bending forward position allows a study of the changes of the measures in back and arm areas, and the squat and step-up postures allow the same in leg and seat zones. In the first posture (see Fig. 11.1A), an individual was standing in an upright position, without hunching shoulders, stretching out the spine and straining the abdomen, and not touching the body with hands or palms. Palms are slightly bent and elbows spread. The upper arm muscles are unstrained. Palms are fisted, thumbs pressed to other fingers. Legs are positioned apart from one another, standing on foot-marks. The gaze is turned straight forward. In the second posture (see Fig. 11.1B), the right leg is placed on the special platform. The gaze is directed straight forward. Hands are bent in and elbows are at an angle of 90 degrees. Forearms and wrists are directed horizontally forward. In the third position, bending (see Fig. 11.1C), the top of the body is tilted forward to the maximum. The chin touches the chest. The arms are held vertically downwards. The legs are not bent at the knees. In the fourth pose, a squat (see Fig. 11.1D), a full squat was made. The gaze is directed straight forward. The hands are bent in and elbows are at an angle of 90 degrees. Forearms and wrists are directed horizontally forward. Common conditions for all the

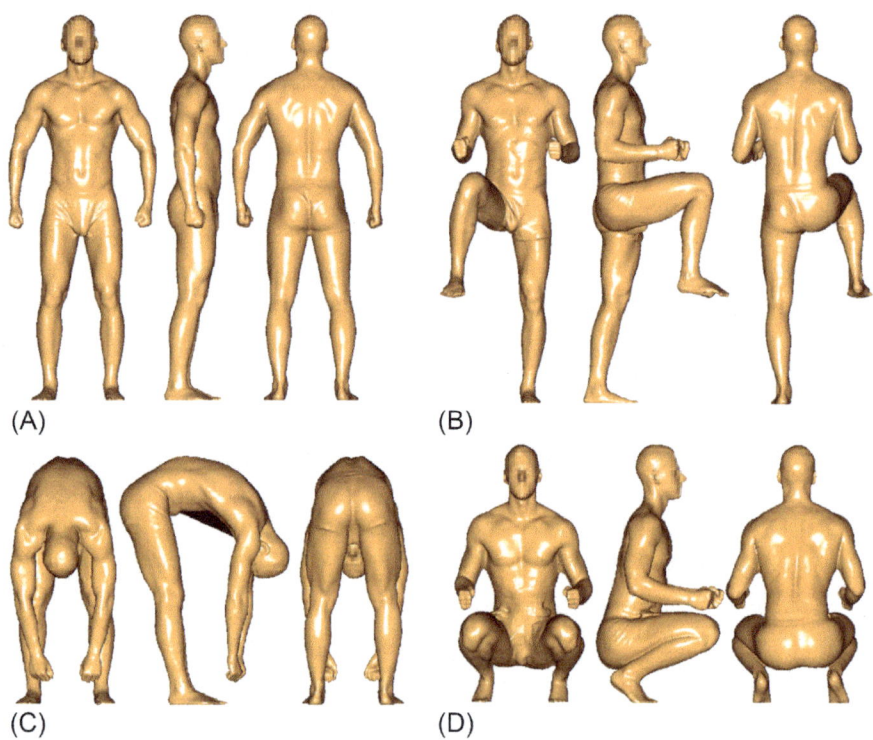

(A)   (B)

(C)   (D)

Fig. 11.1 3D scanning dynamic postures. (A) Standard, (B) step-up, (C) bend, and (D) squat.

postures are palms fisted, holding thumb close to other fingers; standing still during the scan process; and avoiding taking deep breaths.

11.3.4   Sizing for protective clothing

In planning the manufacture of uniforms, decisions must be made as to how many and in what size a particular model series should be manufactured, how they should be labeled, and to what body dimensions the garment sizes should correspond. The purpose of anthropometric parameterization is to introduce garment size classification for mass production clothing, so as to representatively depict wearers' body figure diversity. It is in the interests of manufacturers (design time and costs, logistics, etc.) and buyers/procurement service alike to confine themselves to a minimum number of garment sizes and to use a possibly less complicated garment size classification.

The main dimension for garment sizing is a measurement that introduces one of the classifying dimension scales of clothing sizes by the intervallic division of its value changes, and the clothing products are directly or indirectly (in a coded way) marked with their standardized values. It is also used for the calculation of other measurement values of the standard figure and thereby for the construction of patterns of the correspondingly marked garment. There are control dimensions of head girth (headgear), chest/bust girth, underbust girth, waist girth, hip girth, body height, arm length, leg length, foot length (socks, footwear), hand girth and length (gloves), etc. The determination of these main anthropometric parameters and their accuracy affect all future work in the development of clothing (see Fig. 11.2). If the size is determined incorrectly, the potential wearer is directed to an inappropriate size of clothing in the classification.

According to standards, protective clothing should be marked with its size based on body dimensions measured in centimeters. The size designation of each garment shall compose the control dimensions (ISO 8559-2:2017, 2017). The size designation system is required especially for labeling. Body dimensions for protective clothing sizing according to a type are as follows:

- Jacket, coat, vest: chest or bust girth and height;
- Trousers: waist girth and height;
- Coverall: chest or bust girth and height;
- Aprons: chest or bust girth, waist girth, and height;
- Protective equipment (e.g., knee pads, back protectors, torso protectors): select the relevant measurement:
  - chest or bust girth, waist girth, and height;
  - body height;
  - waist to waist over the shoulder length.

The manufacturer can also designate additional measurements, e.g., the arm length, the inside leg length, or the hip girth for women's garments (ISO 13688:2013, 2013).

It is recommended to describe the closest type of figure to the recipients of the serial manufactured clothing products by means of size titles recorded in or on their pictograms in centimeters. In pictographic markings, the characteristics of clothing

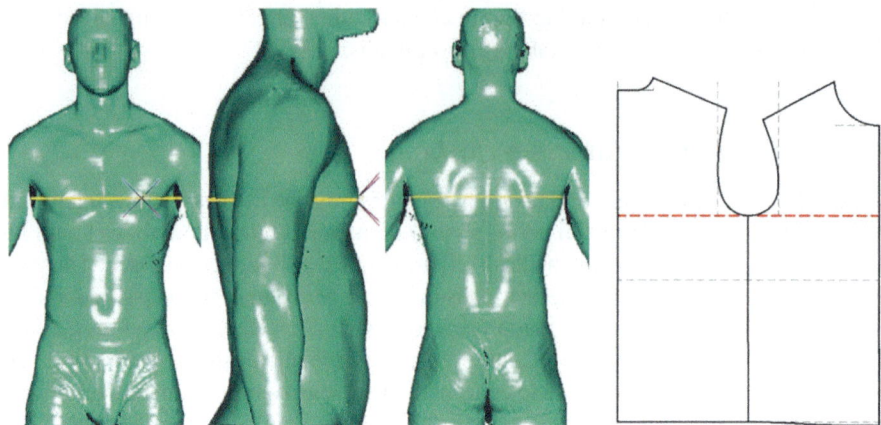

Fig. 11.2 Upon construction of clothing, chest girth is displayed as part of the horizontal width of the structure.

sizes for consumers are less misunderstood because they are not affected by the ignorance of the linguistic or encoding meanings of the labeling.

11.3.5 Usage of 3D scanning and virtual prototyping

There are possibilities to eliminate the deficiencies in PPE design and supply processes, which are associated with body measurements, if body measurement analysis of a large group of people is done using a noncontact measuring method. The significant task is to improve manufacturer data tables of the end user measurements by automated obtaining of a large number of measurements (Song and Ashdown, 2015). It has to be done, considering that so far patternmaking is performed using a body measurement database, which has not been updated for years, and not by taking any additional end-user body measurements.

Anthropometrical data can be acquired with different tools. Traditional methods use different manual tools (measuring tape, anthropometer, a.o.). As the technologies develop, new tools are created, and/or the existent ones are improved.

Photo measuring methods are fast and effective, but the processing of data is time consuming and labor intensive. A relatively new tool (approximately since 1980; Fan et al., 2004) in anthropometry is the 3D scanner. Considering the advantages of 3D scanning, the scanning technologies are being developed and improved. Most of the scanners not only can create a 3D image of the human body, but can also read the *x*, *y*, and *z* coordinates, thereby acquiring precise information about the human body and its volumes (EN ISO 20685:2015). Each method has its advantages and disadvantages. In spite of the fact that laser scanning has been recognized as the most precise method and the gathered results are the most extensive (human body measurement data, a 3D virtual mannequin, a reflection of the actual texture, surface relief measurements, etc.), the light projection method is used more widely in the garment production industry, since the equipment is much cheaper than a laser scanner.

The noncontact measurement method has advantages in obtaining body measurements of a big group of people and further use of the data in different garment design. First of all, it is a reliable data collection device showing sufficiently close correlation with manually gained and standard data. Secondly, a large number of body measurements (153 somatic measure values using the 3D anthropometrical scanner Vitus Smart XXL (©Human Solutions GmbH and VITRONIC GmbH) with Anthroscan software are obtained relatively rapidly (one scanning ~12s) and they are not influenced, for example, by the skills of a measurer, the movements and breathing of a measured person, and the compressibility of soft tissues. 3D body reproductions can be used for further profound studies in different 3D systems, for example, virtual fitting of clothing.

Measurements that are not obvious using manual methods are accessible within 3D methods. For example, inseam length (see Fig. 11.3) measurements may be applicable to review size scales of manufacturers to use them as secondary key dimensions as recommended in the standard (EN 13402-3:2014, 2014), and not body height measurements. This solution is negotiable regarding observations in research, when leg lengths or proportions differ within one body height group, but manufacturers leg length intervals of body height groups are insufficiently wide (currently 6 cm). By actual measurements leg length ranges in one height group occurs from 5 to 11 cm.

Secondly, automatically obtainable crotch length measurements at waist and at waistband (see Fig. 11.4) allow one to analyze and compare the human body surface

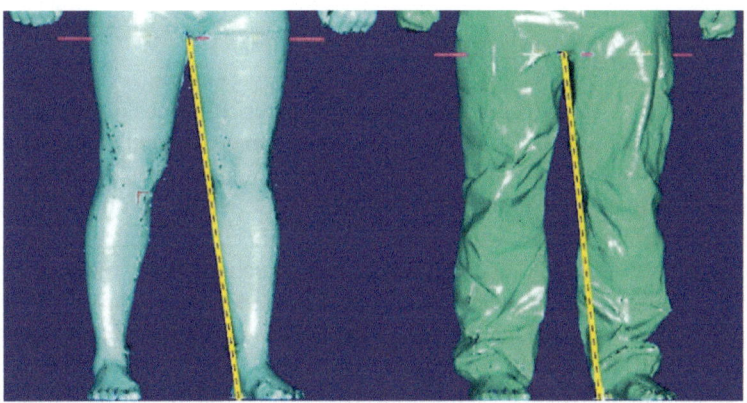

Fig. 11.3 Inseam lengths in underwear and in trousers.

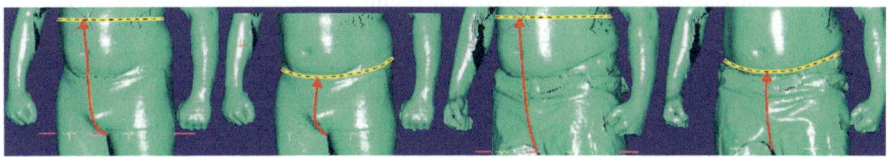

Fig. 11.4 Crotch lengths—at waist and at waistband, in underwear and in trousers.

with finished garments, for example, to determine potential excess volume in the crotch area that occurs as a result of lowering the waistband or belt level. According to measurements taken in the research—crotch lengths of finished products, undressed and dressed scanned bodies—it is concluded that trousers are designed for wearing the waistband at traditional anthropometric waist level, but already dressed body measurements show that all the excess volume of the garment is placed in the crotch area when the wearer chooses to lower waistband level.

The potential of the noncontact method lies in the arrangement of size systems for large target groups, when the obtained information is centrally processed and used further for supply implementation preventing disparities in sizing systems, which can occur if the necessary sizes are indicated by each wearer. For clarity, labeling of clothes in accordance with EN 13402-3:2014 (2014) [Size Designation of Clothes. Part 3: Body measurements and intervals] interval pictograms supplemented by letter designations may be used.

11.4 Design of protective clothing

Construction of garment is a material composition of the garment, which is characterized by the number of cut and other details, by materials, by the package of textile materials, by the configurations, sizes and connections.

11.4.1 Ease allowances

The ease allowance is an algebraic difference of numerical values or (less frequently) its relative expression as a percentage of the control measurement characterizing the outer surface of the finished garment and its decisive standard figure measurement or the corresponding customer's figure measurement. Depending on the location there are chest/bust circumference, waist circumference, hip circumference, upper arm circumference, armhole depth, etc. ease allowances, which numeric values for example in the case of tights, corsets and extensible garments tend to be negative. Functionally distinguishable physiologically minimal, movement comfort, silhouette, etc. ease allowances [ISO/TR 10652].

For example, the analysis carried out by research institute in Moscow (Центральный Научно-Исследовательский Институт Швейной Промышленности) on special clothing structures (male and female) has revealed that the ease allowance of bust is required within the limits of 9–17 cm. This allowance 5–8 times exceeds the size interval for half-bust girth (which is 2 cm). This has allowed to increase the interval of intermediate sizes and, accordingly, reducing the number of sizes without reducing consumer satisfaction with the fit at the same time. As a result, it is concluded that reducing the number of sizes allows for a 40% reduction in the cost of materials and labor costs (Амирова and Сакулина, 1985).

11.4.2   Dynamic allowances

However, the design of protective clothing used in various types of work, where the wearer carries out his characteristic body movements, also requires the use of dynamic ease allowances ensuring the fit of clothing during these movements. If databases of body measurements gained in standard stand position are widely available, there is relatively no common information sources about human body dynamic parameters or changes in body measurements due to movements.

Dynamics can be characterized both by decrease and increase of body measurements, taking into account the wide range of human body movements and uncertainty of displacement directions of the anthropometric points during the movements. In addition, it can be affected by various factors, such as a person's belonging to a particular age-sex group, individual characteristics of body composition, physical fitness, etc.

For motion description and research, it is initially worthwhile to study body size changes (increases and decreases of lengths and circumferences) at the extreme head, torso, and limb states. With such an approach, the patternmaker will be able to apply not only the required mean values of measurements, but also the maximum increments obtained in the extreme position measurements (Preedy, 2012).

Considering the use of dynamic indicators in clothing pattern design, it should be noted that, if in the case of lingerie and sports clothing, knitted fabrics are used to ensure freedom of movements, then in design of protective clothing outerwear made from nonrigid fabrics without the use of appropriate ease allowances, expansions/extensions may be allowed in certain garment areas (McCann and Bryson, 2009a). The obtained dynamic indicators can be used to a reasonable extent on the ease allowances associated with significant body measurements for patternmaking, such as waist and chest circumferences, back width, length and slope, the distance from the back armpit fold to the waist, the upper arm and elbow circumferences, the distance from the back armpit fold to the wrist, the hip/buttock circumference, etc. (Sakharev and Boytsov, 1981).

11.4.3   Design solutions

In addition to ease and dynamic allowances, there can be various constructive and technological solutions, which may be applied in cases, when, for instance, due to various reasons, it is not possible to include dynamic indicators in ease allowance values completely. In order to ensure free movements, one may use, for instance, pleats, inserts (also elastic), wedges, slits, gaps or openings, anatomical seams, and preshaped sleeves and trouser legs. Other additional solutions to ensure easiness of moves are differentiation of the silhouette of the garment, or various types of cut.

However, in the design of work, sport, and special clothing, the use of a loose silhouette considered as a suitable solution is also assessed only in cases involving aspects such as work safety and loads that may be caused by additional weight of clothing and the physical volumes of the products. The design of protective clothing shall guard against parts of the body being uncovered by expected movements of the wearer (ISO 13688:2013, 2013). Clothing of an excessively loose silhouette can also cause undesirable effects due to overlapping shapes, rather than improving the dynamics of movements. For example, while wearing working or special clothing with a free silhouette, it is not permitted by the working snag behind, as well as in addition to the loss of fatigue in the weight of clothing. Similar conditions apply to sports apparel, whose designs must contribute to the dynamics of the athlete's movements, and this depends heavily on the type of materials chosen.

The use of elastic materials may serve as an essential solution for compensating for dynamic indicators in clothing constructions. However, it will not always be allowed due to the application of the clothing and the protective properties prescribed. Working and special clothing aimed at protecting the human body from different affectors, such as dirt, temperature, chemicals, and radiation, is not commonly made of elastic materials. In turn, everyday and also sports clothing, thanks to elastic materials, can combine both the comfort of movement and favorable visual appearance or fashion factors. The comfort of the necessary movements can only be provided with elastic material assistance in products where the values of the dynamic indicators to be used in the structures do not exceed those of elasticity or elongation. If, however, they are exceeded, it is worth increasing the ease allowance for free movements or using any of the additional constructive solutions.

A condition that must not be forgotten is that, by putting on additional protective clothing on top of a basic clothing layer, it must be possible to access the pockets in the lower layer through openings in the top clothing (McCann and Bryson, 2009a). In achieving the desired anthropometric proportionality, maintaining the balance of protective clothing parts and elements already at sizing must not be forgotten. Or, for example, the size and proportional disposition of the pockets must be equally suitable for wearers of different size ranges.

11.4.4   Adjustment solutions

Additional components of protective clothing, which can at the same time serve as components of the adaptation of the exterior shapes of clothing and the fitting of close-fitting elements in certain areas of the body, provide a variety of adjustment solutions. Their use is helpful for both improving external appearance and adapting to human body features and, in addition, to safeguarding protection by tightly loading clothing to potentially endangered areas of the body. For example, the possibility of a regulation at waist level, as well as a cardigan at the downside level with a draw cord, allows the upper-clothing silhouette to be adjusted by the wearer, which is particularly important for female representatives wishing to produce a more feminine appearance when wearing mostly unisex-type clothing. In addition to narrowing and expanding, the possibility of adjustment may be important in areas such as sleeve ends or bracelets, collars, pants and rays of the beams, achieving the desired fitting with a flap, draw cords, or button strip. In addition, in the case of coated protective clothing, these adjustments may be placed only in the inner lining layer, for example by introducing immersed and tightly exposed bracelets, or at the end of a kind of self-propelled bracelet/anklet.

In designing different types of fastenings, it is important to reflect on factors such as the presence of fastening elements in terms of nonloss of resistance and functional properties, as well as restoration or repair options. The various elements of the fastenings, such as zippers, plastic fastenings, push buttons, lines, etc. are designed to be sufficiently robust, for example, to be easily grasped with gloves that are particularly essential when wearing clothing in cold weather and without giving the worker protective gloves (McCann and Bryson, 2009a). Individual solutions may be used to improve the compatibility of protective clothing with PPE products, such as sticking hooks, loops, buttons, saws, belt loops, etc. in the required areas.

The standard of general protective wear requirements states that the design must promote proper positioning of PPE on the user and ensure that it stays in its place for the intended period of use, taking into account environmental factors along with movements and poses that the wearer can take during work or other activities. For this purpose, appropriate means adequate size ranges and adequate adjusting systems, so that protective clothing would be adjustable to a user's morphology (ISO 13688:2013, 2013).

11.5 Fabric parameters affecting the fit

Recent studies show that comfort in wear is the most important property of clothing demanded by users. Comfort in wear is especially important for athletes, physical workers, rescuers, etc. A fundamental understanding of human comfort and knowledge of how to design textiles and garments to maximize comfort for the wearer is therefore essential in the clothing industry (Song, 2011b). In addition, comfort of clothing can be described in several respects, for example, heat balance (increase or decrease in the heat content of the human body caused by an imbalance between heat production and heat loss; EN 13921:2017, 2017), energy metabolism, tactile comfort, body movement comfort, etc. This understanding reflects the demands of a human body: anatomical (morphology, locomotor system, abnormalities), physiological (water regulation, thermal regulation, physical sensation), as well as psychological considerations (duration of use, wearability) (McCann and Bryson, 2009b).

The standard for PPE (EN 13921:2017, 2017) contains guidelines not only for characteristics of ergonomics, associated with anthropomethylene and biomechanical effects between the PPE and human body, but also on thermal works and human sensory interactions with the PPE, such as vision, hearing, smell and taste, and skin contact.

The choice of suitable textiles is largely crucial for best protection, while ensuring the best level of thermal comfort at the same time. General thermal comfort may be defined as total subjective satisfaction with the thermal environment, based on whole body sensation (EN 13921:2017, 2017). The materials chosen for clothing, their design, and the PPE system may make it difficult to exchange heat between the wearer and the external environment, thereby increasing the thermal insulation and/or evaporative resistance (EN 13921:2017, 2017). The choice of materials should be guided by a variety of factors, such as the nature, level, and duration of physical activities to be carried out on a day-to-day basis as well as external environmental conditions.

Given that maximum protection may be worn for clothing covering the largest part of the human body, inappropriate materials may cause comprehensive or local discomfort. Increased thermal insulation and evaporative resistance may result not only in overheating but also increased sweating, which can lead to sweat-soaking and clothing adhering to different areas of skin. This may result in additional unpleasant feelings (such as irritation, tickle, pressure), as well as limited movements, as the clothing can no longer be freely passed over the surface of the skin during movements. The characteristics of the materials to be taken into account for thermal comfort are thermal insulation, water vapor resistance, water vapor permeability, air permeability, water absorption, and desorption (EN 13921:2017, 2017).

11.5.1   Weight and bulkiness

Initially, the use of relatively large values of ease allowances in developing loose clothing without any limitation of movements may seem a simple solution to provide the fit and also the dynamic fit of clothing, but it is essential to look for a balance without making the clothing excessively wide, which may, on the contrary, negatively affect the wearer's comfort. For example, it can result in excessive folds that may limit movements of the body, and the excessive volume can serve as a risk factor to catch on work equipment or, even worse, on moving mechanisms. Heavy, close-fitting, stiff or bulky material can excessively impede the bending of joints and restrict working positions and movements. Materials and products that resist movement elastically and require a continuous muscular effort acting against the elastic recoil to maintain a particular joint position can cause fatigue and injury (EN 13921:2017, 2017).

The design of very wide and loose clothing can increase the total weight, which leads to the fatigue of the wearer carrying it all on the body during the work. In addition, in most cases there is no possibility of choosing lighter material for solving the clothing weight issue, considering the various protective properties that the materials of protective clothing should have, preserving the wearer from different hazards such as hits, cuts, dust, harmful substances, radiation, etc.

PPE designers must have knowledge of *mass distribution*. Peripheral parts of the body are more susceptible to added mass than the trunk, because of the increased moment. Hence, an additional weight is best worn on the trunk and as close as possible to the body center of gravity. This means that the waist is the body location where such a weight is best carried. However, this is often not practical because PPE is normally made to be worn on specific body parts in order to protect against a specific risk. Asymmetric loading should be avoided as far as possible (EN 13921:2017, 2017).

11.5.2   Clothing layer interactions

Frequently protective clothing is worn above the base layers of clothing, and wearers are also provided with multilayer protective clothing systems for more complete protection. In such cases, the selection of appropriate textile materials will be critical, when possible interactions of the interlayers are taken into account. Additional considerable factors are also the vapor resistance and air permeability of the textile materials, which not only guarantee the wearer's thermal comfort but also prevent the creation of conditions where the clothes "stick" to the surface of the skin and no longer slide over it because of the sweating, limiting movements or creating "thickenings" in separate body areas.

The friction of PPE against the body can give rise to redness (erythema) and, if prolonged, to abrasion. Depending upon the duration of wear, compression of the skin can cause unsightly marking. Where blood vessels or nerves pass close to the body surface or are liable to be trapped against hard (bony) body parts, then further adverse effects can occur. For example, straps passing close to the neck/shoulder junction can compress nerves or occlude blood flow. Abrasion or compression might cause irritation or reduce the acceptability of the PPE. Pulse points in the neck, groin, and wrist are obvious possible areas for concern in relation to circulatory effects (EN 13921:2017, 2017).

Low tensile and plowing fabrics have a higher risk of mechanical damage. This may occur in the wear and wash process, when the material is exposed to tensile and deformation. Mechanical damage may be attributable to the design, wear conditions, or inappropriate grooming conditions of the fabric itself (Siegert, 1964; Juodsnukytė et al., 2006).

Clothing textiles in everyday wearing are exposed to a range of tensile strength of different sizes, resulting in a possible deformation of the material. Most frequently, such deformations occur in parts of the clothing exposed to higher loads, in the areas of knees, elbows, and buttocks. In these zones, approximately 10%–12% of the maximum load held by the textile material shall be applied to the material. As uniforms are worn in various weather conditions, this increases or decreases the fabric relaxation period. The permanent deformation deteriorates the aesthetic appearance of clothing by creating prints (Juodsnukytė et al., 2006).

Wear is a mechanical deterioration of the fabric by rubbing the fabric against any other surface. Abrasion is encouraged by excessive rubbing or cleaning of the surface of the material. Many garment parts, such as collars, bracelets, and pockets, are exposed to serious wear during use, which limits their worklife. The life of the garment depends on the individual wear, cleaning, and washing process of each person. The wear of textile materials is influenced by many factors, such as the subtlety of fibers, the type of threads, the fabric weave, etc. The rub creates the scaremongering of the filaments in the thread. Initially, abrasia modifies the top of the fabric, then the internal fabric of the fabric. One of the results of wear is the gradual separation of fibers from the thread. The broken ends of the threads appear on the top of the cloth, which causes a different reflection of light, resulting in a change in color.

During wear, the material is exposed to various types of scrubbing. First, it is the friction of different clothing fabrics against each other or the friction of the cloth against itself. As the wearer simply walks, friction occurs with the trouser beams rubbing against each other, as well as moving the hands along the body during walking or running. There is also a friction between the top of the underlying cloth and the inside of the superseers, the lining or the lining of the casing bag. As a soldier performs a variety of day-to-day activities, when crawling, creeping, sitting, etc., the top of the material is exposed to friction against rough surfaces: land, asphalt, stones, grass, etc. In contact with land, the hard particles of dirt and sand that can cause microscopic damage to the fibers may arrive on the fabric. As a result, dirty material is exposed to faster aging during wear (McLaren et al., 2015).

The effects and properties of the material, such as strength, are deteriorating as a result of wear. Such damage may occur in the wear and care process. In many cases, it is very difficult to determine the exact cause of wear, but usually it depends on the occupation of the wearer. It also depends on the length of the threads in the weave: the longer the covers, the faster the fabric will wear, for example, the satin cloth will wear faster than the twill fabric (Anthesis, 2015; Kaynak and Topalbekiroğlu, 2008; Özdil et al., 2012).

Aging is attributable to all slow and irreversible changes in material properties arising from their instability or environmental effects. These changes may affect the chemical structure, material composition or its physical properties. Factors contributing to the aging of the material: heat movement (exchange), oxygen, sunlight, water, radiation, mechanical stress, biological environment, chemical agents. Material longevity is the time in which certain material quality has deteriorated to a certain critical value.

11.5.3   Clothing care impacts

As the result of clothing care processes, for example, soaking, washing, dry cleaning, or other activities, the shrinkage of textile materials may occur, when linear dimension changes (usually shortening) is observed.

Clothing care (washing, bleaching, ironing, drying, chemical cleaning) and maintenance, such as the removal of ducking, are processes that allow long-lasting clothing but contribute to the inevitable deterioration of the material. How well the consumer understands the grooming label or respects it, and the frequency of washing has a significant impact on the aging time of the garment. Nowadays, clothing is more depreciated in the care process rather than wear, the former taking place by using inappropriate grooming rules, excessive abrasive cleaners, and softeners (Anthesis, 2015; McLaren et al., 2015).

The system produces microscopic material damage, which is not immediately noticeable as a shrinkage or loss of color each time when washing garments in a washing machine. The main cause of fabric wear is mechanical washing in centrifuge. Over time, the tensile strength of the fabrics is not changed, each washing cycle reducing the wear of the garment. Washing and spraying temperature affects the strength and linear size of the material. The contracting of garments is largely encouraged by mechanical washing/drying in centrifuge (ISO 13937-2:2000, 2000).

11.6   Test of PPE conformity

Whether a new uniform for inclusion in the PPE or an improved existing uniform is developed, it is necessary to test its samples for compliance with the task to be carried out and, if necessary, to make improvements. To search functionality and the physical comfort of end-users and the ability to perform tasks in uniform a survey is needed.

In the design of special and work clothing, functional requirements have always been emphasized. For garments under design to meet specific work conditions, materials, and design solutions to protect the user from hazardous working environments, climate conditions and microclimate parameters are assessed, the design elements of the garment and their adjusting possibilities—fasteners, pockets, tightening, etc.—are selected according to the type of work in question, and an end-user training or information provision procedure is planned. In view of functional requirements, workwear designs or elements are very easily united, which in a full-scale production can save resources in both the designing and the production.

The study (see Fig. 11.5) should be started with a survey of end-users, the purpose of which is to identify shortcomings of the current uniform available to the wearers and to determine potential directions of the necessary improvements from the standpoint of the wearers. The study includes analysis of anthropometric data, the obtaining of body measurements. At the end of the study, fit tests are carried out in accordance with standard EN ISO 13688:2013 "Protective Clothing. General Requirements."

The two stages of the study, which are anthropometric studies and interviews, should be conducted during uniform testing. The end-user surveys should result in identification of the most important shortcomings of the PPE uniforms, expressly pointing out the most problematic.

In designing, feedback from end-users is often used to evaluate the dynamic fit. Such a subjective assessment of product properties depends on the assessor's experience, as well as traditions and habits. In fact, ergonomic tests should be carried out by experts, as set by the standard (ISO 13688:2013, 2013), determining that the test or examination is to be carried out by one or more experienced evaluators or experts previously acquainted with the manufacturer's information for further investigation of the protective wear, manually and visually. Such an examination should give answers as to whether the wearer can stand, sit, walk, and climb the stairs; is it possible to raise both arms above the head; can one easily lean down to pick up small items (e.g., a pencil). Also, significant requirements are set out for sleeves and trouser legs, the length of which must not encumber movements of arms and legs; clothing must not be loose enough to flutter and move in a free and cumbersome manner; there must be no points for sudden and unexpected openings between or inside components of the clothing; no unreasonable movement limitation in any of the joints is permissible.

11.6.1 Performance tests

Standard ISO 13688 (ISO 13688:2013, 2013) Annex C (informative) specifies how some basic ergonomic features of protective clothing shall be checked using simple practical tests. Ergonomic assessments are intended to reduce the risk of hazards to the user due to such parameters for example as poor design and fit, poor compatibility with other related items of PPE and poor compatibility with other items of clothing. Assessment points:
- Clothing free from harmful features (sharp or hard edges, protruding wire ends, rough surfaces or other items);
- Appropriate ease of putting on and removing the clothing;

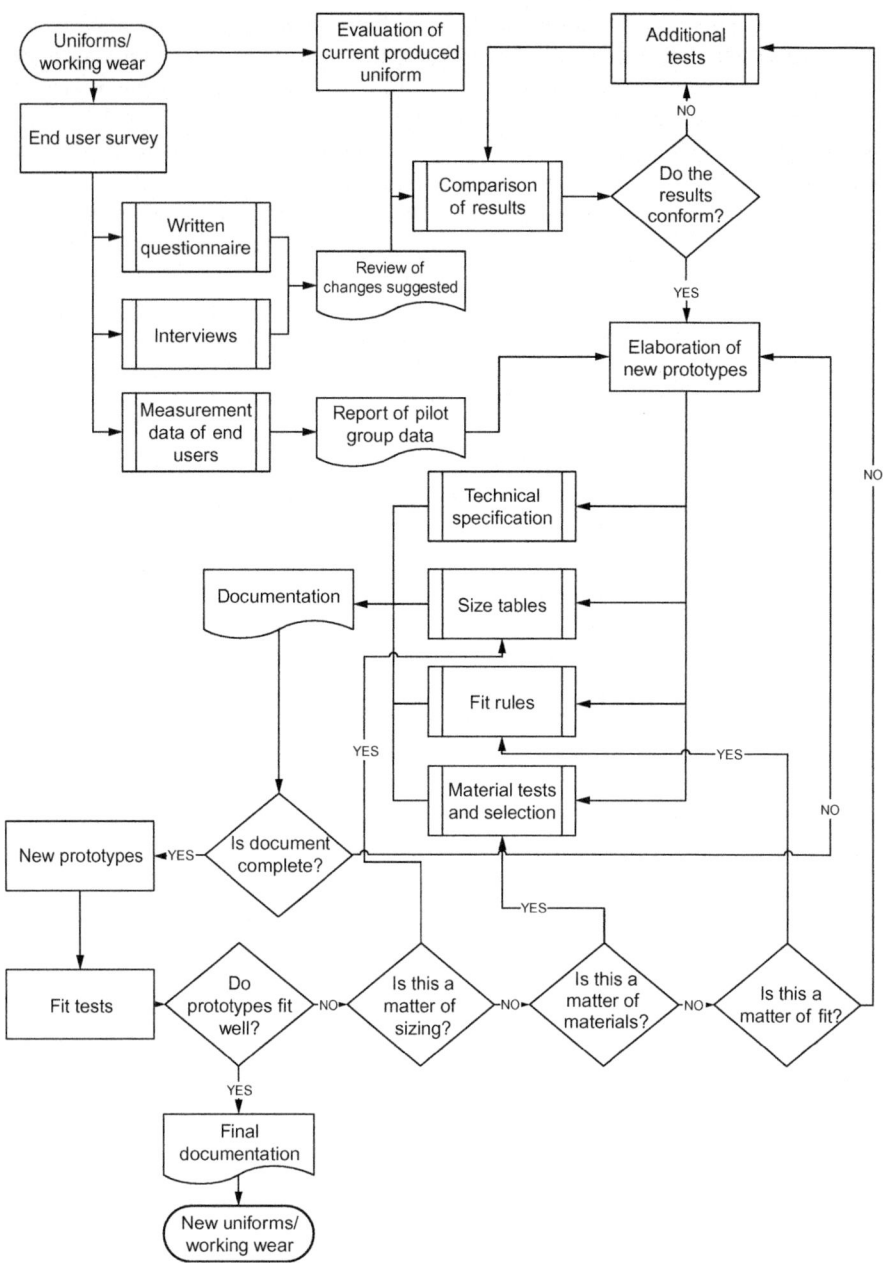

Fig. 11.5 PPE test and enhancement process diagram.

- Appropriate fit: not too tight for comfort, deep breathing is not restricted, no blood flow restriction, appropriately proportioned and positioned clothing design;
- Operation of closures, adjustments, and restraint systems provides adequate range of adjustments available, ease of operation, and security of closures and adjusters, and closures, adjusters and restraint systems are strong enough;
- Coverage of area intended to be protected, coverage maintained during movements;
- Freedom of movements (standing, sitting, walking, stair climbing); raising both hands above the head; bending over and picking up small object;
- Compatibility with other PPE from the same manufacturer.

In order to ensure conformity of size testing in accordance with the standards (ISO 13688:2013, 2013), each surveyor should be given a uniform in line with the size according to size tables; in addition testing should be carried out having season-appropriate underwear. The main points to be covered during the testing, yet potentially complementable depending on the type of garment, are as follows:

1. Visual and tactile assessment of the garment, making sure that it is free of defects that can be detrimental to the user or others (such as sharp or excessively stiff seams or edges; loose threading, ripped up seams, tears, defective fasteners (zippers, buttons, hoop&loop, etc.);
2. Assessment of ease of putting on and taking off;
3. The assessment of the size of the uniform according to the previously defined features of the garment (for example, for the jacket—loosely covers the shoulder and the underarm (rib) area and causes no sensation of tightness; the collar fits well to the neck without creating discomfort; the underarm of the jacket is loose, enveloping waist and pelvis free from strain, the bottom edge reaches beyond the level of the fastening of the trouser cuff; the length of the sleeve (cuff edge) of the stretched-out arm reaches the joint of the base of the thumb, while in the case of bent arm it reaches the joint of the base of the palm. In the pelvic line the trousers are neither excessively loose nor too narrow; the waistband bottom edge of the trousers rests on the upper point of the ilium (above the pelvis); the trousers envelope the pelvis and the stomach loosely enough, without encumbering the placement of arms in the side pockets, the level of the crotch is sufficiently deep for the wearer's comfort; the trouser legs are long enough to form a fall over the top edge of laced boots, etc.);
4. Assessment of freedom of movement (for example, the wearer must be able to perform the following movements: to stand, sit, walk and climb a ladder; lift both arms at the shoulder height in front and to the sides, above the head. To grab shoulders by crossing their arms over the chest; to lean forward and to lift a small object, such as a pencil; squat, spreading knees broadly; to make a deep lunging forward—one leg bent at 90 degree angle and supported on full foot, the other one—stretched out and supported on the fingertip.

11.6.2 Investigation of mobility

Standards already call for the ease of putting on and removing the clothing, with a possibility of unrestricted bending of hands and knees, as well as leaning movements while at the same time leaving uncovered unprotected parts of the body during movements—there is sufficient jacket and trousers overlay (ISO 13688:2013, 2013). Additional conditions that should be taken into account for ensuring movement in clothing worn are, for example:

- the sleeve and trouser legs are not too long without disturbing the movement of the arms and legs;

- the garment is not too loose to cause formation of moving (flowing), disturbing folds of the excessive volume;
- without any places for unexpected tears to appear in the garment or between its elements;
- there is no unjustified movement restriction in any of the joints;
- the jacket loosely covers the shoulder and the underarm (rib) area to cause no perceptible constraints for hand movements in a wide range;
- the trousers cover the pelvis and the shins loosely enough, without limiting the movements of standing, sitting, squatting and leaning;
- the crotch does not restrict movements and does not cause discomfort in the wearer's crotch area;
- all parts of the body to be covered by clothing remain covered during movement.

Mobility, or freedom of movements, is influenced by features of fabric used for the wear as well as the design of the clothes (McCann and Bryson, 2009a), ensuring that they are not too tight or too loose, and do not impair normal body movements (ISO 13688:2013, 2013). Dynamic fit of work wear calls for maximum comfort of movements in response to minimum movements of clothing parts against the wearer's body, because, as defined by the standard (ISO 13688:2013, 2013), the design of clothing must ensure that body parts are not exposed during the foreseeable movements (e.g., the jacket should not rise above the waist by raising hands), thus failing to safeguard the body from surrounding hazards.

Weight and bulk of clothing can also encumber body movement. It has been found that in case of bulkiness, each layer of the outfit may increase wearer's energy consumption by about 4%, caused by the troubled pace and interlayer friction (McCann and Bryson, 2009a).

Interlayer friction is an essential factor in designing each of the separate layers that are to be put atop one another so that the outer layers do not sit too tight and do not impair body movement. In addition, friction may increase due to sweating—thus causing thermal discomfort, as well as limiting move of clothing over the surface of the body parts. For example, a scan experiment with fire proximity suits showed that protective layers can encumber wearer's movements by reducing lifting height of his or her hands by 32 mm (2056–2024 mm) (Ashdown, 2011).

Methods for assessment of dynamic conformity can also be developed in directions regarding to assessment of the volume of deformation in strained parts of clothing, displacement of clothing parts against human body and degree of movement limitations.

References

Anthesis, 2015. Sustainable Clothing (Technical Report). WRAP, Banbury. 47 pp.
Ashdown, S.P., 2011. Improving Body Movement Comfort in Apparel. Woodhead Publishing Series in Textiles, Cambridge, UK.
Chapman, R., 2012. Smart Textiles for Protection, first ed. Woodhead Publishing. 416 pp.
Choi, S., Ashdown, S.P., 2010. 3D body scan analysis of dimensional change in lower body measurements for active body positions. Text. Res. J. 81 (I), 81–93. https://doi.org/10.1177/0040517510377822.
EN 13402-3:2014, 2014. Size Designation of Clothes. Part 3: Body Measurements and Intervals.
EN 13921:2017, 2017. Personal Protective Equipment—Ergonomic Principles.

EN ISO 20685:2015. 3-D Scanning Methodologies for Internationally Compatible Anthropometric Databases.

Fan, J., Yu, W., Hunter, L., 2004. Clothing Appearance and Fit: Science and Technology. Woodhead Publishing Limited, Cambridge, England, ISBN: 0-8493-2594-3, p. 240.

Geršak, J., 2013. Design of Clothing Manufacturing Processes, first ed. Woodhead Publishing. 320 pp.

Hayes, G.S., Venkatraman, P., 2016. Materials and Technology for Sportswear and Performance Apparel. CRC Press, Taylor&Francis Group, p. 370.

ISO 13688:2013, 2013. Protective Clothing—General Requirements.

ISO 13937-2:2000, 2000. Textiles—Tear Properties of Fabrics—Part 2: Determination of Tear Force of Trouser-Shaped Test Specimens (Single Tear Method).

ISO 15535:2012, 2012. General Requirements for Establishing Anthropometric Databases.

ISO 15537:2004, 2004. Principles for Selecting and Using Test Persons for Testing Anthropometric Aspects of Industrial Products and Designs.

ISO 7250-1:2017, 2017. Basic Human Body Measurements for Technological Design—Part 1: Body Measurement Definitions and Landmarks.

ISO 8559-1:2017, 2017. Size Designation of Clothes—Part 1: Anthropometric Definitions for Body Measurement.

ISO 8559-2:2017, 2017. Size Designation of Clothes—Part 2: Primary and Secondary Dimension Indicators.

Juodsnukytė, D., Gutauskas, M., Čepononienė, E., 2006. Mechanical stability of fabrics for military clothing. Mater. Sci. (Medžiagotyra) 12 (3). 243–246 pp.

Kaynak, H.K., Topalbekiroğlu, M., 2008. Influence of fabric pattern on the abrasion resistance property of woven fabrics. Fibres Text. East. Eur. 16(1(66)).

Koblyakova, E.V., Kurshakova, Y.S., Zenkevich, P.I., Dunaevskaya, T.N., 1974. The Size Typology of the Population of the CMEA Member Countries. Light Industry, Moscow, p. 440 (Коблякова, Е.В., Куршакова, Ю.С., Зенкевич, П.И., Дунаевская, Т.Н., 1974. Размерная типология населения стран – членов СЭВ. Легкая индустрия, Москва, с. 440).

Loercher, C., Morlock, S., Schenk, A., 2017a. Design of a motion-oriented size system for optimizing professional clothing and personal protective equipment. In: International Conference on Intelligent Textiles and Mass Customisation, Ghent, 15–18 October. Hohenstein Institut für Textilinnovation gGmbH, Boennigheim, Germany.

Loercher, C., Morlock, S., Schenk, A., 2017b. Motion-oriented 3D analysis of body measurements. In: IOP Conf. Series: Materials Science and Engineering. vol. 254, p. 172016. https://doi.org/10.1088/1757-899X/254/17/172016.

Mattila, H., 2006. Intelligent Textiles and Clothing, first ed. Woodhead Publishing. 528 pp.

McCann, J., Bryson, D., 2009a. Smart Clothes and Wearable Technology, first ed. Woodhead Publishing. 484 pp.

McCann, J., Bryson, D., 2009b. Smart Clothes and Wearable Technology. Woodhead Publishing Limited, p. 484. ISBN: 9781845693572.

McLaren, A., Oxborrow, L., Cooper, T., Hill, H., Goworek, H., 2015. Clothing longevity perspectives: exploring consumer expectations, consumption and use. In: PLATE Conference, 17–19 June. Nottingham Trent University. 229–235 pp.

Naglic, M.M., Petrak, S., Gersak, J., Rolich, T., 2017. Analysis of dynamics and fit of diving suits. In: IOP Conf. Series: Materials Science and Engineering. vol. 254, p. 152007. https://doi.org/10.1088/1757-899X/254/15/152007.

Özdil, N., Özçelik, G., Kayseri, G., Mengüç, S., 2012. Analysis of Abrasion Characteristics in Textiles. Ege University, Turkey. ISBN: 978-953-51-0300-4. 119–146 pp.

Podgórski, D., 2017. A role of smart PPE and IoT technologies in future management. In: Report—Lecture, PPE Conference, Brussels, 22–23 November.

Preedy, R.V., 2012. Handbook of Anthropometry: Physical Measures of Human Form in Health and Disease. Springer Science & Business Media, p. 3107.

Sakharev, M.I., Boytsov, A.M., 1981. Principles of Engineering Designing of Clothes. Light and Food Industry, Moscow, p. 272 (Сахарев, М.И., Бойцов, А.М., 1981. Принципы инженерного проектирования одежды. Легкая и пищевая промышленность, Москва, с. 272).

Siegert, D., 1964. Focus on Fabrics. National Institute of Drycleaning—ASV. 560 pp.

Song, G., 2011a. Improving Comfort in Clothing, first ed. Woodhead Publishing. 496 pp.

Song, G., 2011b. Improving Comfort in Clothing. Woodhead Publishing Limited, p. 496. ISBN: 1845695399.

Song, H.K., Ashdown, S.P., 2015. Investigation of the validity of 3-D virtual fitting for pants. Cloth. Text. Res. J. 1–17. https://doi.org/10.1177/0887302X15592472.

Song, G., Mandal, S., Rossi, R., 2016. Thermal Protective Clothing for Firefighters, first ed. Woodhead Publishing. 242 pp.

Wang, F., Gao, C., 2014. Protective Clothing: Managing Thermal Stress. Woodhead Publishing. 500 pp.

Xiao, P., Ashdown, S.P., 2013. Analysis of lower body change in active body positions of varying degrees. In: Proc. of the 4th International Conference on 3D Body Scanning Technologies, Long Beach CA, USA, 19–20 November, pp. 301–309. https://doi.org/10.15221/13.301.

Амирова, Э.К., Сакулина, О.В., 1985. Изготовление специальной и спортивной одежды: Учебник для кадров массовых профессий. Легпромбытиздат, Москва 256 с.

Further reading

Goldman, R.F., Kampmann, B., 2007. Handbook on Clothing, Biomedical Effects of Military Clothing and Equipment Systems, second ed. 321 pp. Available from: http://www.environmental-ergonomics.org/Handbook%20on%20Clothing%20-%202nd%20Ed.pdf.

Hwang, S.-J., 2001. Three Dimensional Body Scanning Systems With Potential for Use in the Apparel Industry. p. 63.

Foot morphological between ethnic groups

12

Shaliza Mohd Shariff[a,b,c,d], Amir Feisal Merican[a,c,d], Asma Ahmad Shariff[a,c,d]
[a]Centre of Research for Computational Sciences and Informatics in Biology, Bioindustry Environment, Agriculture and Healthcare (CRYSTAL), University of Malaya, Kuala Lumpur, Malaysia, [b]Institute Graduate Studies, University of Malaya, Kuala Lumpur, Malaysia, [c]Centre for Foundation Studies in Science, University of Malaya, Kuala Lumpur, Malaysia, [d]Fashion Department, Faculty of Art and Design, University Technology MARA, Selangor, Malaysia

12.1   Introduction

Human foot shape differs among ethnic groups and individuals (Stanković et al., 2018; Shu et al., 2015). Human foot shape is used as important measure to produce good-fitting shoes that provide comfort (Kouchi, 1995). Previous studies have shown that there exist anthropometric characteristics of adult foot between different ethnic groups (Manna et al., 2001; Ahmed et al., 2013). Some indicated that ethnic or subgroups of adults' foot have differences in foot morphology (Chiroma et al., 2015; Jung et al., 2001). Lee et al. (2015) reported that different ethnic groups tend to have different foot shape characteristics. Since Malaysia is a multiracial and multiethnic country, it is expected to show these differences in characteristics (Karmegam et al., 2011; Joseph, 2014). Malaysian women have been found to have a unique foot shape as the width and length of their feet are greater (Ibrahim and Khedifb, 2004).

Several studies have compared the characteristics of foot shape between ethnicity such as Kouchi (1998), who analyzed foot shape characteristics to identify differences based on ethnic background. These differences arise due to both genetic and environmental factors. As we found significant differences in foot dimensions according to the ethnic groups, it is vital to define Malaysian women's foot dimension data to determine the proper shoe size for the shoe design. Furthermore, small differences in foot shape must be incorporated into designing shoes to meet the requirement for shoe fit comfort.

Previous study on dimension of foot stature by Singh et al. (2013) was performed at the Lady Hardinge Medical College and Asst. Hospital, New Delhi. The results show that foot length measurements are higher in comparison with foot breadth in women (Singh et al., 2013).

A study by Karmegam et al. (2011) displayed evidence that Indian women had longer foot length compared with Malaysian women. This study also determined the differences in anthropometric data among the three ethnic groups in Malaysia in complete body dimension. Meanwhile, among the three ethnic groups, Chinese

Anthropometry, Apparel Sizing and Design. https://doi.org/10.1016/B978-0-08-102604-5.00012-3

women were found to have the shortest foot length, as supported by Onuoha et al. (2013) in their study that compared the foot dimensions between four ethnicities: Nigerian, Malaysian, Chinese, and Indian. The results showed that Chinese have smaller foot length compared with the rest. Onuoha et al. (2013) used traditional anthropometric instruments including anthropometers, steel measuring tape, digital vernier calipers, and digital height gauges.

Several studies identify foot morphology from the 1-D and 2-D measurements. However, there are several flaws for these 1-D and 2-D measurements. It has been noted that the taking of measurements and interpretation of certain parts of the human body is difficult (Telfer and Woodburn, 2010; Novak et al., 2014). This has been supported by Goonetilleke and Weerasinghe (2013) who pointed out that as far as 2-D measurement is concerned, there is a dimensional error between the unconstrained foot and shoe that indicates the quality of fit. The use of three-dimensional (3-D) surface scanning technologies to produce digitized representations of parts of the human anatomy has the potential to help change the way a wide range of products are designed and fabricated (Telfer & Woodburn, 2010). Many footwear designers and manufacturers use an ergonomic form for their products based on 1-D or 2-D measurements, namely, two basic measurements: the length of the foot and girth of waist (Telfer & Woodburn, 2010). Current advanced 3-D surface scanning technologies have the potential to play an important role in the development of customized products (Telfer and Woodburn, 2010).

12.2   3-D INFOOT scanning

In this study, 3-D INFOOT scanner has been used. There are two types of INFOOT three-dimensional foot scanners, which are standard INFOOT scanner and high type foot scanner. The difference between standard INFOOT scanner and the high type INFOOT scanner is that the latter captures the upper part of the leg until calf area, >100 mm higher compared with the standard type. This machine (Fig. 12.1) was due to the accuracy of this foot scanner, which is within 1.0 mm. Additionally, it is a portable machine, weighing only 40 kg (Kouchi and Mochimaru, 2001; Lee et al., 2013).

The advantage of this INFOOT three-dimensional foot scanner is that it can scan a foot's form and its anatomical landmark points, automatically measuring almost 20 measurements using landmarks. This high type INFOOT scanner has 6 lasers points and 12 cameras to capture the upper part of the foot. To run the INFOOT scanner, it must be connected to a personal computer via a USB 2.0 cable.

12.2.1   The specification of an INFOOT scanner

The scanner consists of several parts such as handrail, side cover A, side cover B, scanner cover, top cover, foot cover, foot stage, and switch panel. Each part has its respective function during the scanning process (Fig. 12.2). The detailed specifications of a

Fig. 12.1 3-D INFOOT scanner high type.

three-dimensional INFOOT scanner high type are described in Table 12.1. The machine scans each subject at 30 mm per second. Each foot must be in the correct position in order to obtain the perfect measurement. The advantage of this machine is that the accuracy of measurements is between 1.0 and 2.0 mm. The machine is able to support a person weighing up to 200 kg. Table 12.2 contains a detailed description of the functions of INFOOT high type.

12.3 Participants

A total of 1210 Malaysian women participated in this study between the ages of 20 and 60 years old. The average age is 37.36 ± 12.97. The mean of their stature and body height was 62.54 ± 28.89 kg and 155.97 ± 7.41 cm, respectively. The data were collected for 2 years, from December 2013 to December 2015. All the participants were selected for this study and did not have history of visible foot abnormalities or foot illnesses. Ethical committee approval was obtained by the University of Malaya Research Ethics Committee (UMREC) with an ethical clearance number UM. TNC2/RC/H&E/UMREC-63.

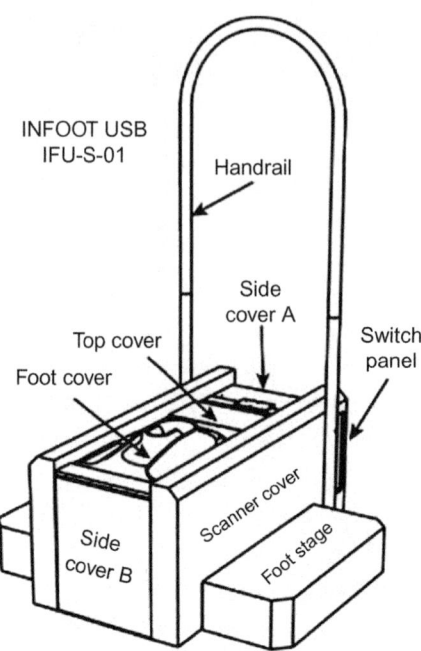

Fig. 12.2 INFOOT high type with eight components.

Table 12.1 The specifications of the INFOOT high type three-dimensional foot scanner

Description	Function
Scanner	Three-dimensional foot scanner
Scanning	30 mm/s
Data processing	20 s/ft
Data accuracy	mm step: Y(W), Z(H) = 1 mm, X(L) = 2 mm 0.5 mm step: Y(W), Z(H) = 1 mm, X(L) = 1 mm
Data format	FBD (original data format)
Scan target	Human foot and landmark points
Weight limitation	200 kg/person
Prohibition	Absorption and reflection of laser
Setting place	In-room use only

12.3.1 The significance of 3-D shoe anthropometric data for the development of shoe sizing

Anthropometry measurement is a necessary tool in developing standardized sizes and applied design for shoes in a specific population. Shoe designs use anthropometric measurements in the foot area (Jamaiyah et al., 2010).

Mostly the existing of standard international shoe sizing system collecting foot anthropometric data with direct measurement instrument is a traditional approach. Among the techniques or devices used to collect the anthropometric measurements or data of the foot size and shape dimensions are caliper and Harris mat print. All these noninvasive, quantitative techniques are either categorized under 2-D type of measurements. It has been noted that the taking of measurements and interpretation of certain parts of the human body is difficult (Telfer and Woodburn, 2010; Novak et al., 2014). This has been supported by Goonetilleke and Weerasinghe (2013) who pointed out that as far as 2-D measurement is concerned, there is a dimensional error between the unconstrained foot and shoe that indicates the quality of fit. Since 3-D scanning technique was found to be more precise and accurate than the manual measuring method, Telfer and Woodburn (2010) used the three-dimensional (3-D) scanner to measure and design the customized foot orthotics and footwear, as an anthropometric measurement and assessment of the foot. The advantage of using the 3-D foot scanner is that it provides an automatic measurement and a detailed picture of the contour of the human foot.

Anthropometric characteristics of the foot in this study have been identified from the ergonomic perspective. The detail measures for marker positions and calculated points using a 3-D foot shape have been conducted.

12.3.2 Marker positions and calculated points

In this study, five bones' landmark positions were chosen: tentative junction point, sphyrion fibulare, metatarsal fibular (MF), metatarsal tibiale (MT), and sphyrion as landmarks to paste the markers (Fig. 12.3). From the five bone positions, the scanner will automatically detect the landmark positions and calculate the 17 ft dimensions used to determine the foot size. The 17 ft dimensions, including foot length, foot breadth, ball girth circumference, instep length, fibulare instep length, instep circumference, heel breadth, height of top of ball girth, height of instep length, toe#1 angle, toe#5 angle, height of sphyrion fibulare, height of sphyrion, angle of heel born, heel girth circumference, calf circumference, and horizontal calf circumference (Fig. 12.4).

12.3.3 Preparation and procedure of the used of 3-D INFOOT scanner

Before the scanning process, participants were required to sign a voluntary consent form with their detailed information. The form consists of details stated in the succeeding text, based on ISO 15535 (general requirements for establishing anthropometric databases):

(a) Biodata participant information (BPI)
(b) Weight and height (in centimeters)
(c) Current shoe size
(d) Comment

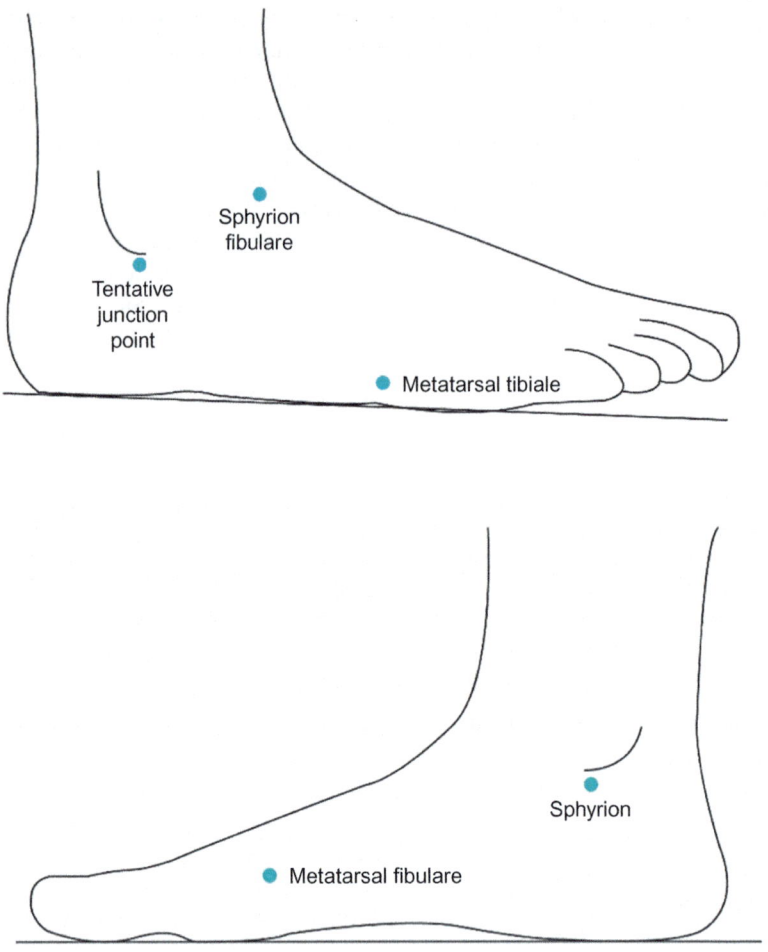

Fig. 12.3 Five points of marker position from five foot bones.

In the comment section, participants were asked to give comments on the challenges when it comes to finding the right shoe size.

Participants must have weight and height taken before the scanning process. The participants were informed of the objectives of the experiment with detailed explanation. The experiment process began with participants taking off their shoes. This was followed by the cleaning of the foot using ethanol sprayed on tissue (suggested by Mr. Kozo from I-Ware Laboratory, Co., Ltd., Japan). Ethanol is used to remove dust from the surface of foot arches to ensure good monitoring and printing quality during the scanning process.

When the participants are in the right position, markers are then placed at the five bones. It is important that the participants stand correctly to ensure that the right

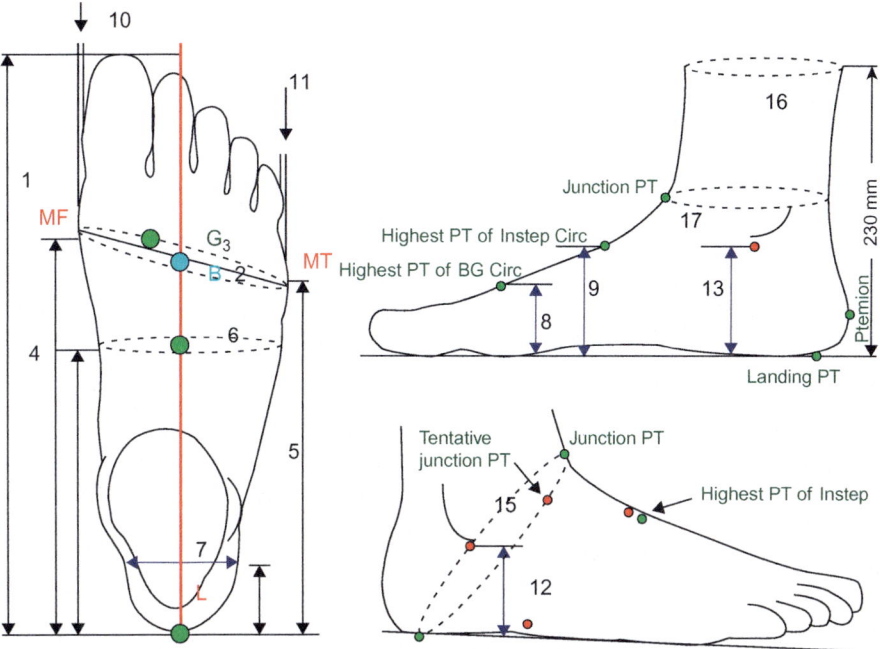

Fig. 12.4 Seventeen measurements of foot dimension: 1. foot length. 2. foot breadth, 3. ball girth circumference, 4. instep length, 5. fibulare instep length, 6. instep circumference, 7. heel breadth, 8. height of top of ball girth, 9. height of instep length, 10. toe#1 angle, 11. toe #5 angle, 12. height sphyrion fibulare, 13. height of sphyrion, 14. angle of heel born, 15. heel girth circumference, 16. calf circumference, and 17. horizontal calf circumference.

pressure is asserted on each foot. The five landmarks define the body dimensions and anatomical correspondence between two different body scans and can also create a homologous model of the foot (Kouchi and Mochimaru, 2011). Markers are designed with circle, velvet material and are blue in color. The blue color reduces the brightness of the laser projection. The material is made out of a very thin fiber and has adhesive elements for easy usage. Participants were required to place one foot (either left or right) onto the scanner while placing the other foot at the same position on the glass plate outside the scanner. Participants stood relaxed and distributed their body weight equally on both feet (Fig. 12.5). Participants were not allowed to move their feet until the scanning process is completed. Before repeating the process with the other foot, the glass plate was cleaned of dust and prints using ethanol. The same scanning procedure was then used for the other foot.

Finally, the data measurement comes from the INFOOT scanner that provides 17 dimensions that are identified. All measurements were in millimeters and are automatically converted into 3-D models.

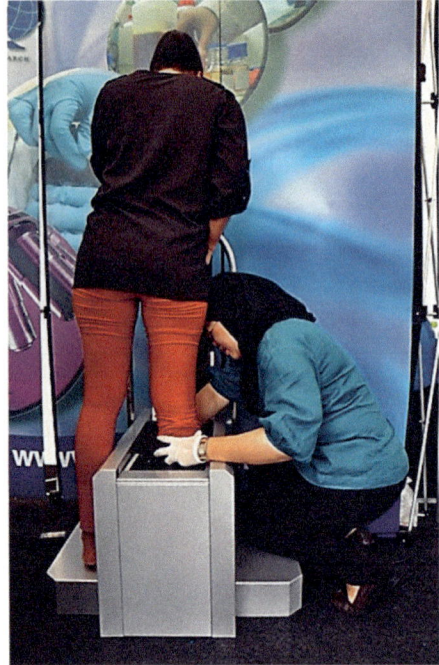 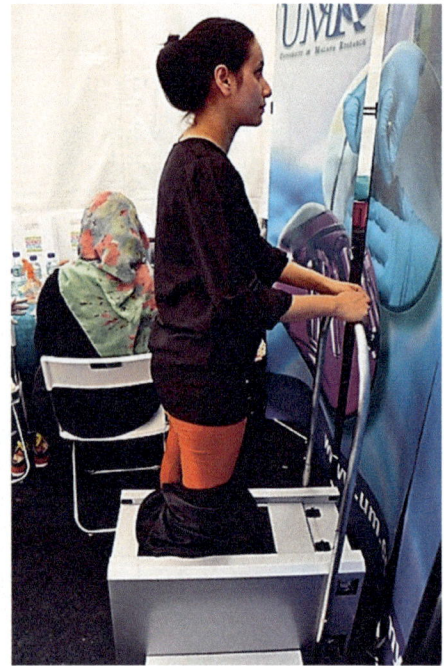

Fig. 12.5 Position during scanning process.

12.4 Foot morphology characteristics between ethnic groups

12.4.1 Comparison FL, FB, and BG for three ethnic groups

The three foot measurements are presented in a polygon mesh image. Foot length (FL) was measured as the distance from the most posterior point of the heel (pternion) to the most anterior point of the second (akropodion) in millimeters by an INFOOT scanner (Dhaneria et al., 2016). The foot breadth was measured from a straight distance from the most medially placed point on the head of the first metatarsal to the most laterally placed point located on the head of the fifth, ball girth circumference was also measured passing over the metatarsal tibiale and the metatarsal fibulae, and the height was measured by vertical distance from the vertex to the bottom of the foot (Shugaba et al., 2013). Three parameters, FL, FB, and BG for the three ethnic groups, were measured using INFOOT 3-D scanner and are shown in Figs. 12.6–12.8. Indians have longer foot length compared with Malay and Chinese at 269.6 mm (Fig. 12.7), while Malay has wider FB (116.45 mm) (Fig. 12.8) and higher BG (280.3 mm) compared with Chinese and Indian. From these images, it can provide clear foot shapes, which are very important in determining foot size.

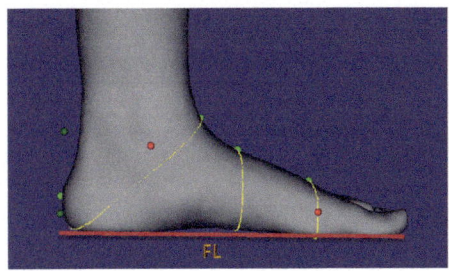
Indian foot length = 269.6 mm

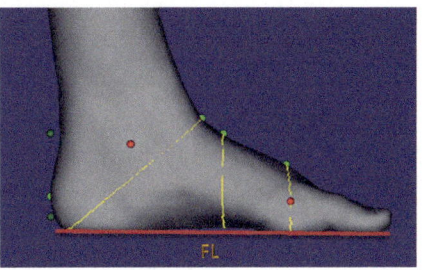
Malay foot length = 268.1 mm

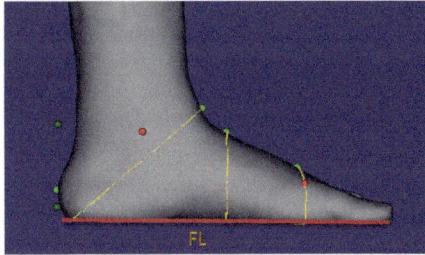
Chinese foot length = 267.1 mm

Fig. 12.6 Comparison of 3-D foot length measurement for three ethnic groups.

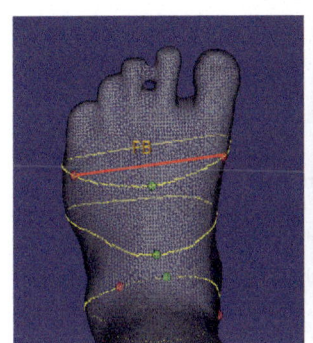
Malay FB = 116.45 mm

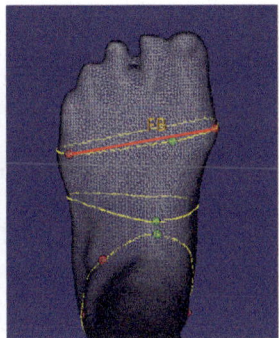
Chinese FB = 107.4 mm

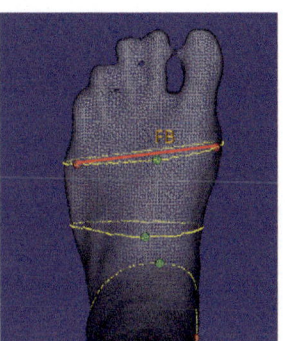
Indian FB = 106. 75 mm

Fig. 12.7 Comparison of 3-D foot breadth measurement for three ethnic groups.

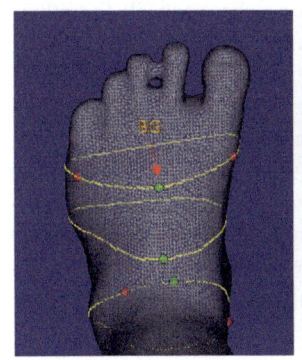
Malay BG = 280.3 mm

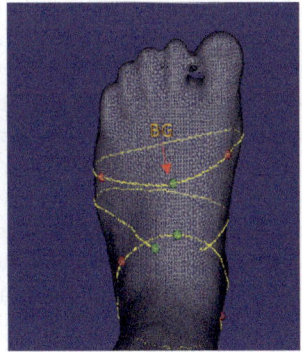
Chinese BG = 270.10 mm

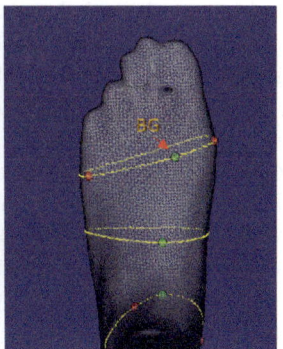
Indian BG = 258.80 mm

Fig. 12.8 Comparison of 3-D BG measurement for three ethnic groups.

Table 12.2 ANOVA test to compare differences between three foot parameters

Foot parameters		Sum of squares	df	Mean square	F	Sig.
Foot length	Between groups	4135.038	2	2067.519	13.254	0.000
	Within groups	188283.066	1207			
	Total	192418.104	1209	155.993		
Ball girth circumference	Between groups	5823.439	2	2911.720	13.099	0.000
	Within groups	268306.217	1207	222.292		
	Total	274129.657	1209			
Foot breadth	Between groups	818.864	2	409.432	10.291	0.000
	Within groups	48020.018	1207	39.785		
	Total	48838.883	1209			

12.4.2   Result in ANOVA and post hoc test for three foot dimension

Table 12.2 shows the results of ANOVA test to compare the differences of the three foot dimensions between the three major ethnic groups. The results clearly show significant differences among ethnic groups with $P < .05$. Table 12.3 also shows the post hoc Tukey test, which shows that only the foot length was significantly different between all the ethnic groups. Concerning ball girth, significant differences were found between Chinese and Malays ($P < .05$), but no significant differences were noted between all three ethnic groups. Meanwhile, there was a significant difference in the foot breadth of Malays compared with Chinese and Indians ($P < .05$). However, there was no significant difference between the Chinese and Indians. The post hoc Tukey test showed that there exists a significant difference between all the ethnic groups for FL, FB, and BG. These results can help the footwear industry to produce shoes that meet customer demands.

The results of ANOVA test and post hoc Tukey test for the three ethnic groups show that Indian women have longer FL compared with other ethnic groups. However, Malaysian women have a wider FB and a higher BG. The results show that Indian women have a slimmer foot shape than Malaysian and Chinese women. Meanwhile, Malaysian women have larger foot shape compared with Chinese and Indian women. The morphology of foot shape is important for the development the new standard shoe sizing system for the multiethnic groups in Malaysia.

Table 12.3 Post hoc Tukey test result

Dependent variable	(I) Race	(J) Race	Mean difference (I-J)	Std error	Sig.
Foot length	Malay	Chinese	2.7590[a]	0.8022	0.002
		Indian	−3.4709[a]	1.1838	0.010
	Chinese	Malay	−2.7590[a]	0.8022	0.002
		Indian	−6.2300[a]	1.2671	0.000
	India	Malay	3.4709[a]	1.1838	0.010
		Chinese	6.2300[a]	1.2671	0.000
Ball girth circumference	Malay	Chinese	4.8348[a]	0.9577	0.000
		Indian	2.8088	1.4132	0.116
	Chinese	Malay	−4.8348[a]	0.9577	0.000
		Indian	−2.026	1.5126	0.374
	Indian	Malay	−2.8088	1.4132	0.116
		Chinese	2.026	1.5126	0.374
Foot breadth	Malay	Chinese	1.7321[a]	0.4051	0.000
		Indian	1.4742[a]	0.5979	0.037
	Chinese	Malay	−1.7321[a]	0.4051	0.000
		Indian	−0.2579	0.6399	0.0914
	India	Malay	−1.4742[a]	0.5979	0.037
		Chinese	0.2579	0.6399	0.914

[a]The Post hoc Tukey test showed that there exists a significant different between all the ethnic groups for FL, FB, and BG.

12.5 Comparison of foot size between ethnic groups

There are three categories of foot sizes: Both feet are full size, both are half size, and either the left or right foot is half size for three ethnic groups shown in Table 12.4. Results show that many Malaysian women have half size in either their left or right foot. <50% of the total number of women had both feet at full size, which means that 66.2% of Malaysian women have a half size of either or both of their feet. The results show that more Malaysian women have more percentage in half size of either one of their feet. Meanwhile, Chinese women have high percentage for full size compared with half full size of their feet or both full size. Indian women were also found to have full-size feet compared with full half size or half full size of their feet. The results show that there are a higher percentage of women with full-size feet, while a small percentage have a half size for their right foot.

12.6 Future trend of shoe sizing

The development of shoe sizing system is different all over the world. Several countries such as the United Kingdom, the United States, Europe, Japan, Taiwan, China, and Australia have their own standard shoe sizing system. Other than that, Mondopoint shoe sizing system is one of the popular standard shoe sizing systems used in many countries. All these standard shoe sizing systems had been developed

Table 12.4 Overall size for three ethnic groups

Ethnic groups	Foot size categories	N	Percentage
Malay	Both full size (left and right foot)	237	33.3%
	Both half size (left and right foot)	203	28.6%
	Either one half size (left or right foot)	270	38%
Total		710	
Chinese	Both full size (left and right foot)	126	34.2%
	Both half size (left and right foot)	118	32.1%
	Either one half size (left or right foot)	124	33.7%
Total		368	
Indian	Both full size (left and right foot)	47	35.6%
	Both half size (left and right foot)	45	34.1%
	Either one half size (left or right foot)	40	30.3%
Total		132	
Overall full size		410	33.8%
Overall half size		366	30.3%
Either half size		434	35.9%
Overall half size or either half foot half size		800	66.2%

manually. Currently, with the rapid development of computer technology such as computer-aided design (CAD) and computer-aided manufacturing (CAM), more options are provided to choose the correct shoe size. Consumers need a comfort and good fitting in buying their shoes.

Several software programs in 3-D to determine the foot and shoe size have been developed such as INFOOT scanner, UCS foot scanner, and Orthobaltic. Based on verities of 3-D foot scanner, this system will recommend the shoe size closest to the foot size and foot shape for perfect fitting. Using a 3-D scanner, it is easier for a customer to find their actual foot size without using a manual equipment to determine the shoe size. Three-dimensional software with the use of tools, such as the Shoe Match Fit System (2018), is an online business that recommends sizes based on what customers know best: the fit of their favorite shoe.

12.7 Conclusion

Feet morphological characteristics between ethnic groups of women in Malaysia have been identified. Indian women were found to have slightly longer foot length than the Malaysian and Chinese women, while Malaysian women were found to have higher ball girth and wider foot breadth compared with the Indian and Chinese women. It was noted that Chinese women have slimmer foot shape compared with Malaysian and Indian women. The results obtained are very useful in providing the information and guidelines in the process of developing the new standard shoe sizing system for Malaysian women. It also can help the footwear industry in producing shoes according to these foot characteristics to meet customer demand.

References

Ahmed, S., Akhte, A.B., Anwar, S., Begum, A.A., Rahman, K., Narayan Saha, C., 2013. Comparison of the foot height, length, breadth and types between santhal and bangalees of Pirganj, Ranpur. Bangladesh J. Anat. 11 (1), 30–33.

Chiroma, M.S., Philip, J., Atta, O.O., Dibal, I.N., 2015. Comparison of the foot height, length, breadth and foot types between males and females Ga'anda People, Adamawa, Nigeria. IOSR J. Dent. Med. Sci. 14 (8), 89–93.

Dhaneria, V., Shrivastava, M., Mathur, R.K., Goyal, S., 2016. Estimation of height from measurement of foot breadth and foot length in adult population of Rajasthan. IJCAP 3, 78–82.

Goonetilleke, R.S., Weerasinghe, W.T., 2013. The Science of Footwear. Taylor and Francis Group, pp. 279–307.

Ibrahim, R., Khedifb, L.Y.B., 2004. Modeling 3D Surface of the Footprint Malaysian Women Perspective. Available from: http://gmm.fsksm.utm.my/~cogramm/cgibin/2004/upload_files/full_paper_backup/144408.pdf. (Accessed 23 February 2013).

Jamaiyah, H., Geeta, A., Safiza, M.N., Khor, G.L., Wong, N.F., Kee, C.C., ... Rajaah, M., 2010. Reliability, technical error of measurements and validity of length and weight measurements for children under two years old in Malaysia. Med. J. Malaysia 65 (Suppl A), 131–137.

Joseph, C., 2014. Growing up Female in Multi-Ethnic Malaysia. Taylor & Francis Groups, p. 2.

Jung, S., Lee, S., Boo, J., Park, J., 2001. Classification of foot types for designing foot wear of the Korean elderly. In: 5th Symposium on Footwear Biomechanics, Zurich, Switzerland, pp. 48–49.

Karmegam, K., Sapuan, S.M., Ismail, M.Y., Ismail, N., Bahri, M.S., Shuib, S., Hanapi, M.J., 2011. Anthropometric study among adults of different ethnicity in Malaysia. Int. J. Phys. Sci. 6 (4), 777–788.

Kouchi, M., 1995. Analysis of foot shape variation based on the medial axis of foot outline. Ergonomics 38 (9), 1911–1920.

Kouchi, M., 1998. Foot dimensions and foot shape: differences due to growth, generation and ethnic o rigin. Anthropol. Sci. 106 (Suppl), 161–188.

Kouchi, M., Mochimaru, M., 2001. Development of a low cost foot-scanner for a custom shoe making system. In: 5th ISB Footwear Biomechanics, pp. 58–59.

Kouchi, M., Mochimaru, M., 2011. Errors in land marking and the evaluation of the accuracy of traditional and 3D anthropometry. Appl. Ergon. 42 (2011), 518–527.

Lee, Y.C., Chao, W.Y., Wang, M.J., 2013. Developing a new foot shape and size system for Taiwanese females. In: The 19th International Conference on Industrial Engineering and Engineering Management, January. Springer, Berlin, Heidelberg, pp. 959–968.

Lee, Y.C., Kouchi, M., Mochimaru, M., Wang, M.J., 2015. Comparing 3D foot shape models between Taiwanese and Japanese females. J. Hum. Ergol. 44, 11–20.

Manna, I., Pradhan, D., Ghosh, S., Kar, S.K., Dhara, P., 2001. A Comparative study of foot dimensions between adult male and female and evolution of food hazards due to using of footwear. J. Physiol. Anthropol. Appl. Human Sci. 20 (4), 421–446.

Novak, B., Babnik, A., Možina, J., Jezeršek, M., 2014. Three-dimensional foot Scanning system with a rotational laser-based measuring head. Strojniški vestnik-J. Mech. Eng. 60 (11), 685–693.

Onuoha, S.N., Okafor, M.C., Oduma, O., 2013. Foot and head anthropometry of 18–30 years old Nigerian polytechnic students. Int. J. Curr. Eng. Technol. 3, 2.

Shoe Match Fit System, 2018. https://www.shoematchfit.com/. (Accessed October 2018).

Shu, Y., Mei, Q., Fernandez, J., Li, Z., Feng, N., Gu, Y., 2015. Foot morphological difference between habitually shod and unshod runners. PLoS One 10 (7), e0131385.

Shugaba, A.I., Shinku, F., Gambo, I.M., Mohammed, M.B., Uzokwe, C.B., Damilola, E.R., Usman, Y.M., 2013. Relationship between foot length, foot breadth, ball girth, height and weight of school children aged 3–5 years old. J. Biol. Chem. Res. 30 (1), 107–114.

Singh, J.P., Rani, Y., Meena, M.C., Murari, A., Sharma, G.K., 2013. Stature estimation from the dimensions of foot in females. J. Anthropol. 9 (2), 237–241.

Stanković, K., Booth, B.G., Danckaers, F., Burg, F., Vermaelen, P., Duerinck, S., … Huysmans, T., 2018. Three-dimensional quantitative analysis of healthy foot shape: a proof of concept study. J. Foot Ankle Res. 11 (1), 8.

Telfer, S., Woodburn, J., 2010. The use of 3D surface scanning for the measurement and assessment of the human foot. J. Foot Ankle Res. 3 (19), 2–9.

Sizing and fit for pressure garments

V.E. Kuzmichev[a], Zhe Cheng[b]
[a]Ivanovo State Polytechnic University, Ivanovo, Russian Federation, [b]Wuhan Textile University, Wuhan, PR China

13.1 Introduction

Nowadays many compression garments having different styles, such as corsets, girdles, pressure and shaping underwear, and so on, can be found on the market in accordance with their different applications, customer demands, and effects provided. Pressure garments are used for body shaping, recovery and protection in daily life, medical uses, and sports applications. The fit is a key feature of these garments, with special meanings that are totally different from other kinds of apparel. The main features of well-fitting pressure garments as a second skin are: the ability to adapt initial human body morphology to a new morphology in static and dynamic postures; the ability of garments to provide the predicted effect; the positive perception of customers in accordance with the length of time the garment is worn; correct construction; and the textile materials. All the features mentioned should be based on a deep understanding and knowledge of three subjects: the soft tissue of the body, textile materials, and garment construction.

According to Gaupta, the compression is the most important functional property of a large group of medical and sports garments and shapewear (Gupta, 2016). Concentrated compression is used in Chinese traditional medicine to improve the functions of many important human organs.

On one side, the general interest in having a beautiful and healthy, well-shaped body is growing; therefore the demands for compression garments and their possibilities for shaping or correcting the body morphology, moving fat, and helping to stay fit are increasing day by day. For example, the ability of a corset to shape human bodies to meet an aesthetic ideal has been known and used for thousands of years.

On the other side, the positive effect of applied compression helps to develop textile and garment technologies and expands such garments in the fields of health and sports applications. In medicine, pressure garments are mainly used in fat suction postoperatively, but pressure therapy can also help to prevent bad scarring, to assist venous return, to reduce peripheral swelling in vascular patients, and in other fields. Some studies have demonstrated and proved the effectiveness of pressure garments in sports (Rider et al., 2014; Hageman et al., 2018), although most of the published research has failed to support any effects of garments for performance capacity in a wide range of exercise tests (Kraemer et al., 2000; Scanlan et al., 2008; Sperlich et al., 2010, 2014; Ali et al., 2011).

Anthropometry, Apparel Sizing and Design. https://doi.org/10.1016/B978-0-08-102604-5.00013-5

The design of pressure garments also needs both traditional and new knowledge about all elements of the system "body-garment" (morphology of human body, sensitiveness of soft tissue, textile materials, structure and construction of garment, technology of production) and the environmental conditions over the lifetime. Traditional databases include the set of body measurements and the properties of textile materials (mechanical, physical, etc.). Additional databases should be built containing special knowledge related to the human body (structure and sensitiveness of soft tissue, morphology and its ability to reshape under applied compression, the main postures and positions, etc.), the textile materials (some abilities to elongate and create pressure, human-friendly properties of the surface, etc.), and IT software for virtual design.

13.2   Brief history of pressure garments

When pressure garments were mentioned from the 16th to the 18th centuries, the corset was similar to a cage and was made of a strip of wood, steel, or whalebone to raise the women's breasts by contracting the limits of the waist as much as possible (Beck, 1995; Cunnington and Cunnington, 1992; Steele, 2001); it was favored by the royal male aristocracy (Ping and Chunhong, 2002; Roberts, 1977). In the 17th century, due to the development of the garment industry, the structure of corsets changed (Lynn, 2014), and the new types of corsets were cone-shaped (Fig. 13.1A) (Salen, 2008). In the 18th century, some producers of corsets began to promote the slogan "good for health" for their corsets, which were easier to wear and more loosely fitting than earlier corsets (Fig. 13.1B). The earliest corsets were very far from human-friendly in design, and they had a strong negative impact on human organs (Fontanel, 1997).

Since the 19th century, corsets have mainly caused discomfort and health issues for women (Lim et al., 2006a). Many scientific studies have focused on the negative

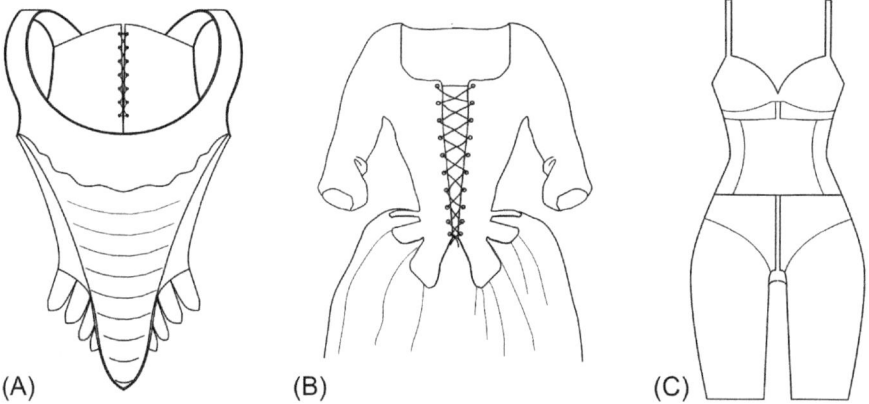

Fig. 13.1 Some pressure garments: (A) multilayer "cone-shape" corset, 17th century; (B) multilayer corset called "jumps," 18th century; (C) one-layer corset, 20th century.

effects of corsets (Kunzle, 2006). In the 20th century, the US conglomerate DuPont came out with Lycra and the resulting revolutionary materials changed pressure garments. A corset, or girdle, worn under a dress and tight pantyhose with a mini-skirt were popular with many women during this era (Fig. 13.1C) (Stewart and Janovicek, 2001).

Later, many people advocated the release of physical restraints by wearing only modified pressure garments and joining them with physical exercises to change their shape fundamentally (Rosoff, 2006; Li et al., 2011). Textile materials also were improved to have better elasticity, moisture permeability, warmth preservation, etc. Socks, sleeves, and some sports protection garments reduce muscle swelling and vibration; increase muscle support; prevent injury and sprains; and improve the recovery and performance of athletes. These positive functions have been proved in many scientific studies (Filatov, 1987; Ivanova, 1998; Ng-Yip, 1993; Hirai, 2002).

From the 20th century, due to the application of pressure garments in the medical field, people began to pay more attention to the pressure created by garments as a positive influence on the human body and their own feelings. For example, pressure treatment prevents venous dilatation of burn scars and aids recovery after liposuction (Klöti and Pochon, 1982; Cheng et al., 1984; Staley and Richard, 1997; Yıldız, 2007). The research in recent years mainly has focused on pressure values, classification, and evaluation of ordinary shaping and functional compression garments (Pickering et al., 2005). To successfully achieve these effects, it is necessary to know the permissible range of garment pressure resulting in influence on physical, physiological, and psychological characteristics of the human body (Surikova et al., 2012).

13.3 Classification of pressure garments

According to the basic characteristics, all pressure garments can be subdivided into five levels. There are several subcategories in each level. Table 13.1 shows the details.

- 1—this is the main level used to differentiate the pressure garment products in the market: daily, sportive, medical, cosmetic.
- 2—the subjective feeling is the human perception of pressure garments by means of skin receptors, determining what is acceptable as a reasonable pressure range for the sensory evaluation: lightly tight, tight, medium tight, extra tight.
- 3—the layer is the number of layers of materials used in pressure garments for performance/construction, manufacturing technology, design purposes, and other aspects: one-layer, two-layers, and so on.
- 4—the objective pressure is the instrumentally measured value of pressure produced by garments and designated as acceptable as a reasonable pressure range for the objective evaluation.
- 5—the coverage is the area of human body skin covered by pressure garments, from 5% to 100%.
- 6—the structure of the pattern block making in accordance with the human body constitution and body measurements needed: top (torso, arm), lower part (below waist), and top+lower (full body, part of body).

Table 13.1 Classification of compression garments

Categories		Subcategories	Description
1	Type	Daily; sportive; medical; cosmetic	
2	Subjective feeling	Lightly tight	The arising feeling is as a second skin; the garments can be worn for a long time (daily underwear)
		Tight	The arising feeling is as a supported shell; the garments can be worn for a long time (functional garments, gymnastics, swimsuit, etc.)
		Medium tight	The arising feeling is as a tight shell; the garments cannot be worn for a long time (sports socks, compressive socks, girdle, etc.)
		Extra tight	The arising feeling is as an obviously very tight shell, the garments can be worn for a short time only (red carpet corset, wetsuits, medical, sports protective, etc.)
3	Layer	$1, 2, 3, 4 \ldots n$	
4	Objective pressure	Low	$<0.5\,kPa$
		Medium	$0.5–3.2\,kPa$
		High	$>3.2\,kPa$
5	Coverage	90%–100%	Overalls, diving suits (professional and amateur), long length sportswear (swimsuit), spacesuits
		70%–80%	Gym suit, tight-fitting trousers (top and leggings), long length sportswear (long sleeve gymnastics, etc.)
		60%–70%	Corset (top), medium length sportswear (track and field, wrestling, weight lifting, etc.)
		30%–60%	Shaping girdles (for bust, waist, abdomen or all upper torso), long johns, stockings, vest
		20%–30%	Tight-fitting underwear (boxers, briefs, etc.), short pants, short length sportswear
		5%–10%	Socks (flight socks, ankle wrap), gloves, body parts protective sportswear (cap, bracer, wrist brace, etc.) Special medical garment (mask treatment for burns, belt for fat suction), short pants, short length sportswear
6	Pattern block	Top part (torso); lower part of body; top +lower part	–

13.3.1 Evaluation of pressure and body reshaping in system "body-garments"

Internationally, the concept of garment pressure was proposed by S.M. Ibrahim (1985) after testing the pressure arising under human dynamic postures (Elder et al., 1985). Subsequently, many explorations have been carried out related to pressure produced by worn garments and the acceptance of subjective perception (Sasaki et al., 1997; Chan and Fan, 2002; Dongsheng and Qing, 2003; Lim et al., 2006b). The skin of different body parts has different receptors (pressure-sensing points located in the deep part of soft tissue) and they generate different feelings of pressure applied, which can be considered as the tolerance to establish the classification of pressure comfort values (Jiang et al., 2016). Whether or not the garment contacts the skin, the pressure evaluation is mainly driven by the evaluation of material in terms of weight, softness, surface, prickle, bondage, etc., which forms a subjective feeling, from very uncomfortable to very comfortable (Liu et al., 2016a).

In recent years, pressure and comfort evaluations were studied in dynamic and static situations for tight-fitting garments (underwear, stockings, etc.) (Nakahashi et al., 2005; Dan et al., 2016; Zhang et al., 2015; Nishimatsu et al., 1998). In general, by using the combination of negative ease in pattern blocks of tight-fitting garments, it is possible to reshape the body in accordance with the material elongation, predicted function, postpartum body sculpting, health care, and so on, and to design different pressures on different anthropometrical levels.

13.3.2 Pressure comfort

In the field of ergonomics, the pressure comfort in the system "body-garment" is an important parameter for evaluating whether the garment will fit the human body in static and dynamic situations (Nishimatsu et al., 1998). The pressure appears for two reasons: the first is due to the garment style, which can cause material deformation (tension, compression, shear, shrinkage) and body skin deformation (compression, lifting, push-up); the second reason is due to human body motion and body measurements changing. The material will be deformed due to the mentioned reasons and produce deformation of soft tissue. So, an acceptable value of pressure will depend on textile materials, body characteristics, and skin location.

The pressure is also related to the warmth, weight, softness, and stiffness of the textile materials (known as the pressure sensitivity) (Xiangling and Weiyuan, 2006; Vuruskan and Ashdown, 2016). Different parts of the human body react differently to the pressure, producing unequal perceptions when acted upon, and different customers can have different acceptability levels in similar positions. For example, a pressure of 3.0 kPa can produce opposite perceptions around the neck (negative) and around the low waist, belt part, and upper iliac bone (positive). Usually, joints and skeletal parts can tolerate higher pressure. Therefore, the objective testing and evaluation of pressure should be aimed at designing different kinds of pressure garments for covering different parts of the body, to set a particular range of pressure acceptability and sensitivity (Cheng and Kuzmichev, 2013; Mengna and Kuzmichev, 2013).

As for the pressure range, the maximum value is approximately <3.50 kPa as applied to the soft parts and approximately <9.50 kPa for skeletal parts (Qiyue and Xin, 2018; Tonghui and Yanbo, 2017). Some studies have shown that, under normal conditions, the range of pressure values is from 1.37 to 5.90 kPa (this value can be increased to >1.97 kPa during physical exercise); the pressure around the bust is from 0.50 to 3.90 kPa; and the comfort range around the waist is from 0.20 to 6.40 kPa, and exceeding 4.00 kPa at the waist is considered harmful to the body (Wu and Yu, 2006; Qi, 2010; Na, 2015). Other results are devoted to the whole body or its parts, the changes of anthropometrical girth (e.g., limbs in one group, bust, waist and hip in another group), the sensitivity of body parts, the influence of body motion, the structural design of garments (Mengna, 2015; Cieślak et al., 2017; Varghese and Thilagavathi, 2016; Xu et al., 2013; Leung, 2010). This schedule represents those factors that should be considered as desirable for the level of comfort pressure.

Nowadays, with the general mainstream of garment improvements, comfort requirements, and popularization of elastic materials, pressure comfort as the core of human physiological and psychological aspects has attracted the attention of many scholars (Wang et al., 2017). Results of considered studies are summarized in Table 13.2.

Table 13.2 Summary of some research results

No.	Locations of pressure applied	Pressure range (kPa)	Effect of compression (reshape or lift)	Researcher
1	Waist, abdomen, hip, thigh	0.72–1.84 3.18–6.37	1. Absolute decreasing of girth, cm: Abdomen 0.07–0.20, Waist 0.05–0.20, Hip 0.92–1.87. 2. Lifting of buttocks 0.59–0.65 cm	L. Yaping, L. Yao, O. Shizue
2	Under bust, waist, hip, thigh, knee	1.10–3.55	Relative decreasing of girth, %: Under bust 0.15–2.5, Waist 0.5–5.5, Waistband 4.21–5.07, Hip 1.5–3.0, Thigh 2.0–4.0	I. Tislenko
3	Abdomen	0.88–1.07	Lifting of abdomen 0.69–0.83 cm	L. lulu
4	Total body	0.16–4.81		M.J. Denton, H.P. Giele, D. Tanaka, J. Ziming, J. Zhennan, L. Xiaoju, Y. Shijia

Table 13.2 Continued

No.	Locations of pressure applied	Pressure range (kPa)	Effect of compression (reshape or lift)	Researcher
5	Bust, waist, hip	0.28–4.98		G. Mengna, M. Sato, W. Yueping
6	Waist, thigh	4.00–5.33		H. Makabe
7	Waist, hip, thigh	0.10–3.89		N. Ito, Y. Xiaohong, W. Haiyan, G. Lei, J. Erfan
8	Waistband, thigh, knee	1.65–5.93		L. Mingxia
9	Hip, thigh, calf	2.40–3.09		T. Tamura, L. Huashan
10	Lower limbs, knee, calf, midcalf, ankle	0.04–4.00		T. Toshiyuki, X. Meiling, Y. Pei, Q. Rui, L. Suzhen

Fig. 13.2A shows the bodies with possible pressure ranges.

In summation, the garments can produce pressure on soft tissue from 0 to 3.7 kPa and on skeletal parts from 3.4 to 9.2 kPa (Giele et al., 1997; Wong and Li, 2004; Yanmei et al., 2014; Guney and Kaplan, 2016; Huashan et al., 2017). Different kinds of garments can produce pressure: swimwear 0.98–1.96 kPa, medical stockings 2.94–5.88 kPa, shaping underwear 2.94–4.9 kPa, and fitness wear at <1.96 kPa (Du and He, 2014; Qu and Song, 2015; Xu, 2016; Xinzhou et al., 2016; Cheng, 2017; Yongrong et al., 2018).

13.3.3   Body soft parts reshape/change range or value

For daily and athletic styles, the pressure garments must squeeze the soft parts of the body. For example, running pants must reduce the muscle oscillations, and shapewear should correct the body shape by changing bust, waist, and hip girth. Few research studies have been done concerning the reshape/lifting (push-up) effects (Table 13.2). Fig. 13.2B shows the collected statistics on how the soft tissue can be displaced under compression (Scurr et al., 2009; Fangyuan and Xiaona, 2013; Milligan et al., 2014; White et al., 2011).

As we can see from Fig.13.2B, the female breast can be lifted from 4.0 to 13.5 cm by means of a bra (the interval is presented of the population of females with average weight 65.6 ± 7.6 kg, height 34 ± 1.8 in., D size of cup). The more delicate area is the male genitalia; its reshaping ability should be taken into account during design of male underwear. The vertical lifting capability of male

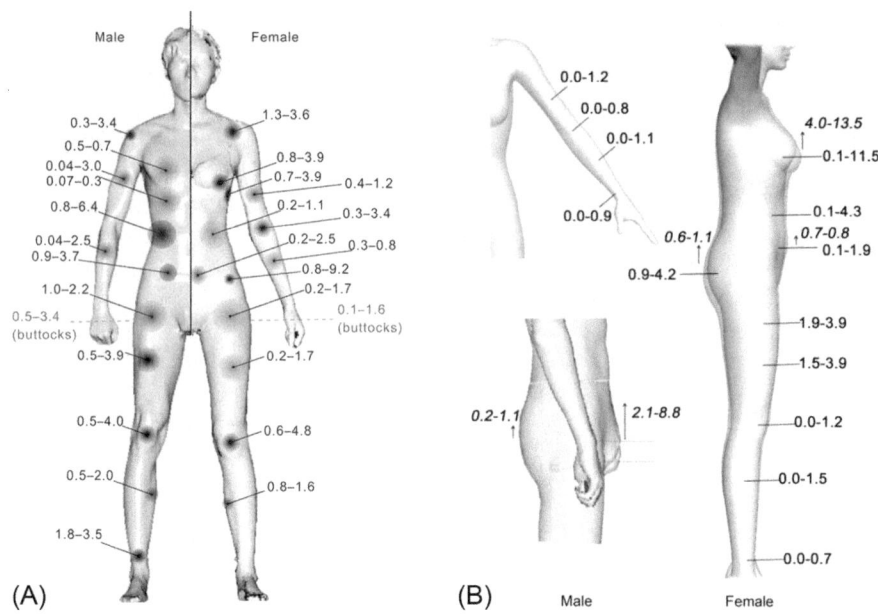

Fig. 13.2 Location of acceptable pressure, kPa (A) and the ranges of shaping around body and lifting of soft tissue, cm (B).

genitalia is 2.1–8.8 cm (the objective results are based on the subjective perception of participants without pressure measurement) (Kuzmichev and Zhe, 2014; Zhe and Kuzmichev, 2014).

13.3.4 Pressure sensors

Today's research methods of pressure measurement make use of sensors and the accuracy of the measurements will directly affect the evaluation results. Objective methods are the basis and foundation of garment design. When the body is dressed in pressure garments, the pressure sensor is located between the body and garment, and the pressure value is directly measured.

The earliest methods used hydraulic U-tubes. At present, pressure tests are mainly focused on airbag methods, optical fiber sensors, and resistance sensors (Nishimatsu, 1988; Wong, 2002). Table 13.3 shows the more popular methods used for testing of compression garments, with the advantages and disadvantages of each.

Most of the scientific researchers prefer to use direct measurement by means of the most popular sensors, such as the AMI airbag (Japan) and the FlexiForce A201 thin-film resistance sensor with the Wireless ELF system (USA) because they are efficient and time-saving equipment.

Instead of a real body, the pressure can be measured on special material simulating real soft tissue, such as cosmetic silicon. Fig. 13.3 shows an experimental stand for

Table 13.3 Pressure sensors

Type	Sensor	Characteristic
Fluid pressure type U-shaped pipe	–	Inaccurate, contact area about 20 cm², convenient for direct reading value, difficult to record results
Resistance	TEKSCAN *FlexiForce Sensors* (Tekscan, 2019a) Interlink Electronics, *Force Sensing Resistor* FlexiForce ELF/ WELF 2 System (Tekscan, 2019b)	High accuracy, wide measurement range, long life, lightweight, can bend. Able to work under harsh conditions, diversification of varieties
Vapor-pressure airbag AMI-3037	AMI-3037 (Japan)	Accurate and convenient
Elastic fiber-optical sensor	Canada FISO FOP-M (FISO, n.d.)	High sensitivity, simple, small, suitable for pressure test, but has position limit

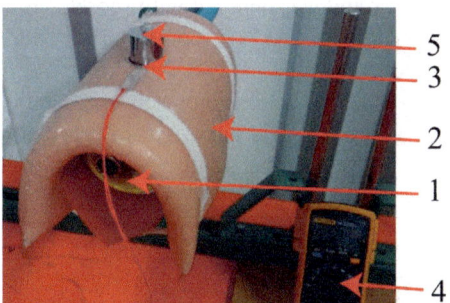

Fig. 13.3 The experimental stand for studying the compression ability of knitted fabrics: *1*—rigid framework ($d = 8$ cm), *2*—elastic substrate made of cosmetic silicone (thickness 2 cm), which imitated the soft body tissue, *3*—pressure sensor, *4*—ohm meter, *5*—calibration load.

simulating real conditions, measuring the pressure between the textile materials and the artificial substrate of soft tissue and testing the compression ability of the textile materials.

The FlexiForce A2013 sensor was fixed on a cylindrical surface of cosmetic silicon substrate *2* (Ziegert and Keil, 1988), which has the same bulk modulus as the soft tissues of the body (GOST 26435-85). This approach saves time during the experiment, and adequate results are obtained.

13.3.5 Expert evaluation

The research methods used to transform human perceptions into digital form are subjective evaluation (sensory analysis), which includes reflecting, joining, and gathering physical, psychological, and other data types. When different participants wear the same clothes, their psychological and physiological responses may be different, for a number of reasons. It is necessary to define the different reactions of the participants by a comfort-content scale, which is a unified reference standard. The scale can be designed from a group of words or corresponding values (this scale is similar to semantic differential). The subjective evaluation method is one of the important methods for studying the compression comfort that results from the wearer's psychological and physiological comprehensive reflection and using subjective intuition as the standard to compare and distinguish feelings during garment wearing. To get the standard score, the group of participants should be gathered in accordance with the common rules and approaches for treating the results obtained.

The special scale can be used to evaluate people's perception process and to eliminate subjective arbitrariness. This approach can be realized by using two kinds of pressure evaluations - objective and subjective (Mengna, 2015). Fig. 13.4 shows the subjective and objective scores of pressures relative to each other and appearing during female dress wearing.

As Fig. 13.4 shows, to clarify the comfort perception, the expert assessment scale was divided into three levels and was denoted with scores: "U" (uncomfortable) with score 2; "E" (endurable) with score 1; "C" (comfortable) with score 0. Both parameters were obtained as average values after testing different female participants in seven postures. Therefore, according to the possibilities and characteristics of both methods, subjective and objective, it is important to use both to study the pressure comfort. The evaluation of pressure must go through subjective evaluation and objective tests that conform with and complement each other.

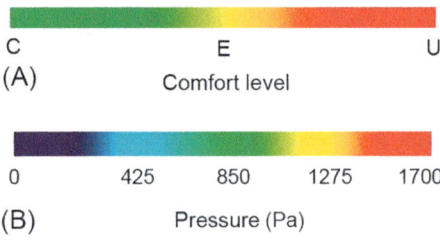

Fig. 13.4 Subjective in comfort scores (A) and objective in Pascal (B) scales for prediction of comfort perception (Mengna, 2015).

13.4 Materials for pressure garments

Pressure garments usually have direct contact and tightly cling to human skin, conforming to the structure of the human body and its dynamic changes. Therefore the elasticity of the material plays a crucial role. Pressure garments also rely mainly on elastic materials, followed by structural design (Liu, 2009). Material properties and garment structure (construction) are the two factors that contribute to or impede the comfort. Therefore, it is necessary to consider the physical properties of materials and the dynamic characteristics of the human body under the condition of health and safety. Generally speaking, the material elasticity of pressure garments will satisfy the dynamic changes of the human body in the covered area.

To investigate the relations existing in the system "body-garment" between the pressure created under the compression garment on one side and the material strain on the other side, the physical model of a cylinder is used. Textile material is considered as a tight-fitted shell covering the body parts. The theoretical pressure P on a cylinder with a radius of curvature r is calculated by the Laplace law (ASTM D2594-04, 2016; Sloan, 1963)

$$P = \sigma/r \tag{13.1}$$

where σ is the stress of textile material.

The human body has a much more complicated shape than the cylinder, and is described in terms of girth, arc, and linear measurements. Besides, the deformable body should be considered after comparison of the surface curvature before and after donning the garment. Thus, for precise calculation of pressure, it is necessary to take into account the deformation of the horizontal sections due to compression of soft tissues in different areas of the body.

So the ability of textile materials to push the human body directly depends on its stress under elongation. Knitted materials have a certain elasticity, with excellent flexibility due to the interlaced structure. In general, the materials are defined as follows: as stretch materials with elongation more than 15%; as rigid materials with elongation smaller than 15%; as power stretch materials with elongation more than 30%. The materials having 15%–30% elongation are called comfort stretch materials (Fan and Hunter, 2009). Close-fitting garments made of such materials will have minimum resistance to body movements. However, the research on the relationship between structural characteristics of high-stretch knitted materials and garment pressure is limited (Sang and Park, 2013). The high-stretch materials are often used for compression garments because the compression garment must fit extremely closely (tightly to body) to apply pressure to the skin (Lee et al., 2017). The pressure level (value) of a high-stretch knitted material depends on the knit structure, yarn composition, and knitting types.

The materials for pressure garments are generally made of synthetic fibers. The garments in daily-wear styles are generally made as a single layer. For production of some sports protective garments, double or multiple layers are used. Typical fiber compositions are complex and contain, for examples:

for functional pressure garment—(72%–92% PET + 8%–28% spandex); (41% nylon + 21% Neoprene + 31% PET + 7% spandex); (55% thermoplastic elastomer TPE + 35% polyester + 10% spandex);

for daily underwear—(viscose fiber + PET + spandex); (cotton/nylon/viscose + spandex); (cotton/viscose + Modal/Nylon + spandex).

Table 13.4 shows the possible combinations of the pressure values and the elongation of materials used in different garments.

To express the abilities of materials, such as to push up the soft tissue of the human body and to create compression, three large databases should be joined: first, the mechanical properties of materials such as an elongation under small loads; second, the pressure that can be applied to soft tissues; third, the human perception of elongated materials by means of skin sensitivity. This last is a complex parameter because it involves both objective and subjective scores and reflects a human-friendly approach (Mengna, 2015).

There are two approaches that allow a calculation of the compression ability of materials.

The first approach uses the results obtained after testing the materials and the soft tissue under conditions with the load of elongation increased constantly. The index of compression performance *CP* shows the relationship between the elongation of knitted materials (%) and their ability to produce pressure on soft tissue (kPa) and can be calculated by the equation (Zhe et al., 2017)

Table 13.4 Parameters of pressure garments

Area of application	Garment function	Pressure (kPa)	Material elongation (%)
Medicine	After-burn therapy	2–3.3 (Macintyre, 2006) (maximum 5,3)	15–20 (Macintyre, 2014)
	Vascular diseases, improving blood circulation by stocking	2–8.0 (http://www.sigvaris.com/usa/en-us/knowledge/compression-levels-and-indications)	
Sports	Reshaping, improving blood circulation, muscle recovery	0.65 (knee)–2.4 (calf) (Troynikova, 2011)	More than 100 (for free movements) (Lim, 2006)
Cosmetic	Reshaping of male bodies	0.65–4.0 (Jennes, 2012)	<40 (Kuzmichev et al., 2016)
	Reshaping of female bodies	0.8–3.6 (Sarah, 2012)	
	Strong correction of breast area of female bodies	<6.6 (http://en.wikisource.org/wiki/Toleration_of_the_corset)	
Daily	Comfort	0.4–1.3 (Boldovkina, 2005; Hang, 2013)	3.5–12.7 (Lim, 2006)

$$CP = \left(P_{\text{warp}} + P_{\text{weft}}\right) / \left(E_{\text{warp}} + E_{\text{weft}}\right), \tag{13.2}$$

where CP is the compression index reflecting the ability of elongated knitted materials to create pressure, kPa/%; P is the pressure measured between the soft tissue of the human body and the elongated materials in warp and weft directions, kPa; E is the elongation of knitted materials in warp and weft directions *or* the design ease allowance in pattern block of a pressure garment, %.

The higher the value of CP, the stronger is the pressure performance of the knitted materials. The lower the value of CP, the weaker the knitted materials' pressure performance. For example, if CP is 0.13 kPa/%, the materials can provide 0.13 kPa to the soft tissue of the body when we decrease the length of sample or increase the negative ease allowance of the pattern block by 1%; when the waist girth is 80 cm, we should decrease the width of the pattern block by 10% to get the new value of 72 cm. The designed garment will provide a pressure of 1.30 kPa in the waist area.

Fig. 13.5 shows the properties of selected knitted materials used for pressure garment production ($T_1 \dots T_9$ have thickness <0.9 mm, T_{10}–T_{18} have thickness of 1.1–2.5 mm). Fig. 13.5 has three axes: the left y-axis is CP, the right upper y-axis is P, and the right lower y-axis is E.

As we can see, the relationship between P and E is complex.

The range of maximum design negative ease is from -15% to -20% and the range of maximum average pressure is from 2.5 to 3.0 kPa (Kuzmichev et al., 2015, 2016). Therefore, the CP is mainly based on the pressure performance of the material, which makes it possible to predict the pressure in advance when the materials are being chosen.

The second approach uses the results obtained after testing the materials and the soft tissue when the elongation of materials was limited and equal to 20%. The compression ability determined as the derivative $dP/d\varepsilon_p$ for strain ε_p is equal to an average value of 20% for many compression garments, as Fig. 13.6 shows. The compression ability $dP/d\varepsilon_\varepsilon = 20\%$ can be presented by Eq. (13.3):

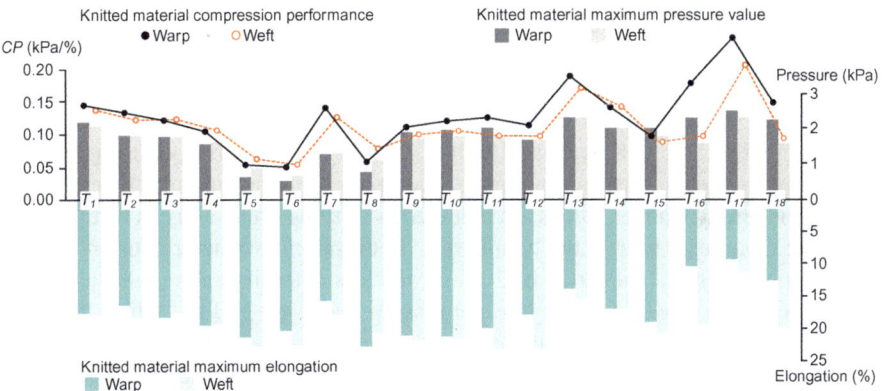

Fig. 13.5 Relations between compression performance CP, pressure P, and elongation (ease allowance) E.

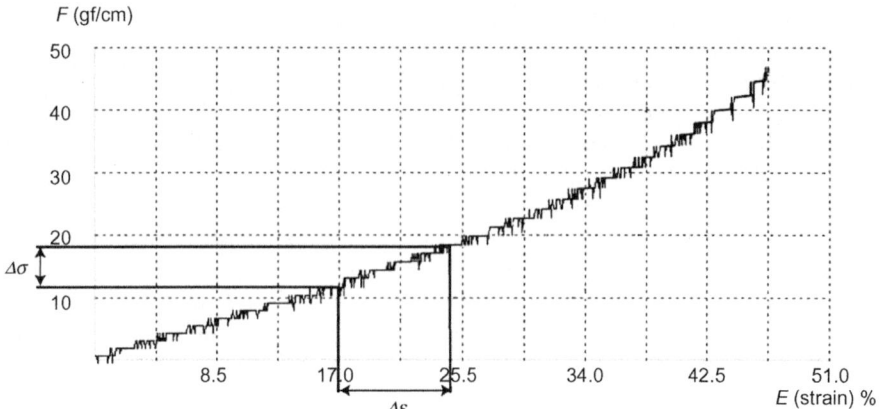

Fig. 13.6 The graphical scheme of finding the values $\Delta\sigma$ and $\Delta\varepsilon$ in the neighborhood of $\varepsilon = 20\%$ by means of Kawabata Evaluation System.

$$dP/d\varepsilon = dP/d\sigma \cdot \Delta\sigma/\Delta\varepsilon, \tag{13.3}$$

where $dP/d\sigma$ is the derivative calculated from the curve $P(\sigma)$, and $\Delta\sigma/\Delta\varepsilon$ are the values calculated from the stress-strain, for example, obtained as curve F-E by the Kawabata Evaluation System KES-FB1. Fig. 13.6 shows the graphical scheme of $\Delta\sigma$ and $\Delta\varepsilon$ calculation.

The module of compression ability M_{comp} can be calculated by

$$M_{comp} = P_{20\%}/20, \tag{13.4}$$

where $P_{20\%}$ is the pressure under the materials after its elongation of 20%. For example, knowing the values found graphically from Fig. 13.6 $\sigma(20\%) = 15$ and $\Delta\sigma/\Delta\varepsilon = 7/6$, and the pressure on soft tissue, the M_{comp} is 33.

The module of compression ability was used for a new classification of compression materials. The boundaries of compression garments were chosen as 0.4, 1.3, 2.0, 3.3 kPa. The values 1.3 and 2.0 kPa correspond to the appropriate and maximum values for daily clothes (Kawabata, 1980). The value 3.3 kPa corresponds to the maximum pressure recommended for sportswear, as well as for vascular diseases and after-burn therapy. Higher values are rendered as a strong effect and can be used only for medical purposes to treat the lymphatic system (http://www.instron.us/en-us/products/testing-systems). The strain of 20% for the less extensible materials could possibly give a pressure exceeding the permissible value of 3.3 kPa, while the most extensible ones did not even produce a minimum pressure of 0.4 kPa. Thus based on the module of compression ability M_{comp} and boundary values of pressure for the garments, the schedule of compression garments and their applications includes four groups, as shown in Table 13.5.

Table 13.5 Groups of knitted materials for compression garments

Compression group	Pressure range (kPa)	Application	M_{comp}
I	0.4–1.3	Permissible pressure for comfortable daily garments. A pressure <0.4 kPa is considered negligible and is not taken into account	20–65
II	1.3–2.0	Critical pressure for comfortable casual wear	65–100
III	2.0–3.3	Permissible pressure for medical garments and shapewear	100–165
IV	More 3.3	Critical pressure for medical and shapewear for daily use (corsets, strong regulation of lymph flow disorders, etc.)	165 and more

A material can have different modules of compression ability along the warp and weft. Knowing both modules, it's possible to cut the pressure garments in different directions in accordance with desirable effects, garment structure, human body morphology, etc.

13.5 Sizing system for pressure garment

13.5.1 Garment labeling

Different styles of garments and different manufacturers have a variety of sets of size/type systems for the human morphology of the populations of different countries and regions. Besides, the different regions have different categories and types of pressure garment distributions, but the general methods of pattern block drafting and labeling are similar (Qiming and Wenbin, 2003; Koszewska, 2004; Hong et al., 2006; Zhe and Kuzmichev, 2017, 2019). These systems were designed to meet the physical characteristics of most people with lower production costs, by reducing the number of garment sizes to satisfy the body characteristics of most people. As usual, these systems are good for the garments produced by pattern blocks with positive ease allowances, when the sizes of ready-to-wear clothes are bigger than the body measurements in total cases.

There are various size charts for women's and men's top and pants (underwear), girdles, socks, etc. using different sizing systems for each category. In the market, the labeling of pressure garments is the same for common garments, but the real sizes are smaller for the needs of fit and compression.

For the top of pressure garments, which cover the torso, the size charts usually are based on the letters "S, M…" or values "46, 48…" (Europe), "36, 38…" (US),

"170/84..." (China), etc. These charts use the traditional body measurements such as height, bust and waist girth, and weight (Numbers.com, n.d.; SIXECHARTER, n.d.). This approach of labeling doesn't include some crucial measurements reflecting the bodies' morphology, which are important to compression garments. Many online stores sell pressure garments with sizes labeled as XS, S, ..., XXL and detail the basic body measurements (girth of bust, waist, hip, height, weight, and thigh) in an additional table.

This simple method of garment size identifications cannot explain the garment's features or help a consumer to compare the garment with their own body measurements. So, consumers usually have to choose a pressure garment based only on their own experience or try it on. However, many pressure garments cannot be tried on in offline/online stores before purchasing, to control the fit and the comfort; it is impossible to determine whether or not to accept the pressure of garments, because some pressure garments are designed for suitable short-time wearing. The consumers with distinctive physiological characteristics and different body measurements in particular should use size charts (Mpampa et al., 2010; Song and Ashdown, 2011). The consumer needs more information about specific features and construction of pressure garments to be confident about their wearing comfort.

To create a pattern block of pressure garments, the patternmaker needs additional information that allows the customer satisfaction to be increased, to produce close-fitting and comfortable garments. Instead of using only the sizes of ready-to-wear garments, such as the bust, waist or hip girth, to calculate the structural values, the compression garments need more detailed and accurate data about human body characteristics. More detailed measurements should be added to describe human morphology.

New body measurements can be obtained by means of body scanning technologies such as the Vitus Smart XXL 3D body scanner with the software Anthroscan for visualization, processing, and evaluation of 3D scan data (Zhe and Kuzmichev, 2016a; Fan et al., 2004; Percoco, 2011; Zhang, 2004; Zhang and Zou, 2003). New research explains how to create the sizing system for male shaping underwear by increasing the number of body measurements. Fig. 13.7 shows all primary and additional body measurements belonging to the lower torso and which can be taken from the bodies scanned. Fig.13.7A shows 14 primary measurements of the lower torso, and Fig. 13.7A (upper) shows the schemes of measuring of the curve and lengths located on the body; Fig. 13.7A (lower) shows the horizontal and vertical distances out of the body. As Fig. 13.7B shows, in order to describe some crucial characteristics of male bodies, 18 additional data points marked as abbreviations should been calculated after data processing.

There are six key measurements that can be used directly for pattern block making, classification of male lower torso, and labeling of underwear:

(1) the additional girth measuring below natural waist and used for a waistband location NW_G,
(2) the distance h_W between two levels—natural waist girth W_G and waistband girth NW_G,
(3) the vertical distance between the point of genital peak and the crotch level h_G,
(4) the horizontal distance in profile view between natural waist front and genital bulge ΔGW,
(5) the horizontal distance between hip peak and natural waist back ΔWH,
(6) thigh girth NT_G as underwear bottom.

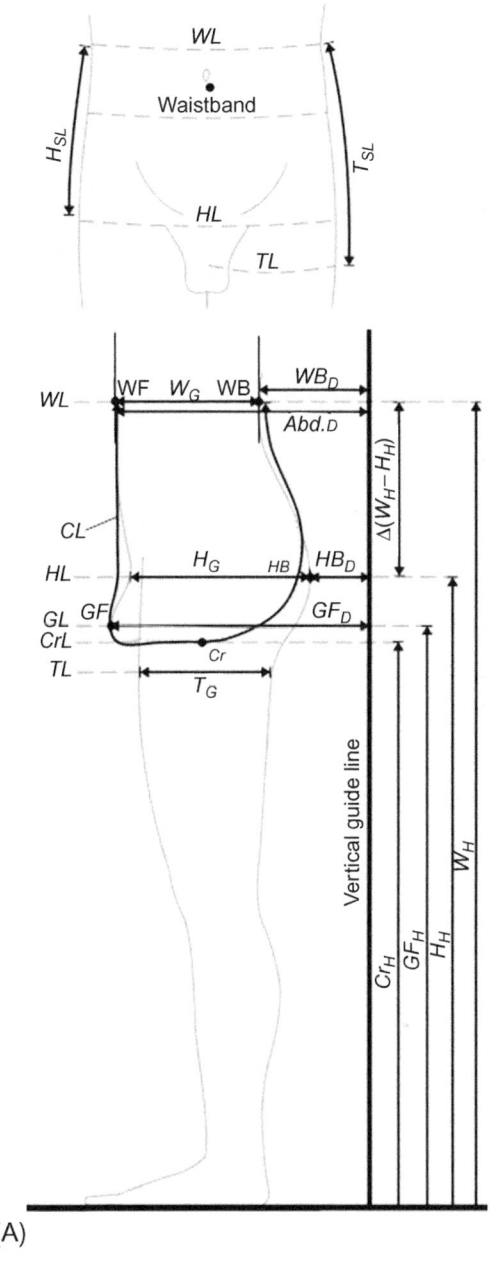

(A)

Fig 13.7 See figure legend on next page.

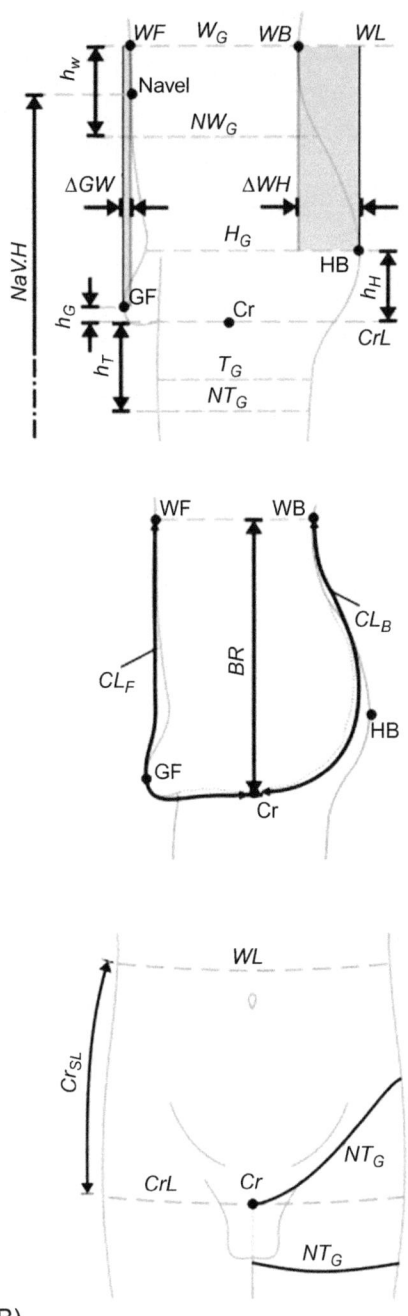

(B)

Fig 13.7 See figure legend in opposite page.

Fig. 13.7 Developed scheme of body measurements to design and label male shaping underwear: (A) primary; (B) additional. (*WL*, natural waist line/level as narrowest girth part; *WF*, *WB*, waist front and back points; *HL*, hip line/level as largest girth part, *HB*, hip back peak point; *GL*, genitals line/level across bulge peak point; *GF*, genital peak point; *CrL*, crotch line/level; *Cr*, crotch point; *TL*, thigh line/level).

(A) Primary measurements horizontally:

W_G, Natural waist girth (narrowest waist girth)
H_G, Hip girth (largest girth)
T_G, Thigh girth (largest girth)
WB_D, Distance from concave waist back to back vertical guideline
HB_D, Distance from hip peak to back vertical guideline
GF_D, Distance from genitals bulge peak to back vertical guideline
$Abd._D$, Distance from front abdomen to back vertical guideline

Primary measurements vertically:

$\Delta(W_H - H_H)$, Natural waist to hip level (vertical distance).
H_H, Hip height
W_H, Natural waist height

Primary measurements in arc length:

CL, Full crotch length (from WF through Cr to WB, close to body)
WL, Length of natural waist to waistband (front, side and back, close to body)
H_{SL}, Side length of natural waist to hip level (close to body)
T_{SL}, Side length of natural waist to thigh level (close to body)

(B) Additional measurements horizontally:

ΔGW, Difference between natural waist front and genitals bulge, $GF_D - Abd._D$
ΔWH, Difference between hip peak and natural waist back, $WB_D - HB_D$
$\Delta(H_G - W_G)$, Difference between hip and natural waist girth, $H_G - W_G$.
NW_G, New waist girth as waistband located below natural waist level
NT_G, New thigh girth as underwear bottom in horizontal and slope directions

Additional measurements vertically:

BR, Distance from natural waist to crotch level, $W_H - Cr_H$
$Nav._H$, Navel level height
GF_H, Height of genitals peak (follow the wearing habits)
Cr_H, Crotch level height
h_G, Distance between genitals peak point and crotch level, $GF_H - Cr_H$
h_T, Distance between crotch and new thigh (underwear bottom) level, $Cr_H - T_H$
h_H, Distance between hip and crotch level, $H_H - Cr_H$
h_W, Distance between natural waist to waistband level, $W_H - NW_H$

Additional measurements in arc length:

CL_F, Front crotch length, from WF front to Cr across genitals peak
CL_B, Back crotch length, from Cr to WB through hip middle groove
ΔF, Value describing genital bulge, $CL_F - BR$
ΔB, Value describing hip bulge, $CL_B - BR$
Cr_{SL}, Side length from natural waist to crotch level, leg inside minus outside length

First new measurement should help to get the comfort in the waistband area, because very often the waistband of daily underwear, excepting some functional compression garments, is located below the natural waist and usually below the navel. So to design the waistband, the girth of the waistband or lower waist NW_G is important.

Second new measurement is the distance h_W between W_G and NW_G, and, following individual preferences, the average distance is approximately 7.7 cm. This level is located close to the top of the anterior and posterior of the superior iliac spine, so it is better to put the waistband at this position so it can support one by both bones and organize a good fixed effect and pressure receptivity. Fig. 13.8B shows the box-plot of NW_G and h_W changing.

The relationship between NW_G and h_W is

$$NW_G = 0.02h_W{}^2 + 0.58h_W + 75.9, \tag{13.5}$$

where NW_G is new waist girth used as the location of the waistband, cm; h_W is decline distance of NW_G from the natural waist, cm.

Third new measurement depends on male genital volume; consumers, when wearing underwear, can feel different tightnesses in the genital area provided by different front pouches of the underwear. h_G describes the genital position put upward or downward according to personal wearing habits and underwear construction. Our research shows that the average value is about 3.1 cm in the range from −4.4 cm to 8.7 cm.

Fourth new measurement ΔGW describes the male genital bulge in the horizontal direction (see small width of the *gray* rectangle in Fig. 13.7B). It will be negative when the abdomen (waist front) bulges larger than the genitals bulge. From statistics, 25% of males have a negative value of ΔGW (mean is −0.68 cm), and 75% of males have a positive value (mean is 0.80 cm).

Fifth new measurement ΔWH presents the buttock volume. Due to this body part, consumers can feel different perceptions of tightness in the hip area provided by

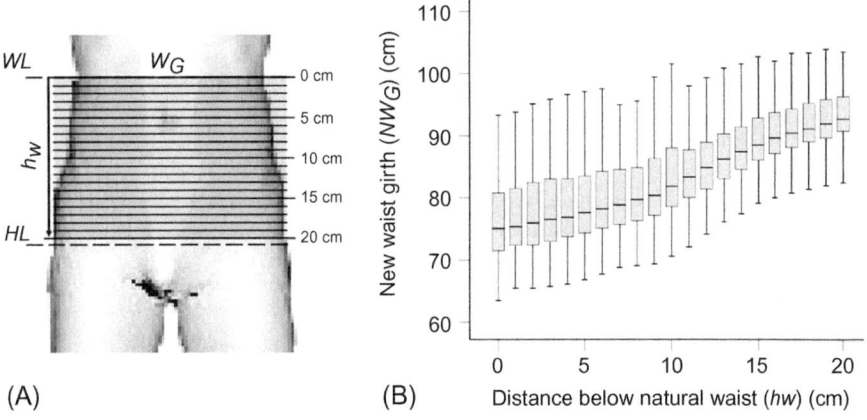

Fig. 13.8 Possible positions of NW_G (A) and (B) relationship between average NW_G and h_W.

different constructions of underwear (with few seams or seamless). From different body characteristics, movements (squatting), or materials properties, the underwear will cause uncomfortable/comfortable feelings appearing on the buttocks.

Sixth new measurement NT_G was called for by various positions of underwear length and bottom. Fig. 13.9 displays six kinds of men's underwear with different bottom lines due to heights and angles. There are some important points located on *HL* and *CrL*: "*a*" on *HL*, "*b*" on underwear bottom, "*c*" on *CrL*.

The first case in Fig. 13.9A with the underwear bottom located upper *CrL* includes different locations of point "*b*" (briefs and trunks), different bottom girth and slope, and also a different length of the side seam. The second case in Fig. 13.9B with an underwear bottom located below *CrL* represents the various underwear styles of tight-fitting boxers (from short to long). Fig. 13.10A shows the possible positions of bottom lines for the first and second cases, Fig. 13.10B is the relationship between the angle and the average values of NT_G, Fig. 13.10C is the side length measured from natural waist level down to *CrL*, and Fig. 13.10D is the relation between horizontal cross-section girth NT_G and h_T.

The one-variable linear equations for calculating new tight oblique girth and horizontal girth NT_G are

$$NT_G = 81.64 - 0.89 \cdot SL, \tag{13.6}$$

$$NT_G = 0.54 \cdot h_T + 54.59, \tag{13.7}$$

where NT_G is new thigh girth measured below or above natural thigh level, cm; SL is side seam length measured from the natural waist, cm; h_T is range of NT_G below *CrL*.

New classifications of lower male torso include new body measurements. Table 13.6 shows the types of male bodies on two levels.

First level is based on W_G and H_G and defined the total sizes of lower torso visually (small or large for slim or obese bodies). The first level has two stages. To identify the *first stage*, the measurements H_G and W_G are needed. Based on H_G, the body can be put to *A*, *B*, *C* types in accordance with the intervals established: smaller than 89 cm,

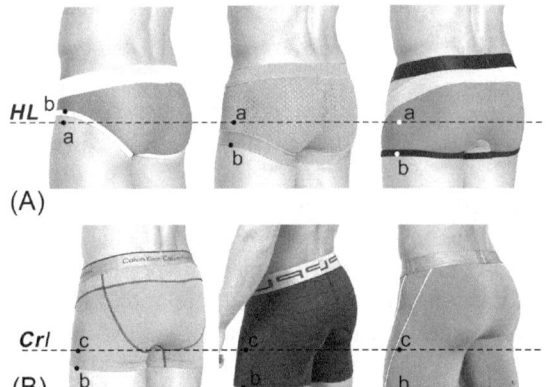

(A)

(B)

Fig. 13.9 Two cases of underwear bottoms: (A) higher than *CrL*; (B) lower than *CrL*.

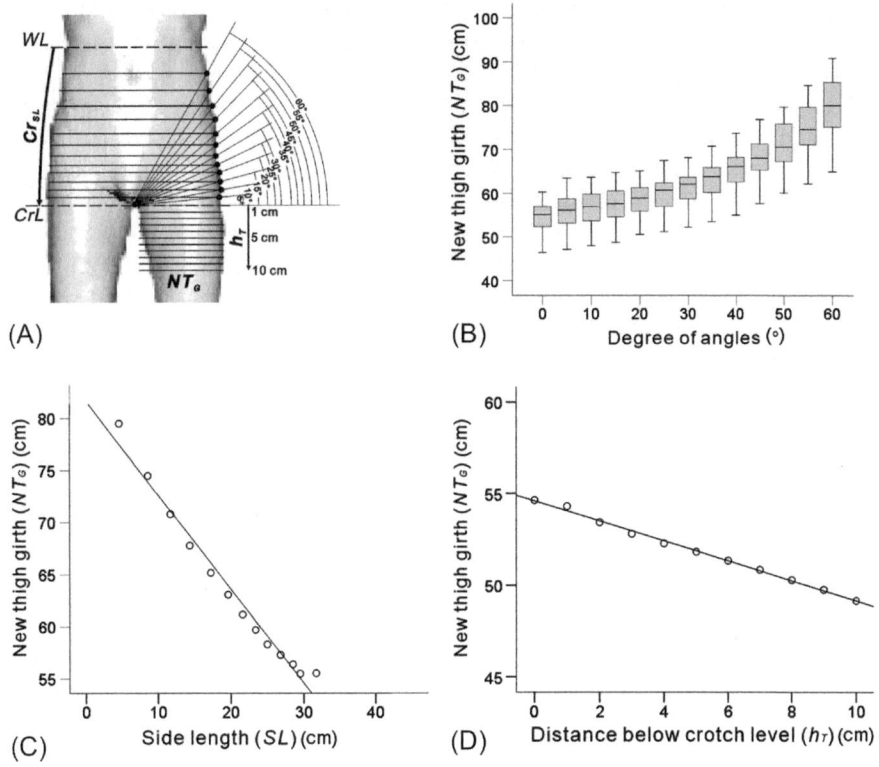

Fig. 13.10 Analysis of new thigh part: (A) angles of thigh cutting; (B) diagram of new thigh angle cutting; (C) diagram of side length with new thigh girth; (D) diagram of thigh girth below crotch level.

between 89...100 cm, and larger than 100 cm. To identify the *second stage*, the difference $\Delta(H_G - W_G)$ should be calculated to describe the contour by marks "++" for the extra-large W_G; "+" for large W_G (or similar values of H_G and W_G); *blank mark* for median of W_G (called normal W_G); "−" for extra-small W_G.

For example, if the body has small W_G and $\Delta(H_G - W_G) > 22.3$ cm, it can be marked as "*A⁻*".

If the body has medium W_G and $\Delta(H_G - W_G) = 95$ cm, it can be marked "*B*".

For equal H_G and W_G (the difference is 0...13.7 cm) and large W_G, the body will be marked as "*C⁺*".

Fig. 13.11 illustrates the body types for the first level.

Second level includes the crucial measurements for describing special characteristics of the male body. The second level characterizes the morphology of the male lower torso and includes indexes of the front and the back and four values have to be processed. The second level has four stages. ΔF and ΔB define the sizes of genitals and buttocks and after combining with ΔGW and ΔWH, the front and the back can be presented as *S, M, L*.

Table 13.6 New classification of male lower torso

Main type	Subtypes algorithms	Intervals (cm)			
		A (small)	**B** (middle)	**C** (large)	
First level	First stage				
	H_G	<89	89...100	>100	
	Second stage	Mark "−" (for small W_G)	"Blank mark" (for middle W_G)	Mark "+" (for large W_G)	Mark "++" (for extra large W_G)
	$\Delta(H_G - W_G)$	>22.3	13.7...22.3	0...13.7	<0
Second level	Third and fourth stages	**S** (small)	**M** (middle)	**L** (large)	
	$\Delta F = CL_F - BR$	<7.9	7.9...10.9	>10.9	
	$\Delta GW = GF_D - Abd_D$	<0	0...1.2	>1.2	
	Fifth and sixth stages	**S** (small)	**M** (middle)	**L** (large)	
	$\Delta B = CL_B - BR$	<6.1	6.1...9.1	>9.1	
	$\Delta WH = WB_D - HB_D$	<2.3	2.3...5.4	>5.4	
Proportions rates (%)		12...20	64...73	11...17	

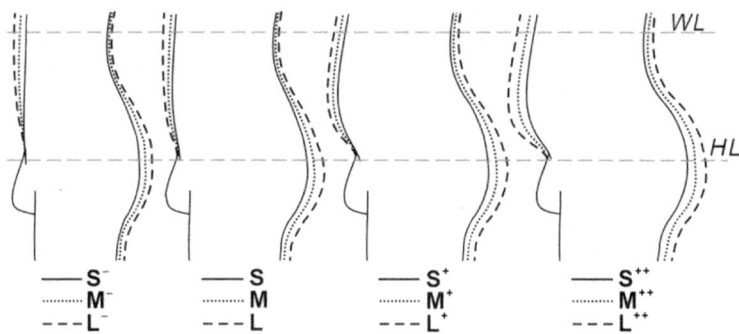

Fig. 13.11 Body types on first level of classification.

The third stage relates to the body's front and needs ΔF and ΔGW to identify the types of *S*, *M*, *L*. E.g., if the body has $\Delta GW = -0.1$ cm and $\Delta F = 7.5$ cm, it will be marked as *S*.

The fourth stage relates to the buttock shape by means of ΔB and ΔWH also for *S*, *M*, *L* types. It's a special case where four measurements of *third and fourth stages* and *fifth and sixth stages* cannot match the same intervals completely. E.g., for ΔF (or ΔB) in *S* types, ΔGW (or ΔWH) in *M* types, we comply with the principle of ΔF (or ΔB) priority to define ΔGW (or ΔWH) as *S*.

Finally, the lower torso and underwear could be marked as **B^-/LM** and so on. Here, **B^-** describes the total type of the body based on W_G and H_G, and **LM** describes the front (ΔF, ΔGW) and hip (ΔB, ΔWH) characteristics. To use this method, it is easy to describe the lower torso to help in the sizing of the underwear.

Fig. 13.12 shows three examples of how to mark the lower torso of the bodies.

This method of body type classification is therefore based just on girth and differences and allows expressing of the main body features.

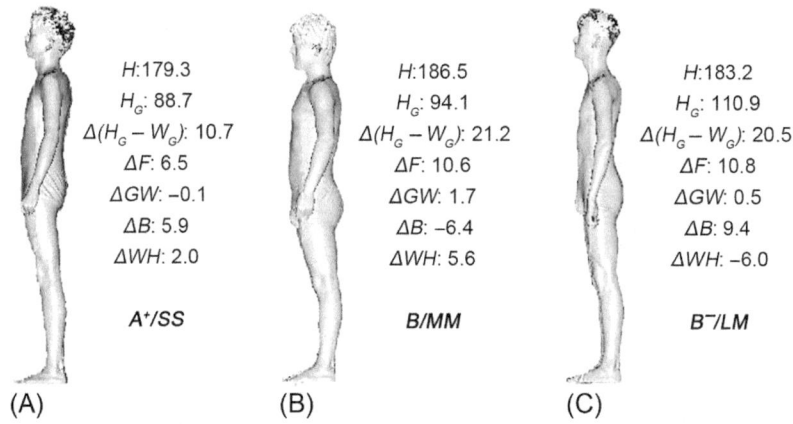

Fig. 13.12 Examples of male body identification, cm: (A) *A^+/SS*, (B) *B/MM*, (C) *B^-/LM*.

13.5.2 2D design of pressure garments

In the patternmaking process, the ease allowance is equal to the difference between the corresponding garment dimension and the body measurement. First, when the dimensions of the compression garment are smaller than the corresponding body measurements the pattern block should be designed with "negative ease." Second, the dimensions of loose and semiloose fitted garments are larger than the body measurements, so the ease has to be "positive." Third, for close-fitted garments that look like a copy of an avatar shape, the pattern blocks should be designed with "zero ease."

Pressure garments are mostly designed from knitted materials with negative ease. Because of their good elasticity, it is easy to increase the length in one direction and to shrink it in the opposite one. For example, in order to maintain the pressure distribution around the human body, the horizontal girth needs to be shortened and the vertical length increased at the same time. Fig. 13.13 shows the shape of knitted materials

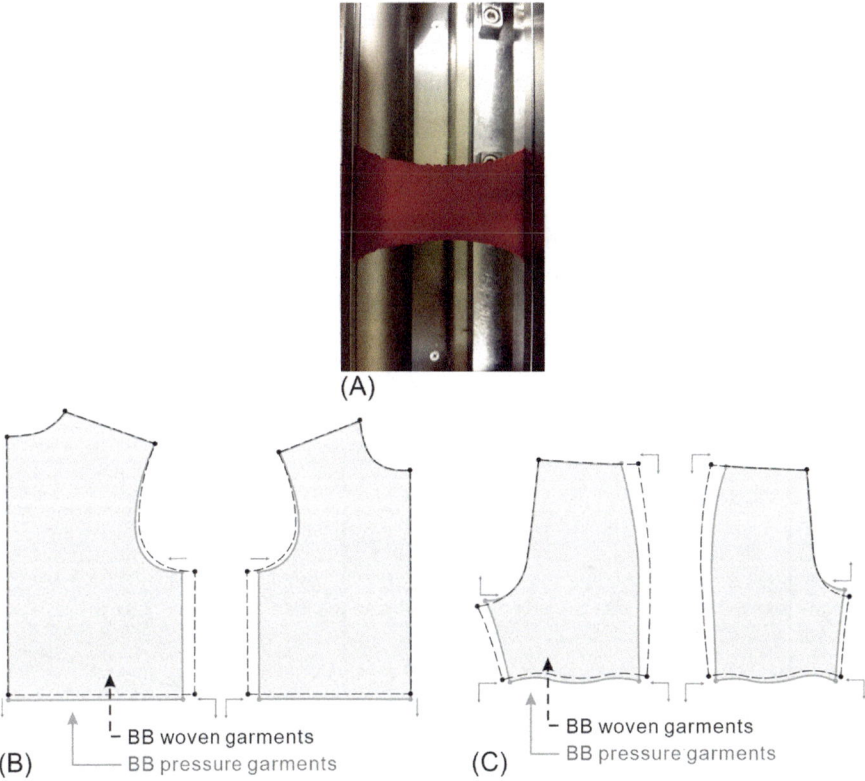

Fig. 13.13 Sample of knitted materials under an elongation test by KES FB-1 (A) and the scheme of transforming basic blocks of woven loose garments into compression garments: (B) top; (C) pants.

under elongation and methods of correction of the pattern blocks from loose shapes into tight-fitting shapes.

Knitted materials have shrinkage (or edge curl). The amount of shrinkage is determined by the material type, structure, processing technology, and finishing methods. Due to the laddering characteristic of the knitted materials, the seam allowance (including sewing loss) is usually 0.12–2.50 cm. In order to obtain the necessary pressure on the shoulder area, the pattern block should be reduced by an average of 10% compared to the dimension, calculated for a rigid cylinder of the same size (Ng, 2001).

When knitted material has good draping and little shrinkage, the length and width of the finished pressure garment will become longer and narrower. To get a desirable shape, the pattern block should be drawn with negative ease in the horizontal. When knitted material has poor draping and large shrinkage properties, we need to increase the length in the vertical in the pattern block. To sum up, the pattern drawing rules should include edge size (seam allowance), sewing loss, horizontal ease, vertical shrinkage and other various shrinkages, etc. (Anbumani, 2007).

Due to the characteristics of the yarn coil structure, the shrinkage rate of most knitted materials is from 2% to 7% (Zhang, 1990; Cui and Wu, 2018). Fig. 13.14 shows the relationship between the elongation and the shrinkage of knitted materials used for daily underwear, tight-fitting garments, shapewear, etc.) measured in two perpendicular directions.

For example, the basic pattern block of a T-shirt is designed with bust girth 84 cm and an ease value of 6 cm (total width is $84 + 6 = 90$ cm) and the full length is 50 cm. To modify this basic block into a block for a tight-fitting T-shirt, and in order to prevent the impact of tensile and shrinkage of the material after wearing, the horizontal size is reduced ($84 - 8 = 76$ cm), but the amount of shrinkage is needed to be designed in the longitudinal direction, that is, the length of the tight-fitting T-shirt becomes $50 + 3 = 53$ cm.

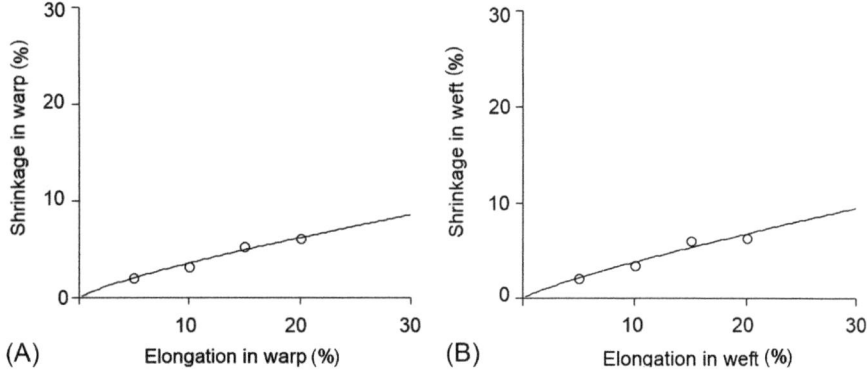

Fig. 13.14 Relationships between the elongation and the shrinkage of knitted materials in warp (A) and in weft (B).

13.5.3   Difference between common and pressure garment structures

Design of pressure garments mainly depends on the elastic properties of materials, and, according to the structural design, is different from design of the more common garments. Some thin or one-layer pressure garments are mostly designed as one-piece, seamless, or with a small number of seam lines, in order to avoid discomfort. Therefore to design pressure garments, it is necessary to attend to a structure in which simple seam lines are usually used; besides, the dart design and the common garment modeling process are not usually adapted to human body morphology. For functional pressure garments, some seam lines are necessary. To be certain of the process technology, material selection, and design of seam lines, comfort tests, including body pressure sensitivity, wearing comfort, functionality, material breathability/permeability, hygienic performance, and other aspects, should be carried out (Kuzmichev and Zhe, 2015; Xizhou et al., 2015).

13.6   Ease allowance

As mentioned before, the pattern blocks of pressure garments are designed with negative ease allowance and the dimensions of the pattern blocks are smaller than the corresponding body measurements. The ease should be determined according to the design purpose and the comfort range. The design ease of pressure garments usually is in the range from −40% to 0% in accordance with the elasticity of the knitted materials (Lu, 2013). To design pressure garments with similar functions, the ease allowance range depends upon the properties of the knitted materials and should be (in %): 0–19.5 for low elastic materials, 19.5–25.5 for medium elastic, larger than 25.5 for high elastic.

Fig. 13.15 shows the maximum elongations measured after reaching the acceptable pressure for male and female bodies; the full range of elongation is about 5.5%–38.9% with 3D samples around the body parts (Zhe and Kuzmichev, 2016b).

Fig. 13.15 shows that female bodies can accept the pressure with very large elongations of 24.3%–35.9% around the full torso (shapewear, corrective garments, functional pressure garments). Because male bodies are more muscular than female ones, the compression is more balanced, about 18.2%–20.4% for body-fitting underwear.

13.7   3D design of pressure garments

Through the use of software such as CLO 3D, Marvelous Designer, or Vidya, it's easy to make the 2D → 3D interactive design of pressure garments. There are several important factors of 3D pressure garment design, such as the accuracy of body measurements, 3D body (avatar) building technology, 3D virtual try-on simulation technology, the technique of 2D graphics to 3D garments, etc. At present, there are a large number of research papers or reports on 3D try-on, testing of the pressure garments

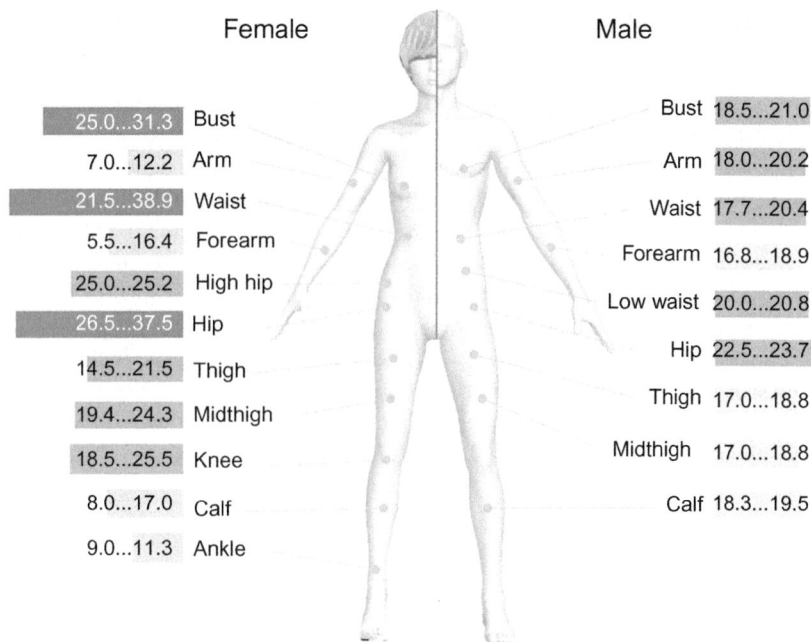

Fig. 13.15 Recommended maximum ease ranges, −%.

(Abtew et al., 2018; Kuzmichev et al., 2018; Mengna et al., 2015), and software (Abtew et al., 2018; Wang and Tang, 2010; Liu et al., 2017; Nakashima et al., 2016).

To design 3D garments, the connection between the virtual avatar and the clothes should be established by means of the ease allowances. The calculation of "negative ease" of a pattern block for compression garments is a complex scientific problem in the case of the 2D→3D method, as was shown before.

For positive ease, it is possible to generate the patterns by flattening the parts of the 3D garment in a 3D-to-2D approach. The ease allowances are created like a gap between the avatar and the inner surface of a woman's dress (Mengna, 2015). CAD Lectra DC3D and LooksTailorX were used as the methods for design of loose and semiloose fitted garments, with positive eases on the base of a solid avatar. To create a 3D garment with positive ease directly on the torso and the subpart of the scanned body, the horizontal cross-sections can be obtained after a relative increase of the original cross-sections of the individual avatar (Thomassey and Bruniaux, 2013; Tao and Bruniaux, 2013). Garments for diving and cycling with zero ease allowances are designed by flattening the parts of the avatar surface (Naglic et al., 2016; Liu et al., 2016b).

Reviewed workflows of 3D designs cannot be used for creation of compression garments with "negative ease," since they do not consider the differences between the garment dimensions and body measurements to obtain the required compression pressure. In order to take this fact into account, the following three approaches can be used:

(1) an inclusion of special procedures in a flattening algorithm;
(2) a transformation of pattern blocks taken from an avatar on the 2D surface after flattening (Ziegert and Keil, 1988; Watkins, 2011);
(3) flattening of the 3D surface which has "negative ease" compared to the surface of the avatar.

The existing implementations of the first approach don't consider the anisotropic properties of textile materials and require special software.

The second approach, as a kind of traditional method to design the basic blocks, is not appropriate for 3D design.

The third approach is based on the creation of a compression garment surface in a form appropriate for 3D-to-2D flattening of its parts. For its implementation, it is necessary to build a 3D model of the compression garment in a "relaxed state" (when the shell of the garment is not elongated and locates inside the avatar), and then to generate its flattened pattern blocks. The technical capability of constructing and flattening the parts of a 3D garment with "negative ease" exists in CAD LooksTailorX. However, the body measurements taken from the typical solid avatar that presents the standard body might be not enough to create a more complex and realistic pressure profile. Besides, there is no opportunity to calculate an ease based on the requirements of pressure comfort. The 3D shape of the compression garment based on the cross-sections of the human head and pressure comfortable data can be calculated (Balandina, 2007).

The most convenient and common way of representing the surface of 3D virtual garments is a polygonal mesh. It allows reproduction of complex surface topography and achieving of a high accuracy level of the shape. There are two main methods for flattening of 3D polygonal mesh objects: physical and geometrical. The physical method is based on the calculation of minimal tension energy of the deformable elements of the polygonal net; the advantages of deformable elements are an isomorphism and an allowance for the anisotropic properties, but elements are iterative and may be unstable (Hsiao, 2014; Liang and Bin, 2004; Wang et al., 2002; Li et al., 2005; Zhang and Luo, 2003).

The geometrical method is based on geometrical values of deformable elements, such as edge lengths, angles, areas of faces, etc. (Wang and Tang, 2010; Sheffer et al., 2005; Azariadis and Aspragathos, 2001). Geometrical constraints help to determine proper positions of mesh vertexes.

Usually, the methods for calculating the pattern blocks of garments made of knitted materials are physical or combined and give an opportunity to directly use the equations for tensile force and strain of real materials.

Most CAD applications allow 2D-to-3D virtual try-on, so the generated pattern blocks could be used for virtual simulation, such as stitching of the parts and dressing the garment on the avatar (Browswear, n.d.; Lectra Modaris 3DFit, n.d.; Optitex 3D Runway, n.d.).

According to Eq. (13.1), the tensile force determines the pressure on the body, at the same time, the extensibility of the textile material must correspond to the garment pressure. The appropriate range of knitted fabric deformation can be calculated by the function $\sigma = f(\varepsilon)$ (Ibrahim, 1968). Since the materials for compression garments exhibit deformations in a wide range, this approach will be difficult to apply for an inactive part of the body (Liu et al., 2007).

The process of compression garment design has the following main steps:

(1) the creation of an avatar by the body scan data,
(2) the choosing of knitted materials,
(3) the exploration of system "body-compression garments" to find which dimensions of both objects are changing,
(4) the creation of a 3D surface of the compression garment in a "relaxed state,"
(5) the design of pattern blocks by a 3D-to-2D flattening procedure.

To take into account the corresponding deformations of the soft tissues and the textile materials and to calculate the parameters of the pattern blocks, a soft deformable avatar is needed, different from a solid avatar. After deformation, the morphology of the soft avatar will present the shape of the compression garment in a "relaxed state," on the basis of which the pattern blocks can be generated by the flattening procedure. This approach allows consideration of the morphology of the body, the compression abilities of soft tissues, and the properties of knitted materials used. Fig. 13.16 shows the flowchart of the 3D-to-2D method.

To create digital avatars of female bodies, a body scanner such as Vitus Smart XXL (Human Solutions GmbH and VITRONIC GmbH) can be used. Fig. 13.17A shows the example of the lower subcorpus part of the scanned body. The surface of the 3D scan consists of a huge number of points, so it is a complex object to specify seam lines and flattening its parts. To simplify the surface and reduce the number of polygon vertexes, the scanned image can be presented as the combination of *B*-splines for horizontal cross-sections. Since the method was explored for a compression garment

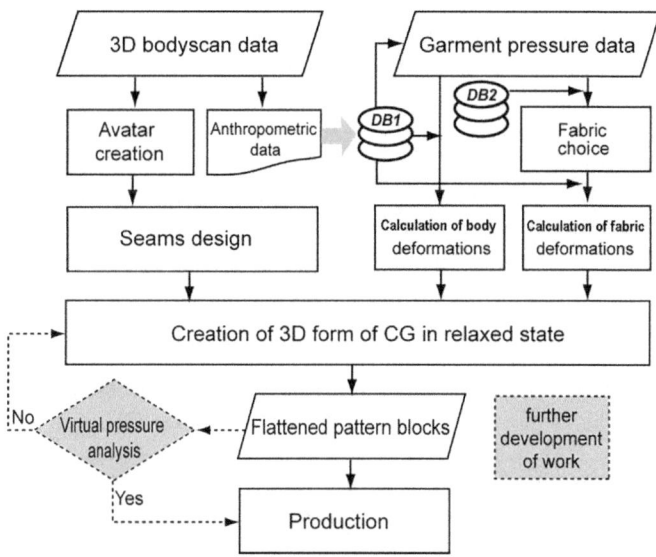

Fig. 13.16 Flowchart of 3D-to-2D approach for compression garment design.

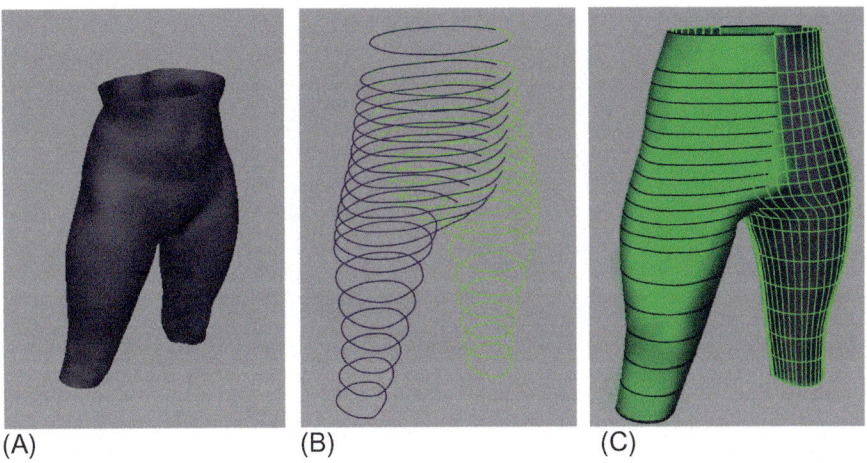

(A)   (B)   (C)

Fig. 13.17 Algorithm for processing the scanned body: (A) polygonal surface of scanned body, (B) cross-sections as *B*-splines, (C) 3D spline of NURBS surface.

covering the subcorpus part of the torso, the avatar was not designed completely. Fig. 13.17B shows that the cross-sections were built by means of Maya Autodesk. To construct the avatar on the basis of the curves framework, the left and the right were mirrored by nonuniform rational B-spline (NURBS) surfaces, saving the main body measurements as Fig. 13.17C shows.

The surface of the avatar designed from NURBS curves is the base of the structure of the garment by means of the seam lines of the proposed garment. Fig. 13.18B shows

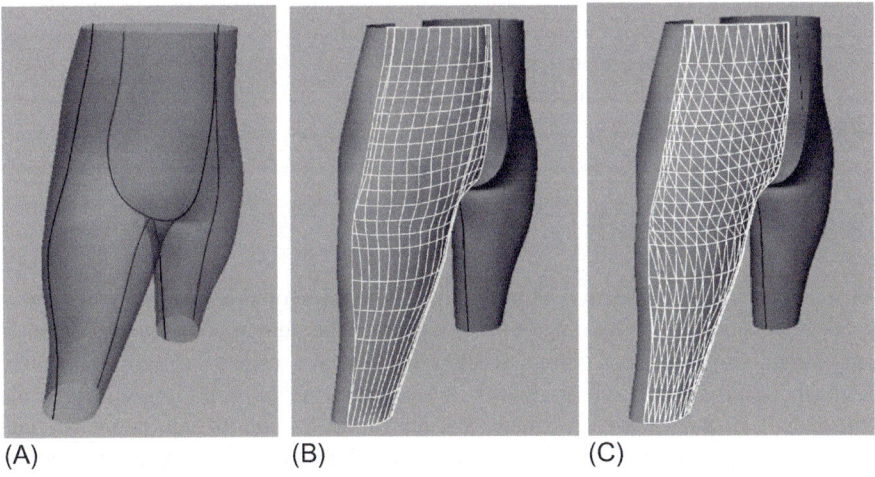

(A)   (B)   (C)

Fig. 13.18 Specifying the blocks of the compression garment: (A) seams on the avatar, (B) blocks of compression garment as a trimmed NURBS surface, (C) blocks as the polygonal mesh objects.

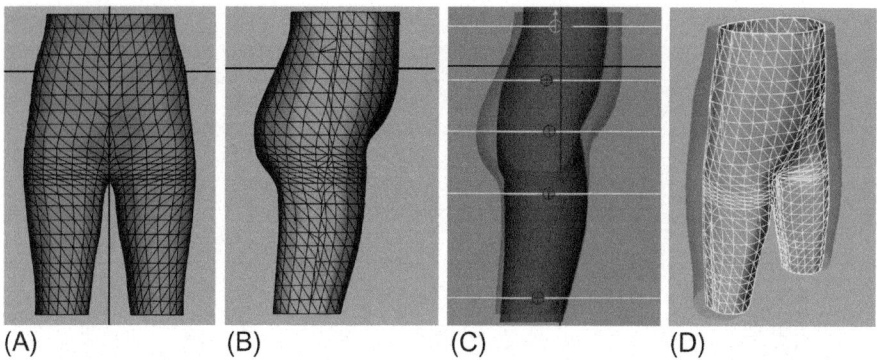

(A)   (B)   (C)   (D)

Fig. 13.19 The avatar and the compression garment: (A, B) the avatar surface covered with the compression garment in the frontal and lateral projections; (C, D) lateral projection and the general view of the system "avatar (outside)—compression garment (inside)."

the position of the crotch, side, and inseam lines. Then, the resulting sections, located between the seam lines, were trimmed and converted to the polygons (Fig. 13.18B and C).

To create the 3D form of the pressure garment in a "relaxed state," the 3D avatar should be scaled in accordance with the deformation of soft tissue to get the desirable pressure. Fig. 13.19A and D shows the projections of the avatar and the compression garment in the "relaxed state."

It is clear from Fig. 13.19C and D that the sections of the compression garment in the "relaxed state" are significantly reduced relative to the initial avatar. The elongation of patterns in the vertical direction corresponds to the contraction of fabric in a direction transversal to the tensile force. To obtain the flattened pattern blocks, the 3D-to-2D approach was implemented. Flattened pattern blocks made by this approach are shown by the triangulated grid in Fig. 13.20.

Virtual reality brings remarkable new benefits to the modern design of pressure garments.

13.8 Future trends

As seen in the previous sections, through recent research on pressure garments, a reasonable pressure sensitivity range of many parts of the human body and the range of body remodeling have been obtained. These studies have made a basic contribution to the further study of pressure garments in the future. At the same time, through the future study of new innovative materials, the rational value of ease allowance can be designed for a reasonable pressure range. The influence of pressure garments on comfort has always been a matter of special concern to people.

In recent years, the design of pressure garments has also been mainly based on human body scanning technology. At the same time, virtual simulations can be carried

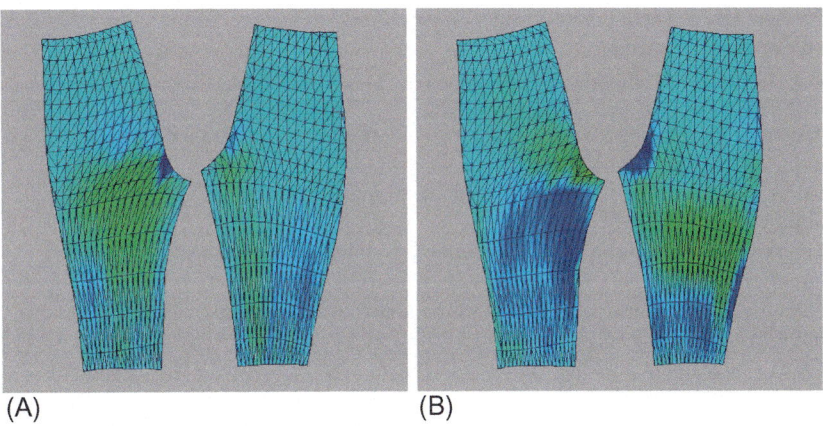

Fig. 13.20 Flattened pattern blocks with the presentation of deformation inside: (A) strain in vertical direction and (B) shear.

out by using digital twins of bodies, textile materials, clothes after certain virtual verification, measurements, final design, actual experiments, evaluation and detection. According to developing trends, the future research will focus on the design and evaluation of pressure garments from a 3D simulation perspective.

Future development involves comprehensive scientific research on the "avatar of human body-pressure garment" system. The human body or its virtual digital avatar is the starting point to create the pattern block through the technology promotion of materials, visualizing the final product through the new technology of virtual simulation. Emerging technologies are facilitating this movement and include 3D virtual simulation, 3D body scanning, and 3D garment design. On the one hand, the sizes of pressure garments need to be further optimized according to the characteristics of human body morphology for more accurate design of details. On the other hand, the designers can design pressure garments in accordance with real human body shape and customized preferences. We can also reclassify the sizing system for pressure garments. For consumers, this would make it easier to find the clothes with the best fit for their body characteristics.

13.9 Sources of further information and advice

Some brief introductions to the history of pressure garments, most of which were published on the Internet site, also include defects in measurement items described in anthropometric textbooks. Information on medical research on pressure garments can be found on the Internet.

https://www.ncbi.nlm.nih.gov/pmc
The US national library of medicine national institutes of health.
http://xueshu.baidu.com

Pressure test research report on stress apparel.
https://www.clo3d.com/
A virtual garment simulate and design software.
http://www.tc2.com
A company providing 3D body scanning equipment and software for measurement.
http://www.gerbertechnology.com
Gerber, a provider of software and manufacturing systems for the garments.
http://www.lectra.com
Lectra, a French company offering technology solutions for garments.
http://www.nike.com
Nike Inc., the pressure garment and sizing system reference is provided.
https://www.taobao.com
China's largest Internet Mall; can view a large number of pressure garment photo materials, with the size of the table, as well as wear evaluation.
https://www.isobar-compression.com/
Isobar, a company for pressure garments made to order.
German (VDI/VDE 2634-1, -2, -3) and Japanese (JIS B7441) standards are the only standards that establish a protocol for evaluating the accuracy of 3D scanners.
Japanese Standards Association (1997), JIS L 4007:1997 Sizing System for Hosiery, Japanese Standard Association, Tokyo.

References

Abtew, M.A., Bruniaux, P., Boussu, F., Loghin, C., Cristian, I., Chen, Y., 2018. Development of comfortable and well-fitted bra pattern for customized female soft body armor through 3D design process of adaptive bust on virtual mannequin. Comput. Ind. 100, 7–20.

Ali, A., Creasy, R.H., Edge, J.A., 2011. The effect of graduated compression stockings on running performance. J. Strength Cond. Res. 25 (5), 1385–1392.

Anbumani, N., 2007. Knitting Fundamentals, Machines, Structures and Developments. New Age International, p. 274.

ASTM D2594-04, 2016. Standard Test Method for Stretch Properties of Knitted Fabrics Having Low Power.

Azariadis, P.N., Aspragathos, N.A., 2001. Geodesic curvature preservation in surface flattening through constrained global optimization. Comput. Aided Des. 33 (8), 581–591.

Balandina, G.V., 2007. Investigation of the impact of a corset on the female silhouette. Sewing Ind. 4, 52–53.

Beck, T., 1995. The Embroiderer's Story: Needlework From the Renaissance to the Present Day. David & Charles, p. 160.

Boldovkina, O.S., 2005. Garment Design. Vladivostok, Russia, p. 160.

Browswear, n.d. http://www.browswear.com.

Chan, A.P., Fan, J., 2002. Effect of garment pressure on the tightness sensation of girdles. Int. J. Garment Sci. Technol. 14 (2), 100–110.

Cheng, H., 2017. Research on compression on legs in the process of hose top constraint failure. Mod. Text. Technol. 25 (5), 38–42.

Cheng, Z., Kuzmichev, V.E., 2013. Artistic and constructive database for the design of men's underwear (part 1). Sewing Ind. (6), 26–29.

Cheng, J.C.Y., Evans, J.H., Leung, K.S., Clark, J.A., Choy, T.T.C., Leung, P.C., 1984. Pressure therapy in the treatment of post-burn hypertrophic scar—a critical look into its usefulness and fallacies by pressure monitoring. Burns 10 (3), 154–163.

Cieślak, M., Karaszewska, A., Gromadzińska, E., Jasińska, I., Kamińska, I., 2017. Comparison of methods for measurement of the pressure exerted by knitted fabrics. Text. Res. J. 87 (17), 2117–2126.

Cui, J., Wu, Y.-M., 2018. Current status of research on shrinkage between different linings and knitted fabrics. Liaoning Silk (2), 25–28.

Cunnington, C.W., Cunnington, P., 1992. The History of Underclothes. Courier Corporation, p. 272.

Dan, R., Fan, X.R., Shi, Z., Zhang, M., 2016. Finite element simulation of pressure, displacement, and area shrinkage mass of lower leg with time for the top part of men's socks. J. Text. Inst. 107 (1), 72–80.

Dongsheng, C., Qing, Z., 2003. A study on garment pressure for men's suit comfort evaluation. Int. J. Garment Sci. Technol. 15 (5), 320–334.

Du, Y.D., He, T.H., 2014. The influence of fabric property on the pressure comfort of ladies swimsuit. Prog. Text. Sci. Technol. 6, 029.

Elder, H.M., Fisher, S., Hutchison, G., Beattie, S., 1985. A psychological scale for fabric stiffness. J. Text. Inst. 76 (6), 442–449.

Fan, J., Hunter, L., 2009. Physiological comfort of fabrics and garments. In: Engineering Apparel Fabrics and Garments. Woodhead Publishing, pp. 201–250.

Fan, J., Yu, W., Hunter, L., 2004. Clothing Appearance and Fit: Science and Technology. CRC Press, p. 260.

Fangyuan, W., Xiaona, C., 2013. Preliminary study on the relationship between chest displacement and bra comfort. J. Text. 34 (1), 106–109.

Filatov, V.N., 1987. Elastic Textile Shells. Legpromizdat, Moscow, p. 248.

FISO, n.d. Fiber Optic Sensing Solutions. http://www.fiso.com/

Fontanel, B., 1997. Support and Seduction: A History of Corsets and Bras. Harry N. Ambrams, New York, p. 160.

Giele, H.P., Liddiard, K., et al., 1997. Direct measurement of cutaneous pressures generated by pressure garments. Burns 23 (2), 137–141.

GOST 26435-85. Russia. Knitted materials. Test of elongation.

Guney, S., Kaplan, S., 2016. Parameters affecting pressure comfort performance and measurement of pressure comfort. J. Text. Eng. 23 (102), 152–163.

Gupta, D., 2016. Functional clothing—definition and classification. Indian J. Fibre Text. Res. 36 (4), 321–326.

Hageman, D.J., Wu, S., Kilbreath, S., Rockson, S.G., Wang, C., Tate, M.L.K., 2018. Biotechnologies toward mitigating, curing, and ultimately preventing edema through compression therapy. Trends Biotechnol. 36 (5), 537–548.

Hang, T.W., 2013. The Evaluation of Pressure and Tactile Comfort of Girdles. (Thesis Submitted for the Degree of Bachelor of Arts)Institute of Textiles & Clothing, The Hong Kong Polytechnic University.

Hirai, M., 2002. Effect of elastic compression stockings in patients with varicose veins and healthy controls measured by strain gauge plethysmography. Skin Res. Technol. 8, 236–239.

Hong, S., Yonggui, L., Airong, B., et al., 2006. Discussion on size standard of knitted underwear. Knitting Ind. (9), 58–59.

Hsiao, S.W., 2014. Surface flattening assisted with 3D mannequin based on minimum energy. Int. J. Comput. Inf. Syst. Control Eng. 8 (8), 87–94.

Huashan, L., Yuxiu, Y., Mengyuan, L., Shuyan, X., Zimin, J., 2017. The effect of clothing pressure on lower extremity muscle fatigue in running. J. Text. 38 (7), 118–123.

Ibrahim, S.M., 1968. Mechanics of form-persuasive garments based on spandex fibers1. Text. Res. J. 38 (9), 950–963.

Ivanova, Z.T., 1998. Development of Compression Garment Design (Ph.D. thesis). MGUDT, Moscow, p. 262.

Jennes, L.K.P., 2012. Study on Shapewear Preferences for Women in Hong Kong Intimate Apparel Market. (Thesis Submitted for the Degree of Bachelor of Arts)Institute of Textiles & Clothing, The Hong Kong Polytechnic University.

Jiang, L., Cheng, H., Zhang, P., 2016. Subjective evaluation of failure performance of socks and mouth restraints. Wool Text. J. 44 (10), 70–73.

Kawabata, S., 1980. The Standartization and Analysis of Hand Evaluation. Textile Machinery Society of Japan, Nishi-ku/Osaka, Japan.

Klöti, J., Pochon, J.P., 1982. Conservative treatment using compression suits for second and third degree burns in children. Burns 8 (3), 180–187.

Koszewska, M., 2004. Outsourcing as a modern management strategy prospects for its development in the protective garment market. AUTEX Res. J. 4 (4), 228–231.

Kraemer, W.J., Volek, J.S., Bush, J.A., Gotshalk, L.A., Wagner, P.R., Gomez, A.L., 2000. Influence of compression hosiery on physiological responses to standing fatigue in women. Med. Sci. Sports Exerc. 32 (11), 1849–1858.

Kunzle, D., 2006. Fashion & Fetishism: Corsets, Tight-Lacing and Other Forms of Body-Sculpture. The History Press, p. 400.

Kuzmichev, V.E., Zhe, C., 2014. Analysis of pressure distribution in system "body-men's underwear" In: Book of 14th AUTEX World Textile Conference. Uludag University Publishing House, Turkey, p. 45.

Kuzmichev, V.E., Zhe, C., 2015. Men underwear design-main problems and solutions. In: 45th International Conference on Computers & Industrial Engineering 2015 (CIE45)vol. 4(1). pp. 256–263.

Kuzmichev, V.E., Tislenko, I.V., Zhe, C., et al., 2015. Investigation of compression ability of knitted materials. Bull. Kazan Univ. Technol. 18 (20), 179–180.

Kuzmichev, V.E., Zhe, C., Mengna, G., 2016. Experimental basement of compression pressure prognosis. News of higher educational institutions. Technol. Text. Ind. 364 (4), 91–96.

Kuzmichev, V.E., Tislenko, I.V., Adolphe, D.C., 2018. Virtual design of knitted compression garments based on bodyscanning technology and 3D-to-2D approach. Text. Res. J. https://doi.org/10.1177/0040517518792722.

Lectra Modaris 3DFit, n.d. http://www.lectra.com

Lee, H., Eom, R., Park, S., et al., 2017. Effects of knit fabric layering and flat seam direction on stretchability and garment pressure. Kor. J. Living Environ. 24 (4), 533–540.

Leung, W.Y., 2010. Evaluation of Compression Garment Design Factor and Prediction of Garment Pressure on Wearer. Hong Kong Polytechnic University.

Li, J., Zhang, D., Lu, G., Peng, Y., Wen, X., Sakaguti, Y., 2005. Flattening triangulated surfaces using a mass-spring model. Int. J. Adv. Manuf. Technol. 25 (1–2), 108–117.

Li, Y.N., Lu, A.M., Dai, X.Q., Chen, J.A., Zhao, X.W., 2011. Effect of garment pressure on lower limb muscle activity during running. Adv. Mater. Res. 175, 832–836.

Liang, W., Bin, H., 2004. Optimal flattening of freeform surfaces based on energy model. Int. J. Adv. Manuf. Technol. 24 (11−12), 853–859.

Lim, N.Y., 2006. Innovation and Technology of Women's Intimate Apparel. Woodhead Publishing, pp. 114–131.

Lim, N.Y., Zheng, R., Yu, W., Fan, J., 2006a. Assessment of Women's Body Beauty. Innovation and Technology of Women's Intimate Apparel, pp. 1–27.

Lim, N.Y., Ng, S.P., Yu, W., Fan, J., 2006b. Pressure evaluation of body shapers. Innovation and Technology of Women's Intimate Apparel, pp. 151–168.

Liu, D., 2009. Tight-garment structure design and grading. J. Text. Res. (8), 113–116.

Liu, R., Kwok, Y.L., Li, Y., et al., 2007. Quantitative assessment of relationship between pressure performances and material mechanical properties of medical graduated compression stockings. J. Appl. Polym. Sci. 104 (1), 601–610.

Liu, L., Liu, X., Li, Y., 2016a. Young women's subjective evaluation of pressure comfort in body suits. Shanghai Text. Sci. Technol. (9), 39–43.

Liu, K., Wang, J., Zhu, C., Hong, Y., 2016b. Development of upper cycling clothes using 3D-to-2D flattening technology and evaluation of dynamic wear comfort from the aspect of clothing pressure. Int. J. Cloth. Sci. Technol. 28(6).

Liu, K., Wang, J., Hong, Y., 2017. Wearing comfort analysis from aspect of numerical garment pressure using 3D virtual-reality and data mining technology. Int. J. Cloth. Sci. Technol. 29 (2), 166–179.

Lu, L., 2013. Research on the Structure of the Trousers Based on the Hip Pattern of Young Women. (Doctoral Dissertation)Donghua University, Shanghai, p. 92.

Lynn, E., 2014. Underwear: Fashion in Detail. V&A Publishing, p. 224.

Macintyre, L., 2006. Pressure garments for use in the treatment of hypertrophic scars—a review of the problems associated with their use. Burns 32, 10–15.

Macintyre, L., 2014. New design tool for delivering graduated compression. In: Paper Presented at NED University of Engineering and Technology's International Textile Conference, Karachi, Pakistan.

Mengna, G., 2015. Development of Virtual Garment Design With Comfort Simulation (Ph.D. thesis). ISPU, Ivanovo, p. 215.

Mengna, G., Kuzmichev, V.E., 2013. Pressure and comfort perception in the system "female body-dress" AUTEX Res. J. 13 (3), 71–78.

Mengna, G., Kuzmichev, V.E., Adolphe, D.C., 2015. Human-friendly design of virtual system "female body-dress" Autex Res. J. 15 (1), 19–29.

Milligan, A., Mills, C., Scurr, J., 2014. Within-participant variance in multiplanar breast kinematics during 5 km treadmill running. J. Appl. Biomech. 30 (2), 244–249.

Mpampa, M.L., Azariadis, P.N., Sapidis, N.S., 2010. A new methodology for the development of sizing systems for the mass customization of garments. Int. J. Cloth. Sci. Technol. 22 (1), 49–68.

Na, Y., 2015. Clothing pressure and physiological responses according to boning type of non-stretchable corsets. Fibers Polym. 16 (2), 471–478.

Naglic, M.M., Petrak, S., Stjepanovic, Z., 2016. Analysis of tight fit clothing 3D construction based on parametric and scanned body models. In: D'apuzzo, N. (Ed.), International Conference on 3D Body Scanning Technologiespp. 302–313.

Nakahashi, M., Morooka, H., Nakamura, N., Yamamoto, C., Morooka, H., 2005. An analysis of waist-nipper factors that affect subjective feeling and physiological response—for the design of comfortable women's foundation garments. Sen'i Gakkaishi 61 (1), 6–12.

Nakashima, M., Hosoya, T., Shimana, T., 2016. Musculoskeletal simulation of sports motion considering tension distribution in a whole body compression garment. Procedia Eng. 147, 252–256.

Ng, S., 2001. Pressure model of elastic fabric for producing pressure garments. Text. Res. J. 71 (3), 275–279.

Ng-Yip, F., 1993. Medical clothing. A tutorial paper on pressure garments. Int. J. Cloth. Sci. Technol. 5 (1), 17–24.

Nishimatsu, T., 1988. Fabric stiffness. Text. Inst. 75, 307–313.

Nishimatsu, T., Ohmura, K., Sekiguchi, S., Toba, E., Shoh, K., 1998. Comfort pressure evaluation of men's socks using an elastic optical fiber. Text. Res. J. 68 (6), 435–440.

Numbers.com, n.d. Size Conversion—Shoes, Clothing, Ring Size Charts. https://www.asknumbers.com/ClothingMensConversion.aspx

Optitex 3D Runway, n.d. http://www.optitex.com

Percoco, G., 2011. Digital close-range photogrammetry for 3D body scanning for custom-made garments. Photogramm. Rec. 26 (133), 73–90.

Pickering, T.G., Hall, J.E., Appel, L.J., Falkner, B.E., Graves, J., Hill, M.N., Roccella, E.J., 2005. Recommendations for blood pressure measurement in humans and experimental animals: part 1: blood pressure measurement in humans: a statement for professionals from the Subcommittee of Professional and Public Education of the American Heart Association Council on high blood pressure research. Circulation 111 (5), 697–716.

Ping, J., Chunhong, S., 2002. Brief History of Clothing. China Textile & Apparel Press, Beijing, p. 152.

Qi, C., 2010. Study on Bra Comfort Comfort. (Doctoral Dissertation)Tianjin University of Technology, p. 89.

Qiming, W., Wenbin, Z., 2003. Research on men's prototype theory and technology based on human body characteristics in Zhejiang province—men's prototype structure design principle and technology. J. Donghua Univ. 29 (6), 30–34.

Qiyue, Z., Xin, L., 2018. Analysis of the status quo of garment pressure comfort. Liaoning Silk (1), 29–30.

Qu, C., Song, X., 2015. Effect of Women's one-piece swimwear pressure on human physiology. J. Shanghai Univ. Eng. Sci. 3, 022.

Rider, B.C., Coughlin, A.M., Hew-Butler, T.D., Goslin, B.R., 2014. Effect of compression stockings on physiological responses and running performance in division III collegiate cross-country runners during a maximal treadmill test. J. Strength Cond. Res. 28 (6), 1732–1738.

Roberts, H.E., 1977. The exquisite slave: the role of clothes in the making of the Victorian woman. J. Women Cult. Soc. 2 (3), 554–569.

Rosoff, N.G., 2006. "A glow of pleasurable excitement": images of the new athletic woman in American popular culture, 1880–1920. In: Sport, Rhetoric, and Gender. Palgrave Macmillan, New York, pp. 55–64.

Salen, J., 2008. Corsets: Historic Patterns and Techniques. Anova Books, p. 128.

Sang, J.S., Park, M.J., 2013. Knit structure and properties of high stretch compression garments. Text. Sci. Eng. 50 (6), 359–365.

Sarah, C.S.W., 2012. Development of a Men's Body Re-Shaper. (Thesis Submitted for the Degree of Bachelor of Arts)Institute of Textiles & Clothing, The Hong Kong Polytechnic University.

Sasaki, K., Miyashita, K., et al., 1997. Evaluation of foundation comfort based on the sensory evaluation and dynamic garment pressure measurement. J. Jpn. Res. Assoc. Text End Uses 38, 53–58.

Scanlan, A.T., Dascombe, B.J., RJ Reaburn, P., Osborne, M., 2008. The effects of wearing lower-body compression garments during endurance cycling. Int. J. Sports Physiol. Perform. 3 (4), 424–438.

Scurr, J., White, J., Hedger, W., 2009. Breast displacement in three dimensions during the walking and running gait cycles. J. Appl. Biomech. 25 (4), 322–329.

Sheffer, A., Mogilnitsky, M., Bogomyakov, A., 2005. Abf++: fast and robust angle based flattening. ACM Trans. Graph. 24 (2), 311–330.

SIXECHARTER, n.d. American Apparel Size Charts. http://www.sizecharter.com/brands/americapparel/womens

Sloan, W., 1963. Cotton stretch fabrics by slack mercerization: part 1: the effects of yarn and fabric construction. Text. Res. J. 33, 191–199.

Song, H.K., Ashdown, S.P., 2011. Categorization of lower body shapes for adult females based on multiple view analysis. Text. Res. J. 81 (9), 914–931.

Sperlich, B., Haegele, M., Achtzehn, S., Linville, J., Holmberg, H.C., Mester, J., 2010. Different types of compression clothing do not increase sub-maximal and maximal endurance performance in well-trained athletes. J. Sports Sci. 28 (6), 609–614.

Sperlich, B., Born, D.P., Zinner, C., Hauser, A., Holmberg, H.C., 2014. Does upper-body compression improve 3×3-min double-poling sprint performance. Int. J. Sports Physiol. Perform. 9 (1), 48–57.

Staley, M.J., Richard, R.L., 1997. Use of pressure to treat hypertrophic burn scars. Adv. Wound Care 10 (3), 44–46.

Steele, V., 2001. The Corset: A Cultural History. Yale University Press, p. 208.

Stewart, M.L., Janovicek, N., 2001. Slimming the female body? Re-evaluating dress, corsets, and physical culture in France, 1890s–1930s. Fash. Theory 5 (2), 173–193.

Surikova, G.I., Surikova, O.V., Kuzmichev, V.E., 2012. CAD. Infra-M Publishing House, Moscow, p. 476.

Tao, X., Bruniaux, P., 2013. Toward advanced three-dimensional modeling of garment prototype from draping technique. Int. J. Cloth. Sci. Technol. 25 (4), 266–283.

Tekscan, 2019a. Pressure Mapping, Force Measurement Tactile Sensors. https://www.tekscan.com/store/category/force-sensors-flexiforce.

Tekscan, 2019b. FlexiForce WELF 2 System. https://www.tekscan.com/products-solutions/systems/flexiforce-welf-2-system.

Thomassey, S., Bruniaux, P., 2013. A template of ease allowance for garments based on a 3D reverse methodology. Int. J. Ind. Ergon. 43 (5), 406–416.

Tonghui, Z., Yanbo, Y., 2017. Status quo and development trends of clothing pressure comfort. Shandong Text. Technol. (2), 37–41.

Troynikova, O., 2011. 3D body scanning method for close-fitting garments in sport and medical applications. In: HFESA 47th Annual Conference.

Varghese, N., Thilagavathi, G., 2016. Handle, fit and pressure comfort of silk/hybrid yarn woven stretch fabrics. Fibers Polym. 17 (3), 484–494.

Vuruskan, A., Ashdown, S.P., 2016. Fit analyses of bicycle clothing in active body poses. In: International Textile and Apparel Association (ITAA) Annual Conference Proceedings. vol. 73. pp. 1–2 Vancouver, BC.

Wang, C.C., Tang, K., 2010. Pattern computation for compression garment by a physical/geometric approach. Comput. Aided Des. 42 (2), 78–86.

Wang, C., Smith, S., Yuen, M., 2002. Surface flattening based on energy model. Comput. Aided Des. 34 (11), 823–833.

Wang, Y., Liu, Y., et al., 2017. Pressure comfort sensation and discrimination on female body below waistline. J. Text. Inst. 1–9.

Watkins, P., 2011. Designing with stretch fabrics. Indian J. Fibre Text. Res. 36 (4), 366–379.

White, J., Scurr, J., Hedger, W., 2011. A comparison of three-dimensional breast displacement and breast comfort during overground and treadmill running. J. Appl. Biomech. 27 (1), 47–53.

Wong, S.W., 2002. Prediction of Garment Sensory Comfort Using Neural Networks and Fuzzy Logic. (Doctor Thesis).

Wong, A.S., Li, Y., 2004. Influence of fabric mechanical property on clothing dynamic pressure distribution and pressure comfort on tight-fit sportswear. Sen'i Gakkaishi 60 (10), 293–299.

Wu, J., Yu, W., 2006. A review of the evaluation of the compression comfort of knitted fabrics. J. Wuhan Text. Univ. 19 (3), 1–4.

Xiangling, M., Weiyuan, Z., 2006. Research progress of clothing pressure comfort. J. Text. 27 (7), 109–112.

Xinzhou, W., Kuzmichev, V.E., Adolf, D.C., 2016. Modeling of the deformation of figures under water environment and water pressure for underwater sports. In: Innovative Development of Light and Textile Industry: A Collection of Materials of the All-Russian Scientific Student Conference.vol. 191. pp. 32–36.

Xizhou, W., Jinsong, D., Yunxiang, L., 2015. Analysis of motion mechanics and design points of wet diving pants. J. Beijing Inst. Cloth.: Nat. Sci. Ed. (3), 33–39.

Xu, C., 2016. Evaluation and Study of Physiological Comfort of Women's Swimwear. (Doctoral Dissertation)Shanghai University of Engineering Science, p. 86.

Xu, Y., Liu, D., Wu, Z., 2013. Study on pressure comfort of female one-piece swimsuit. Knitting Ind. (6), 65–67.

Yanmei, L., Weiwei, Z., Fan, J., Qingyun, H., 2014. Study on clothing pressure distribution of calf based on finite element method. J. Text. Inst. 105 (9), 955–961.

Yıldız, N., 2007. A novel technique to determine pressure in pressure garments for hypertrophic burn scars and comfort properties. Burns 33 (1), 59–64.

Yongrong, W., Shengli, L., et al., 2018. The test and research on the pressure comfort threshold of women's clothing. J. Text. (3), 132–136.

Zhang, P., 1990. Geometric structure and dimensional properties of flat knitted fabrics woven with air-spun cotton yarn. Part V: coil length. Text. Technol. Overseas (15), 1–3.

Zhang, Y., 2004. Application research of TC2 non-contact three-dimensional measurement system. J. Donghua Univ. (Nat. Sci. Ed.) 30 (3), 93–96.

Zhang, Q.L., Luo, X.Q., 2003. Finite element method for developing arbitrary surfaces to flattened forms. Finite Elem. Anal. Des. 39 (10), 977–984.

Zhang, Y., Zou, F.Y., 2003. Principle and application of 3D human measurement technology. J. Zhejiang Univ. Technol. 20 (4), 310–314.

Zhang, M., Dong, H., Fan, X., Dan, R., 2015. Finite element simulation on clothing pressure and body deformation of the top part of men's socks using curve fitting equations. Int. J. Cloth. Sci. Technol. 27 (2), 207–220.

Zhe, C., Kuzmichev, V.E., 2014. Analysis of pressure test of men's underwear. J. Zhejiang Text. Fash. Technol. Coll. 13 (1), 42–45.

Zhe, C., Kuzmichev, V.E., 2016a. Consumer preferences about men's underwear. In: Information Environment of the University-International Science and Technology Conference vol. 2. pp. 192–195.

Zhe, C., Kuzmichev, V.E., 2016b. Prediction of pressure and looseness in underwear knitted fabrics. Knitting Ind. (3), 62–64.

Zhe, C., Kuzmichev, V.E., 2017. Classification of male lower torso for underwear design. In: IOP Conference Series: Materials Science and Engineering. vol. 254(17). IOP Publishing, p. 172007.

Zhe, C., Kuzmichev, V.E., 2019. Research on the male lower torso for improving underwear design. Text. Res. J. 89 (9), 1623–1641.

Zhe, C., Kuzmichev, V.E., Adolphe, D.C., 2017. Development of knitted materials selection for compression underwear. Autex Res. J. 17 (2), 177–187.

Ziegert, B., Keil, G., 1988. Stretch fabric interaction with action wearables: defining a body contouring pattern system. Cloth. Text. Res. J. 6 (4), 54–64.

Sizing and fit for plus-size men and women wear

14

Simone Morlock, Andreas Schenk, Anke Klepser, Christine Loercher
Hohenstein Institute for Textilinnovation gGmbH, Boennigheim, Germany

14.1 Introduction

The demand for plus sizes has increased sharply in recent years. However, the clothing industry has not been able to react competently to the increasing demand, as no substantiated body data have been available from people with large measurement and volumes above the standard clothing sizes. The reason for this is that the proportion of men and women with large sizes has always been too low in all series measurements carried out in Germany since 1960. For this reason, no reliable size charts for plus sizes have been available, although it has been well known that "the anthropometric data will generate meaningful size distribution which in turn will provide accurate pattern cutting and satisfactory fit for the consumers" (Zakaria and Gupta, 2014) and will allow cost-effective garment production as well.

Due to the lack of real body data, the clothing industry has assumed an idealized customer image. The body measurements for the large sizes were extrapolated from the proportions of the medium clothing sizes. Clothing was produced on this basis. However, this approach led to significant fitting problems, as the body shapes of people with large dimensions do not correspond to the ideal image of proportionally balanced "normal sizes." Morphological studies prove that human body proportions do not change linearly with increasing chest girth and that the variation of individual body shape characteristics increases considerably in the subsequent sizes. This fact must definitely be taken into account in the development and manufacture of well-fitting clothing.

For this reason, two series measurements were initiated in Germany especially for men and women of large sizes (Morlock et al., 2009, 2015) to, for the first time, provide the clothing industry with reliable body data as a foundation for an optimized fit development of plus-size clothing.

14.2 The importance of sizing systems for plus-size men and women

Comparing series measurement results confirms the increasing demand for large sizes on the market. Juxtaposing German men's sizing survey from 1980 (Forschungsinstitut Hohenstein et al., 1980) and the latest representative series measurement in Germany, SizeGERMANY 2009 (Hohenstein and Human Solutions,

Anthropometry, Apparel Sizing and Design. https://doi.org/10.1016/B978-0-08-102604-5.00014-7

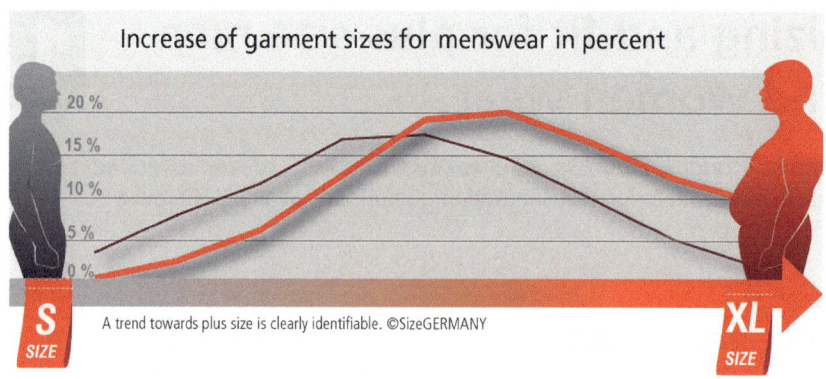

Fig. 14.1 Comparing the German sizing surveys results from 1980 and 2009.

2007–2009), the results indicate that the market share of large menswear has increased significantly. Fig. 14.1 shows a clear movement from small and middle sizes to large sizes. Chest girths increased 7.3 cm on average. This corresponds to a difference of roughly two German sizes.

In women, too, a comparison of the series measurement results from 1994 and 2009 shows that the percentage shares of large sizes have risen in the market in recent years (Verband der Damenoberbekleidungsindustrie e.V, 1994; Hohenstein and Human Solutions, 2007–2009).

The challenge faced by clothing manufacturers is seeing that the body shapes of people with large dimensions do not correspond to the ideal image of proportionally balanced "standard sizes." Therefore it is not possible to build on the patterns of small and medium sizes when developing patterns for the large sizes. Extrapolating body dimensions without considering the changing proportions causes fit problems in this size segment. People simply do not grow at a consistent rate. Certain parts of the body, such as the waist area, increase significantly more than the hip area, for example. This must be taken into account when grading the pattern pieces. In addition, there is the problem of the homogeneity of individual body shapes decreasing significantly with increasing body volume and the body shape variance increasing within the sizes. This is shown in Fig. 14.2 using women as an example.

All women in this figure were randomly selected from the sampling of 3-D scan data available in Hohenstein based on bust girth. While women with a bust girth of 92 cm have a generally homogeneous overall image with fewer physical differences, the body characteristics of women with a bust girth of 140 cm differ significantly. These women cannot be dressed with one and the same clothing product. Women's body shapes differ especially in the ratio of waist-hip proportions. This is different in men. The focus here is on the individual stomach characteristics, as shown in Fig. 14.3. In this figure, all men have a chest girth of 120 cm. However, each of them looks different. Here, too, it is not possible to dress all men with the same clothing product due to the conspicuous variance in body shape with the same chest girth.

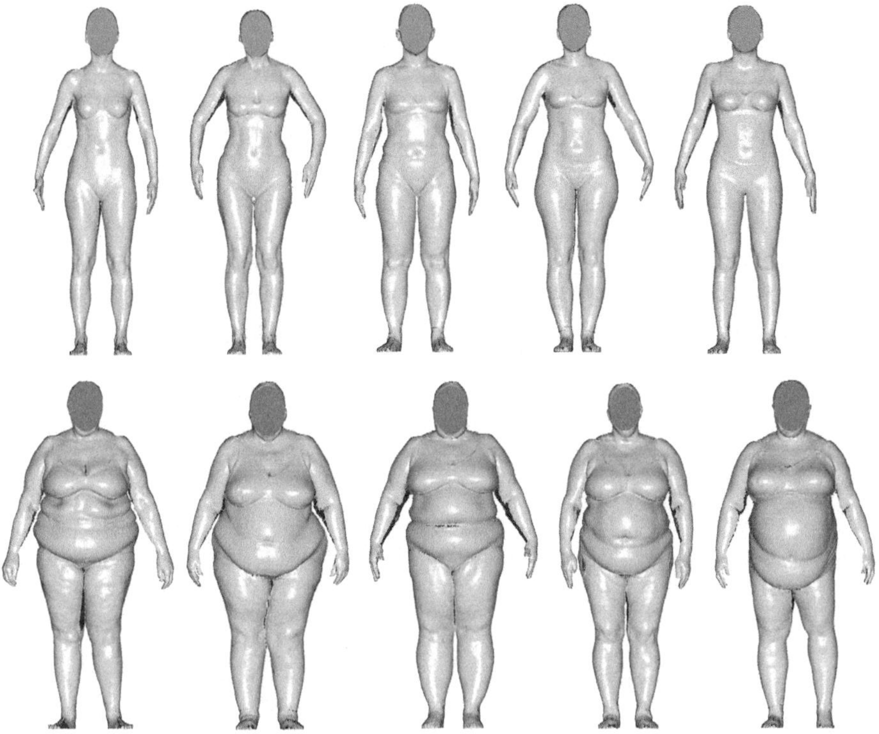

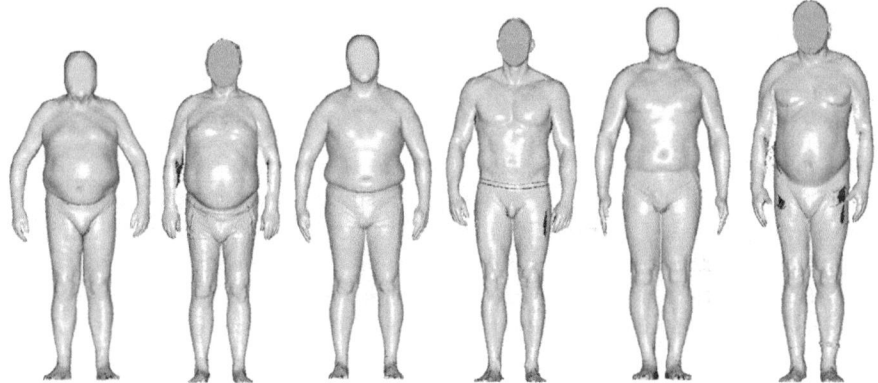

Fig. 14.2 Increasing body shape variance—comparison between women with bust girth 92 and 140 cm.

Fig. 14.3 How does a man with chest girth of 120 cm look like?

This shows that it is important not only to know the average body measurements but also to know which body type represents the largest market shares. Which body type, at which height and age group, is most strongly represented? This is the only way to develop pieces of clothing suitable for the widest possible range of potential customers, and that thus enables a wide market coverage. This is fundamental to the success of a mass product.

Nonconsideration of the morphological characteristics of large sizes inevitably leads to fitting defects when developing clothing. However, both the body data as the basis for perfect pattern development and the market shares for developing and planning optimal size ranges with a suitable model silhouette can only be collected by carrying out series measurements.

Due to the lack of body data, two significant problems are arising in the industry: Firstly the fitting problem arises because the dimensions and body shape characteristics in the large sizes are not being taken into account in clothing development. On the other hand an enormous variety of sizes have manifested itself in the market over the last few decades. Due to the lack of size standards, each clothing company has developed its own individual size tables, which differ greatly from one another. For this reason, there is no uniform size designation on the market in the large-size segment. Intensified by the internationalization of the markets, this leads to a strong customer irritation in retail, as the sizes differ significantly from brand to brand and the customer must always choose a different size. This leads to high returns, especially in the fast-growing online retail sector, where it is impossible to try things on before buying.

To close the data gap and make body data available to the plus-size industry, two series measurements were initiated in Germany, carried out as part of publicly funded research projects. In both projects, all persons were recorded using a 3-D body scanner, their dimensions were digitally derived from the 3-D scan, and size tables were developed for large sizes. Thanks to 3-D technology the specific body shapes of people with large clothing sizes were able to be visualized in addition to their height and circumference dimensions, which enabled significant important insights to be gained for the development of clothing with a secure fit.

14.3   Realization of the sizing surveys for plus sizes

The aim of the "large-size men and women" series measurements was to generate meaningful samples with which to derive well-founded body measurements and average body proportions. The first step was the series measurement for women conducted from 2007 to 2009 (Morlock et al., 2009). The series measurement for men took place from 2013 to 2015 (Morlock et al., 2015, 2016). Because these are two different projects and some differences in the requirements and implementation must therefore be named, the two studies are described in the following together but if necessary also separately from each other.

Subject recruitment in both series was random and without age restrictions. Different media were used for the call, for example, press releases, newspaper ads, and flyers and posters. Furthermore the survey was supported by manufacturers and retailers for large sizes with store gift cards. Despite extensive recruitment measures the search for suitable study participants turned out to be much more difficult and lengthy than expected at the beginning of the project. In contrast to small and medium sizes, it is very difficult to motivate men and women of large sizes to let themselves be measured wearing only their underwear. Here, recruiting test subjects required considerably more effort than is the case with measuring standard sizes. The reason for this is that, in general, it is very difficult to motivate people with large body measurements to take part in standard series measurements.

In both measurement studies, it was not possible to record persons throughout Germany within the framework of publicly funded projects for cost reasons. Nevertheless, the random samples of men and women do cover regions from all over Germany. On the one hand, this is due to the increasing blending of the population due to relocation within Germany. On the other hand, scan data from previous studies, within the framework of which measurements were taken at several locations in Germany, were also integrated into the samples. The entire data pool of scan data available in Hohenstein now comprises approx 20,000 individual data records. At this point the question of the representativity of the samples is always posed. It must be stressed here that, for target group-specific body data analysis, no representative data with regard to the regional distribution need to be available. Findings on body shape differences in terms of people's regional origins could be comprehensively obtained within the representative series measurements and can be transferred to the target group sample if required. However, a reliable distribution of clothing sizes and body shapes in a sample is important. These criteria are met for both women and men.

The aim of the "plus-size women" series measurement was to measure women with a bust girth larger than 104 cm. In men, subjects were sought with chest girths greater than 116 cm. The subjects were recorded using the Vitus Smart XXL 3D body scanner (Vitronic, 2018). Scanning was done inpatient in Hohenstein (Stuttgart, Baden-Württemberg, Germany). In addition to the scanning, a test subject survey was carried out. All men and women were questioned on their socioeconomic background, fit requirements, clothing preferences, and buying behavior.

The subjects were scanned according to SizeGERMANY standard in two scanning poses, the so-called standard posture and the relaxed posture. These postures are also defined by ISO 20685:2010 (ISO—International Organization for Standardization, 2010). These postures differ in arm and leg positioning. When relaxed the arms and hands hang down loosely to the side and do not touch the body. The feet are pelvis width apart and facing forward. The relaxed posture corresponds to the posture that people would adopt when wearing clothing and when measuring manually. In the standard posture the arms hang down and are slightly angled at the elbow joint. They should not touch the body. The feet are aligned at shoulder width, so the thighs touch each other as little as possible. The standard posture ensures surface coverage of the human body with as little shadowing in the thigh and upper arm area as possible. Shadowing is the term used when a 3-D scan shows data gaps because body areas have not been detected. A body scanner can only record the surfaces that are visible to the cameras. Put simply a scanner can only depict what people can see with their eyes. Therefore concealed body areas cannot be detected. It is therefore important to scan a person in different positions to obtain the most complete image possible in the scan results (point cloud). Fig. 14.4 shows the two scan postures for men and women.

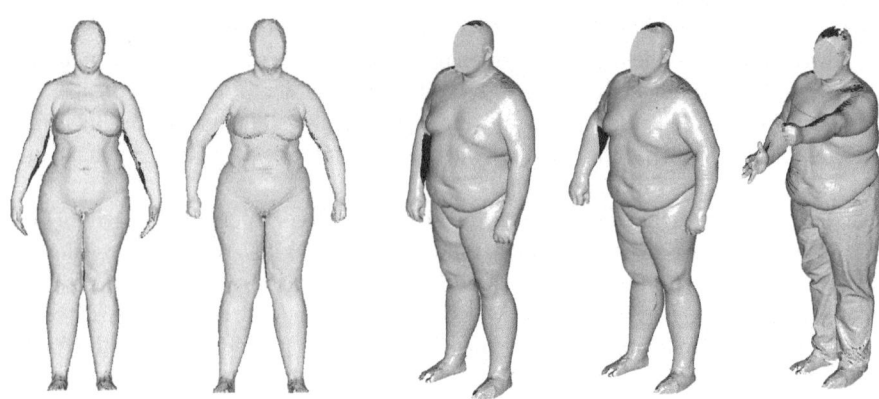

Fig. 14.4 Scanning positions "plus-size women and men."

For the men a third scan position that is likewise shown in Fig. 14.4 on the right side was introduced. This is the so-called "reach posture," which SizeGERMANY has integrated into the measurement report for measuring ergonomic dimensions. For men of large sizes, this scanning position was used to scan all subjects in their own pants they happened to be wearing on the day of the measurement. The goal was to perform a 3-D-based fit analysis of their pants.

To ensure the comparability of the body measurement results, it is important that all test subjects wear uniform measurement clothing during the scanning process. Therefore suitable scan underwear was made available to all test subjects if the underwear they were wearing did not meet the requirements and, for example, protruded far away from their bodies. To ensure the comparability of the bust girth measurements, all female test subjects had to wear a special measurement bra in the correct cup size, which they also received. The measurement bra is a soft cup bra that supports the breasts without heavily changing their natural shape. Different bra shapes such as minimizers, push-up bras, and bras with or without underwire significantly influence the bust girth and other dependent measurements such as neck, shoulder, and chest points and lead to nonreproducible and noncomparable measurement results. This can be reliably excluded with the uniform measurement bra.

As a result of the series measurements, male and female random samples are available. The sample of women comprises 2265 individual scan data pieces with bust girths from 104 to 169 cm and covers age groups from 18 to 87 years. The sample of the men consists of 664 3-D scan data pieces with chest girths from 120 to 175 cm and covers the age range from 18 to 81 years. In addition to the 664 scans, further 150 body scan data pieces in size 58 from previous measurement studies were integrated into the sample and used for statistical body data analyses. The final sample therefore consists of 814 individual data records. The men's size tables do not begin until size 60, but on this basis the transition from the standard size tables to the large sizes was able to be optimally evaluated.

14.4 Development of new sizing systems for plus sizes

14.4.1 Fundamentals of the sizing chart

The basis for the creation of the size tables are all individual personal data records from the scanned male and female subjects. The body measurements were measured digitally on the scan using the software Anthroscan (Human Solutions GmbH, 2018). The definition of the anthropometric measuring points for measuring the body dimensions in 3-D scans was carried out according to the measuring methods of ISO 8559 (ISO—International Organization for Standardization, 1989) and the SizeGERMANY standards to guarantee homogeneous and reproducible measurement results. No international specifications exist for some target group-specific body measurements, such as the natural waist girth (women) or waistband girth (men), as they were specially developed for the target group within the scope of the project.

Influences caused by changes in posture during the measuring process are excluded by the scanner technology and digital body measurement derivation. The body measurements were digitally measured by qualified personnel on the 3-D scan data. Due to the possibility of fixing important anthropometric measuring points—such as waist and hip height on which further elementary body measurements are based—clearly on the 3-D scan, in contrast to manual measurement with measuring tape, errors in deriving the body measurements are largely excluded. In both scan postures (relaxed and standard), numerous body measurements that describe the human body in great detail were measured on each person and were available for the evaluation of the body measurements. A total of 144 body measurements were derived from the 3-D scan for women and 123 for men. In addition, body height was recorded manually as a comparative measurement. Head girth was only measured manually to exclude errors resulting from subjects' hair. Only some of the men wore special scan caps to minimize the influence of their hair and to be able to record head dimensions and shape.

All measured body measurements were transferred to a common basic value table. Within a plausibility check process, outliers in these value tables were determined and corrected after the body dimension derivation on the scan. An exact and consistent data basis was therefore available for the evaluation and development of the size tables.

While developing size tables the individual body measurements are never isolated, but are always evaluated multidimensionally, taking into account the length- and circumference-dependent correlations. Fig. 14.5 demonstrates the multidimensional evaluation, using the example of the evaluation of the men's waist girths. In this scatter plot, each point represents the individual body measurement of a single subject. The diagram shows the different dependencies between the body measurements of waist girth, bust/chest girth, and body height.

The body dimension values in the size table do not always represent the mean values of all body dimensions. Figures with real body dimensions and proportions are reproduced, corresponding to the majority of the persons. In addition, technical aspects of clothing play a decisive role in the evaluation. Clothing that is a little too long and too wide causes fewer problems than clothing that is too short and

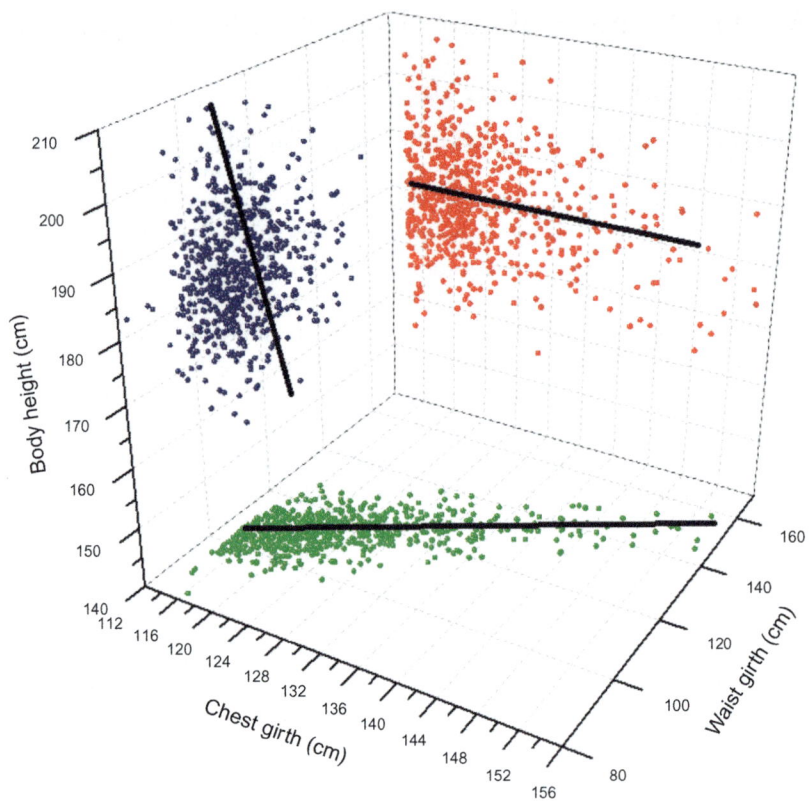

Fig. 14.5 Evaluation of body measurements: body height, chest, and waist girth.

too tight. Shortening or narrowing clothing is easier to implement than lengthening or extending. For this reason the measurement values tend to increase, especially when the individual values are widely dispersed to achieve better market coverage.

The statistically evaluated body measurements were verified by the parallel development of the virtual 3-D body models. This ensures that the measurements in the size tables correlate with each other and can be traced on a representative, average body.

14.4.2   Classification of men's and women's outerwear

There are different possibilities to create sizing systems (Ashdown, 2007). In general the structure of sizing systems is based on the division of the population into groups with similar body measurements and body shapes. The clothes sizes of a size system are fundamentally determined based on the body shape and body height series that have the largest market shares. To develop the "plus-size" size tables, the question of the appropriate size system had to be considered first. For women the existing and proven system could be used, but for men, more detailed questions had to be answered in the first step.

14.4.2.1   Size system of women's outerwear

The "plus-size women" size table system is based on the standard size table system for women. These have remained unchanged in this form for decades and have proven their worth. The validity of the system was most recently confirmed by the body data results from SizeGERMANY in 2009 (Hohenstein and Human Solution, 2008). The proven size system was also able to be used for women of large sizes. The size tables comprise the three size ranges of short, normal, and long sizes. For each of the size ranges, the three body shapes are shown as narrow hips, normal hips, and wide hips. The body shape is calculated from the difference between bust girth and hip girth. The bust girth increases by 6 cm per clothing size at the same height, as you can see in Table 14.1.

14.4.2.2   Size system of men's outerwear

A central point of discussion at the beginning of the project was the selection of the right sizing system, as a sizing table was to be developed that was to be widely accepted in the market. Due to the historical development of the men's size tables and their systematics in Germany, however, the course of action at the beginning of the project was unclear.

In Germany the assignment of garment sizes is based on chest girth. For men's outerwear the size designation is half the value of the chest girth, for example, a chest girth of 100 cm is size 50. Determining the measurement interval is therefore of fundamental importance. Market practice showed differing business solutions relating to plus sizes. At the time of the project start, a sizing standard was missing for the plus sizes. The men's size tables ended with the German size 58 (chest girth 116 cm) and 60 (chest girth 120 cm). For the standard sizes from 42 to 60, a chest girth interval of 4 cm applies. In the market, intervals of 4 cm, 6 cm, and up to 8 cm are used and sometimes even alternate. Therefore, for the new size tables, it was necessary to define the appropriate chest girth interval per clothing size in the large-size sector. The jump in size for the chest girth as finally set at 4 cm. The 4-cm interval guarantees a detailed evaluation of the individual sizes but can be converted to the 6- or 8-cm interval if required.

A further question arose when determining the body heights in the individual clothing sizes. Since 1960 a total of three series measurements have been carried out in Germany to measure men: 1960, 1980, and 2007–09 SizeGERMANY (Verband der Herrenbekleidungs-Industrie e.V. (HG.), 1960; Forschungsinstitut Hohenstein et al., 1980; Hohenstein and Human Solutions, 2008). However, the results of the measurements from 1980 were never put into practice. The point of contention was and still is the increase in height of men with increasing chest girth. In 1960 the body height was increased by 3 cm for small and medium sizes and by 2 cm for larger sizes. This means that the more corpulent a man gets, the taller he has to be, which does not

Table 14.1 Size system of women's outerwear

Bust girth (cm)	110	116	122	128	134	140	146	152	158
Standard sizes (normal hips)									
Short sizes	**24**	**25**	**26**	**27**	**28**	**29**	**30**	**31**	**32**
Body height (cm)	160	160	160	160	160	160	160	160	160
Normal height	**48**	**50**	**52**	**54**	**56**	**58**	**60**	**62**	**64**
Body height (cm)	168	168	168	168	168	168	168	168	168
Long sizes	**96**	**100**	**104**	**108**	**112**	**116**	**120**	**124**	**128**
Body height (cm)	176	176	176	176	176	176	176	176	176
Sizes for narrow hips									
Short sizes	**024**	**025**	**026**	**027**	**028**	**029**	**030**	**031**	**032**
Body height (cm)	160	160	160	160	160	160	160	160	160
Normal height	**048**	**050**	**052**	**054**	**056**	**058**	**060**	**062**	**064**
Body height (cm)	168	168	168	168	168	168	168	168	168
Long sizes	**096**	**0100**	**0104**	**0108**	**0112**	**0116**	**0120**	**0124**	**0128**
Body height (cm)	176	176	176	176	176	176	176	176	176
Sizes for wide hips									
Short sizes	**524**	**525**	**526**	**527**	**528**	**529**	**530**	**531**	**532**
Body height (cm)	160	160	160	160	160	160	160	160	160
Normal height	**548**	**550**	**552**	**554**	**556**	**558**	**560**	**562**	**564**
Body height (cm)	168	168	168	168	168	168	168	168	168
Long sizes	**596**	**5100**	**5104**	**5108**	**5112**	**5116**	**5120**	**5124**	**5128**
Body height (cm)	176	176	176	176	176	176	176	176	176

Notes: The sizes are determined by breast/chest circumference and body height. The normal height sizes show the largest market shares. Therefore, these are the most provided sizes in the German market.

correspond to reality. In 1980 the sizing experts wanted to adapt the men's height system to that of women and not increase the body height with increasing chest girth. This meant a fundamental change in the existing size system that was consistently rejected by the men's clothing industry at the time, so the results of this series measurement were never applied. For this reason the men's sizes in German retail are still largely based on the size designations in the table from the 1960s.

A new, solid size base has been created with SizeGERMANY, but it has not yet fully penetrated the market. The reason for this can still be seen in the size system. In SizeGERMANY a sensible compromise between history and reality has been created. Body height now only increases by 1 cm per clothing size. SizeGERMANY represents the current state of the art, however, and it seemed sensible to build on this from the very beginning. After the first statistical analyses of body measurements from the present sample "plus-size men," it was possible to prove that the SizeGERMANY system can also be used for the new size table "plus-size men" that is to be developed. Table 14.2 shows the size systematics for men of large sizes using the example of the normal, slim, and heavy body shapes. Other body shapes are extra slim and extra heavy.

Figure type classification—Men

The difference between chest girth and waist girth determines the body shape. The body shape designation is taken from SizeGERMANY as represented in Fig. 14.6 extra slim, slim, normal, heavy, and extra heavy. The body shape is differentiated within the body height classifications.

Body height classification—Men

One widely discussed issue in the men's outerwear sector is whether a constant or an increasing body height is more reasonable, while chest girth is increasing from size to size. Up to SizeGERMANY 2009, this was common practice. But it is well known that increases in body height cannot be extended without end. Following this rule a person in size 82 would have a body height of 210 cm. Obviously, this does not reflect reality. In the "plus-size men" project, the question of the perfect increase in body height was finally answered. Fig. 14.7 is a scatter diagram showing chest girth on the x-axis and body height on the y-axis. Each point represents the body measurements of one individual subject. Apparently, there is no significant correlation between the two dimensions. The trend line is horizontal, showing no increasing or decreasing tendency. The results clearly prove that the average body height does not increase with increasing chest girth. In addition, Fig. 14.7 emphasizes the necessity of five body height types due to the large variation in individuals. The body height classification—extra short, short, normal, long, and extra long—was taken from SizeGERMANY.

On the basis of these results, the body height of the standard size tables is increased by 1 cm per size, but starting from German size 60 (chest girth 120 cm), it remains constant for all sizes. Differences can be represented by the various size ranges of extra short, short, normal, long, and extra long.

Table 14.2 Size system of men's outerwear

Chest girth (cm)	120	124	128	132	136	140	144	148	152	156
Figure type normal										
Extra-short sizes	K30	K31	K32	K33	K34	K35	K36	K37	K38	K39
Body height (cm)	169	169	169	169	169	169	169	169	169	169
Short sizes	30	31	32	33	34	35	36	37	38	39
Body height [cm]	177	177	177	177	177	177	177	177	177	177
Normal height	60	62	64	66	68	70	72	74	76	78
Body height (cm)	185	185	185	185	185	185	185	185	185	185
Long sizes	118	122	126	130	134	138	142	146	150	154
Body height (cm)	193	193	193	193	193	193	193	193	193	193
Extra-long sizes	L118	L122	L126	L130	L134	L138	L142	L146	L150	L154
Body height (cm)	201	201	201	201	201	201	201	201	201	201
Figure type slim										
Extra-short sizes	K30S	K31S	K32S	K33S	K34S	K35S	K36S	K37S	K38S	K39S
Short sizes	30S	31S	32S	33S	34S	35S	36S	37S	38S	39S
Normal height	60S	62S	64S	66S	68S	70S	72S	74S	76S	78S
Long sizes	118S	122S	126S	130S	134S	138S	142S	146S	150S	154S
Extra-long sizes	L118S	L122S	L126S	L130S	L134S	L138S	L142S	L146S	L150S	L154S
Figure type heavy										
Extra-short sizes	K30B	K31B	K32B	K33B	K34B	K35B	K36B	K37B	K38B	K39B
Short sizes	30B	31B	32B	33B	34B	35B	36B	37B	38B	39B
Normal height	60B	62B	64B	66B	68B	70B	72B	74B	76B	78B
Long sizes	118B	122B	126B	130B	134B	138B	142B	146B	150B	154B
Extra-long sizes	L118B	L122B	L126B	L130B	L134B	L138B	L142B	L146B	L150B	L154B

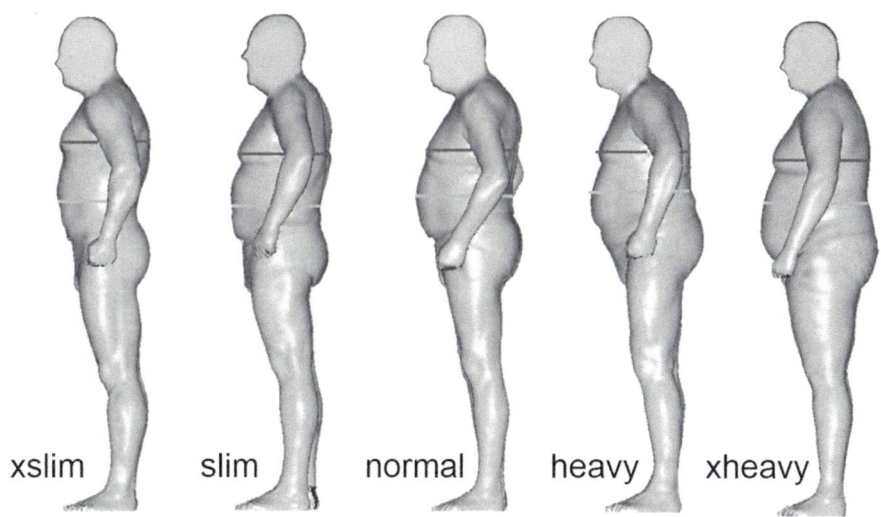

Fig. 14.6 Presentation of the five figure types, each one with chest girth 128 cm.

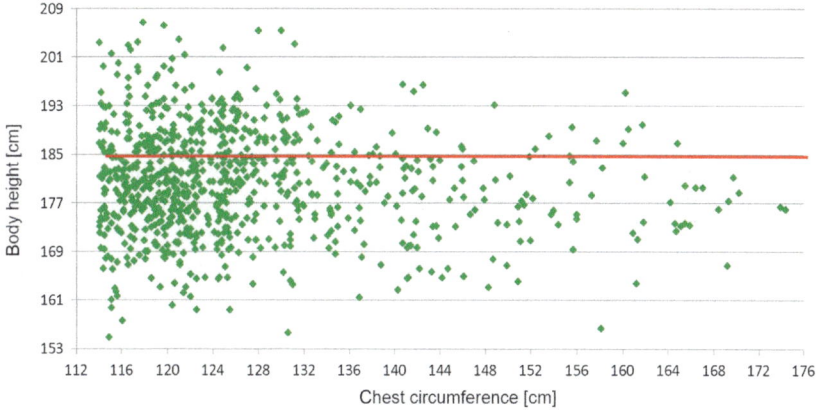

Fig. 14.7 Correlation analyses of body height and chest girth starting from size 58 (chest 116 cm).

14.4.3 Definition of target group–specific measurements

The waist girth shown in the standard size tables is measured horizontally on the body according to specified measurement standards, such as ISO 8559: "the girth of the natural waistline between the top of the hip bones (iliac crests) and the lower ribs, measured with the subject breathing normally and standing upright with the abdomen relaxed." However, 3-D body shape and posture analyses have made it clear that for the larger-waist sizes, which are required as a measure for fit-oriented clothing development, it is no longer horizontal. Rather the horizontal waist directly below the chest

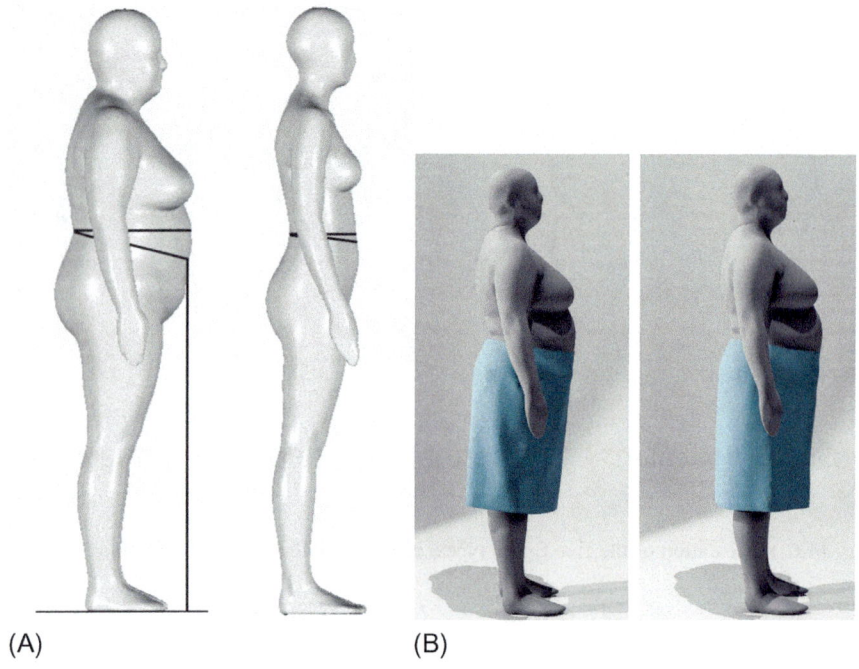

(A)   (B)

Fig. 14.8 Comparison of horizontal and natural waist girth—Part (A) and (B).

is almost at the level of the underbust girth, as can also be seen in Fig. 14.8A for the
example of women. Apart from the position of the measurement, the circumferential
value also changes significantly. Both have a significant influence on the fit of clothing. For people with heavy corpulence, the standard waist size cannot be used as a
basis for pattern construction, because these people wear neither pants nor skirts at
the measured body position.

The horizontal waist dimension is important as a base value for the description of
the bodies and for the construction of jackets and other tops, but is not suitable for
integration into the pattern construction for pants as described earlier. For this reason,
in addition to primary measurements such as chest, waist, and hip girth; inner leg
length; and arm length, the "waist measurement" was redefined especially for the target group of large sizes and measured on the 3-D scan. This measurement was designated as "natural waist girth." Fig. 14.8B visualizes the problem of horizontal
waist girth in comparison with the natural waistline of women. If the natural waistline
is incorrectly implemented in the technical development of pants and skirts, the balance of the products will be incorrect, and they will typically slip forward when worn,
resulting in significant fitting defects.

The "natural waist girth" is measured starting from the rear waist point (anthropometric measuring point) along the natural waistline. This line is not based on anthropometric measuring points but in most women is created over the years by a
permanent change in the connective tissue and becomes visible as an
"indentation," so to speak, in the abdomen. This is where most women's pant or skirt

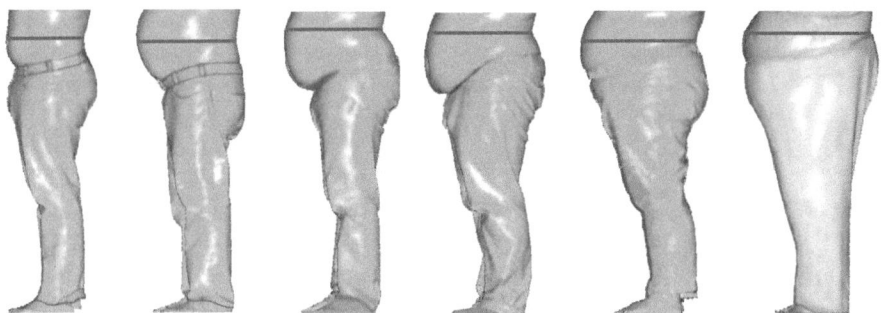

Fig. 14.9 3-D-based fitting analysis of pants—comparison of waist height and waistband height—side view.

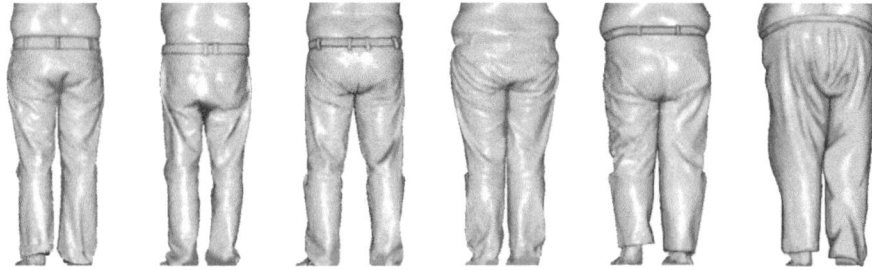

Fig. 14.10 3-D-based fitting analysis of pants—back view.

waistbands sit in their everyday lives. In addition to the circumference, the vertical distance from the front waist point to the standing surface was measured as "distance of waist girth-natural waist" to determine a comparison with the classic waist height (standing surface to the horizontal waist). This measurement forms the elementary basis for the technical deepening of the waistband line of pants and skirts.

Although there are no anthropometric measuring points and only the individually pronounced change in connective tissue, both measurements were able to be determined very reliably. The values for the "natural waist girth" can be transferred 1:1 into the pattern for pants and skirts, which has been proven by extensive fit tests in the further course of the project. For the German size 64, with a bust girth of 158 cm and a body height of 168 cm, a depression of approx 8 cm in the waist is to be implemented in the development of the pattern.

There is a similar problem with men as with women. The larger the size, the lower the front waistband girth in the belly area. To evaluate this problem a 3-D-based fit analysis of pants was carried out on the basis of 3-D scans to analyze the individual waistband shape and the fit of the pants on the body. During the series measurement, the men were scanned not only in their underwear but also in their own pants they happened to be wearing that day.

Figs. 14.9 and 14.10 each show six test subjects of different sizes, all of whom are wearing their own pants in their usual way. Fig. 14.9 shows the men in side view. The

Fig. 14.11 Target group-
specific measurements "plus-
size men."

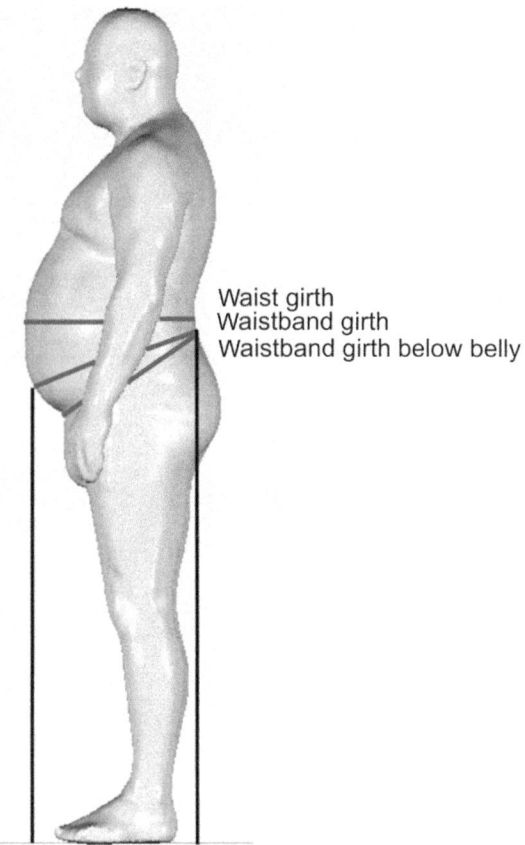

Waist girth
Waistband girth
Waistband girth below belly

illustration shows that the horizontal waist girth does not have any noteworthy relation to the waistband girth. The pants all sit significantly lower. However, this fact was obviously not taken into account for any of these pants in terms of pattern. The result is that all pants that are insufficiently recessed at the front will slide down when worn, which causes typical fitting errors or wrinkles. The exception is the man who wears his pants above his stomach. Here the pattern geometry of the pants corresponds to the man's wearing habits.

The fit problems are most obvious in the rear view shown in Fig. 14.10. The fact that men of large sizes tend to have flat buttocks and narrow thighs was not taken into account in the design of these pants. Both the buttocks and thighs are too pronounced in the pattern. This results in significant fit defects. The leg position of the pants is also not adapted to the target group.

Not only the six pairs of pants shown but also almost all pairs of pants of the scanned 664 test subjects show these typical fit problems. This shows the great need for reliable body data and its correct conversion into the pattern for the target group described.

As with the women the men's waistband girth was measured in the 3-D scans to depict these in the size tables (see Fig. 14.11). However, the men's waistband girth could not be determined nearly as reliably as was possible for the women.

The waistband girth is measured from the rear "waistband point" perpendicular to the torso axis. The digital measuring tape does not run below at the front, but rather over the belly. From there the waistband height at the front is determined in comparison with waist height. It is important to note that this measure of length cannot be determined from the "waistband girth below the belly." Measured in this way the pants would lack the extra length required by men in the front area.

No anthropometric measuring points are available for measuring the waistband girth and its front position height. In contrast to women, men have a deeper back waist position than the anthropometric back waist point. Furthermore the waistband girth of men on the human body can only be determined approximately. Men have no change in connective tissue like that in women. Therefore the fit of the pants depends exclusively on individual wearing preferences. However, these differ heavily among men. They wear their pants both below and above their stomachs. Some men also wear their pants just above the belly. Individual wearing habits are influenced by the shape of the belly itself. Due to men's highly individual stomach expression and the habit of wearing pants below the belly, the waistband position is hardly reproducible for a representative average. Fitting uncertainties must therefore always be planned on when implementing these values in patterns.

It must also be taken into account that the waistband width is changed by a belt. Most men of large sizes typically have a large belly, but not large hips. The pants therefore have no hold and will slide down. The majority of men use belts. To prevent the trousers from slipping, the belt must be pulled so tightly that the body tissue is compressed. How strong this compression needs to be or is desired to be is unknown and certainly different among individuals. However, this means that the waistband girth, which is measured on the body surface, differs from the real, compressed circumference. Due to the reasons mentioned earlier, the waistband girth cannot be converted 1:1 into the pattern and is therefore only a guiding measurement.

For men the "waistband girth below the belly" was also determined. This measurement follows the belly. This measure is also not based on any anthropometrically reproducible measuring point. If the stomach droops very far, the measurement on the scan cannot be measured. This waistband size is certainly interesting for underwear manufacturers, but as a measure for pant development, it is usually too small, because the belly is completely ignored.

With regard to the fit of pants, the "waistband height" at the front is one of the most important measurements. Like the waistband girth the values can only be determined approximately, as men's wearing behavior varies greatly with regard to the waistband. Nevertheless, the measurements in the size table give a clear indication of the extent to which the front pattern of the waistband must be recessed if the pants are to be worn below the stomach. As an example for the value, for the German size 70, with a bust girth of 140 cm and a body height of 185 cm, a depression of approx 12.5 cm in the waist is to be implemented in the development of the pattern.

In summary, it can be said that it will always be very difficult to develop pants with a secure fit for men of large sizes and a large belly, due to the specific morphological characteristics and individual wearing habits.

14.4.4   Definition of morphotypes

Especially for the target group of large sizes, it is of elementary importance to describe the three-dimensional body shapes in detail, because the larger the clothing size, the greater the inhomogeneity of the bodies. Three-dimensional body shape data are crucial for well-fitting pieces of clothing, as body shape cannot be precisely defined with the help of simple height and circumference measurements. People of a clothing size with almost identical dimensions can have different body characteristics. This results in differentiated requirements for the fit of clothing products within a size that must be taken into account in pattern construction and development.

Morphotypes are an important supplement to the classic body shapes (e.g., narrow, normal, or wide hips), which are based only on the classical height and girth measurements. A morphotype, on the other hand, is used to identify size-specific characteristics and designates a specific body type whose definition is based on 3-D dimensions and clustered 3-D body characteristics. As the many-body variants in the large sizes make it impossible to serve all of them with one and the same clothing product, the market shares by morphotype ensure the greatest possible market coverage in product development by knowing the most commonly represented body shape within one size.

The basis for defining the morphotypes is the ratio of the body dimensions "shoulder diameter" to "hip diameter" in frontal view and "waist diameter" in frontal view to "shoulder diameter" and "hip diameter." Fig. 14.12 shows the diameter

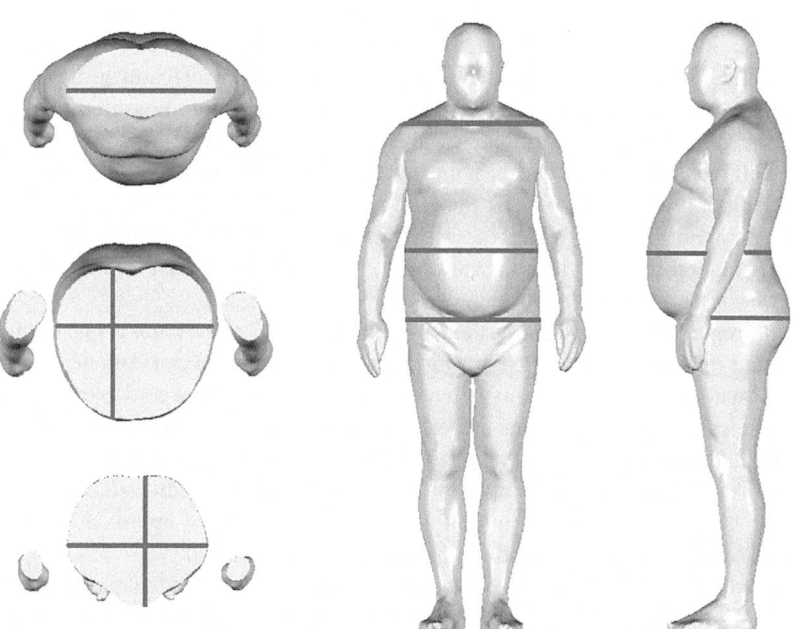

Fig. 14.12 Body cross sections shoulder, waist, and hip from front and side view.

dimensions using the example of men. The body diameters allow conclusions to be drawn about the actual shape of the torso. They show whether the bodies are more oval or round in shape at the same circumference. This way, important findings relevant to fit can be derived. For example, different abdomen types were identified based on the "waist diameter."

The information on morphotypes represents a significant supplement to the traditional size tables, which describe the conventional body shapes based on height and girth measurements. Morphotypes are therefore key in ensuring that products are developed with the right fit.

14.5   Generation of virtual 3-D body models

Based on the statistical evaluation of body measurements and the 3-D-based body shape analysis, average 3-D body models can be developed that visualize the typical body shapes of a size. On the one hand, these serve as the foundation for verifying the size table. The parallel development of 3-D body models and size tables ensures that the tables' dimensions correlate with each other and can be traced on a representative, average body. On the other hand the 3-D models can be used to design perfectly fitting basic patterns based on 3-D. The scan data, which are available as so-called point clouds, are connected to closed polygon surfaces. This polygon model is the basis for all further modeling steps. The software Geomagic Studio 2012 was used to process the digital data and create the 3-D body models. The basic procedure for generating 3-D body models is as follows:

Step 1: Selection of a sufficient number of 3-D scans to be assigned to the same defined size cluster

The selection parameters are the primary measurements of body height, bust girth, and hip girth (women) or waist girth (men) and then the secondary measures in descending order of priority. Depending on the determined scattering of the individual measurements, the areas of application of the body measurements should be narrowed or broadened.

Step 2: Editing and data conversion of the selected 3-D scan data

The selected data sets are converted into ASCII format. Before the polygon surface can be created, the point cloud must undergo a series of processing steps to achieve optimal results in the triangular surfaces that represent the body surface. This includes reducing noise and an even and curvature-based reduction of the points. In the following polygon phase, the points are connected to triangular surfaces by triangulation, and the normals of the surfaces are determined so that all are oriented in the same direction. The surface created in this way must then be subjected to a series of further work steps to achieve an even result. Among other things the number of triangles is decimated, and the polygons are smoothed and cleaned.

Step 3: Calculation of the average bodies based on the appropriately selected 3-D scans

Based on the processed 3-D scans, Geomagic calculates the average bodies. For the normal sizes, 20–30 individual 3-D scans are used as a basis to calculate a clothing size, depending on availability.

Step 4: Preparation of the determined 3-D bodies to create watertight surfaces

This is followed by the work process of further modeling the polygon models. Changes or additions are made to fill existing holes and correct or model inadequate shapes, such as the body areas under the arms or between the legs.

Step 5: Determination of the permissible degree of idealization of 3D body models

Depending on the desired purpose—simple visualization of the body shapes, as a basis to develop fitting busts or for 3-D design and simulation—the degree of idealization of the models must be implemented, for example, the shaping of the abdomen and the beginning of the abdomen. The body details must then be shaped differently. The legs and arms in particular are difficult to convey due to different postures while scanning, and the postprocessing effort is extremely high here.

Step 6: Modification of the average 3-D body models through modeling in relation to size- and shape-specific body characteristics and body dimensions

The 3-D body models are verified in terms of their dimension and shape conformity based on the size table and adapted or scaled to ensure that the created 3-D models match the developed size table.

Step 7: Final preparations of the models

All triangles in the mesh must have the same orientation and cannot have any holes, gaps, or overlaps. Extreme edge angles, self-overlaps, and spikes must also be machined and removed. Once all errors have been corrected, the data record can be saved in the desired format depending on further use and the software used.

14.6   3-D-based development of basic patterns for plus-size jackets and pants

With increasing growth of the bust/chest girth, the body proportions change considerably. The changes in the body geometry have a decisive effect on the pattern design. In some cases, they lead to major changes in the pattern geometries. The new body data were therefore converted into optimized body-consistent basic patterns for women and men of large sizes.

14.6.1   Plus-size women's outerwear

Within the "plus-size women" project, the focus was on the development of guidelines and rules for optimal, functional design. Based on the new body data, basic patterns for the lower and upper halves of the body were developed, and test clothing was produced. Differentiation was made according to increasing body volume and the resulting special body proportions to determine the boundaries between ideal silhouette design and functional requirements for clothing comfort for the range of examined sizes. The results were evaluated and verified by comprehensive fit and comfort tests on suitable test subjects of various sizes. Different model characteristics were evaluated, regarding elements such as the sleeve/armhole construction, the different gradations of the back and shoulder widths, and the angle of the seats of pants. Test samples were manufactured, and fit tests were carried out. From this, optimized patterning guidelines were derived and formulated. A special focus of the study was the technical conversion of the natural waist into the pattern of pants and skirts.

14.6.2 Plus-size men's outerwear

Based on the analyzed body measurements and average body models, basic patterns were developed for jackets and pants for men of large sizes. In doing so, new paths were taken. The traditional approach of 2-D pattern development has been supplemented by a new 3-D-based solution that has already been developed in previous projects for women. (Kirchdörfer et al., 2011) This procedure ensures the perfect fit of basic patterns, as the human 3-D body geometry is incorporated directly into the pattern pieces.

The principle of 3-D-based design is shown in Fig. 14.13 using pants as an example. The 3-D body models (left) serve as the basis for the transformation of the body surface into the plane. The first step is preparing for surface unwinding. The 3D software Rhinoceros 4.0 is used for the implementation. Only nonuniform rational B-spline (NURBS) surfaces with one-way curvature radii can be unwound. The polygon meshes cannot be transformed. For this reason, 3-D surfaces are constructed on the meshes that simulate the body surface and idealize the body like clothing (see Fig. 14.13 second from left). To achieve uniform results when unwinding different body surfaces, the polygon meshes must be segmented horizontally and vertically (see third from left). Decisive criteria for segmentation are the changes in the

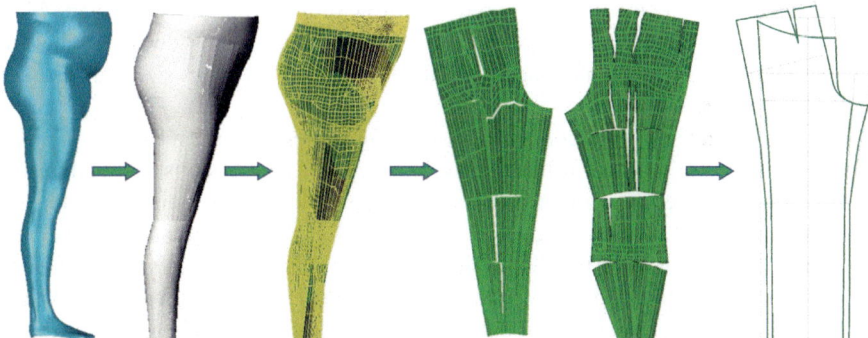

Fig. 14.13 Unwinding and rearranging of constructed NURBS planes—for example, pants.

curvature radii on the body, which do not need to be reproduced exactly and instead bridged, such as the areas abdomen-thighs or buttocks-thighs. The edges of the segments form the boundary for the individual unwinding areas. The transformation of the 3-D surfaces into the plane is done according to the segmentation of the polygon meshes. The unwound surfaces should not yet be used for the development of patterns in this form. By separating the segments and rearranging the individual elements, the first basic shape is achieved (see fourth from left). The outer contour of the rearranged surfaces, including internal gaps and overlaps, is used as a DXF file in the CAD system as a basis for developing the basic pattern design (see right) or the patterns for test clothing.

14.7   Results and discussion

14.7.1   Size charts: "Plus-size women and men"

For the first time, body measurements of plus-size women and men are available to the clothing industry in Germany but naturally for all interested countries worldwide as well. The size charts are related to SizeGERMANY systematics.

14.7.1.1   Size charts: "Plus-size women"

For a long time, there was speculation as to the size at which women could reasonably be considered to be of large sizes. Body shape analyses have shown that this is the case from size 48 with a bust girth of 110 cm. The sizes beneath this still have homogeneous body shapes. The visualization of the bodies using 3-D technology proves that, from size 48 and up, the variation in body shape increases significantly in the sizes with the same bust girth.

Therefore the "plus-size women" size tables cover the range of German sizes 48–64. This corresponds to bust girths of 110–158 cm. The tables contain 7 primary measurements and 43 secondary measurements, plus the average body weight of the test subjects in the individual sizes. The size system is adapted to the SizeGERMANY size tables and describes the three size ranges of short, normal, and long and, for each of the size ranges, the three body shapes of narrow hips, normal hips, and wide hips. This size system for women has been established in Germany for decades. The following body dimensions are shown in the tables for all nine size ranges (see Table 14.3).

The body measurements contained in the size tables correspond to the average body dimensions of women of large sizes. However, these measures do not describe statistically calculated average bodies, but rather bodies that correspond to the ideal of the individual sizes based on their proportions and allow the greatest market coverage. Here the waist girth is to be mentioned as an example. The maximum span is between the smallest and largest waist value with the same bust girth, sometimes up to 35 cm. This is due to the large variation in body shape in the large sizes. Therefore the tables do not contain simple calculated statistical averages, but rather values that have been formed according to clothing principles as required.

Table 14.3 Body measurements of size chart: "plus-size women"

Body measurements of size chart: "Plus-size women"			
1	Bust girth	26	Arm length
2	Hip girth	27	Upper arm length (acromion to elbow)
3	Body height	28	Upper arm girth—at base of armscye
4	Waist girth	29	Biceps girth
5	Natural waist girth	30	Elbow girth
6	Distance horizontal to natural waist	31	Forearm girth
7	Underbust girth	32	Wrist girth
8	Waist height (distance)	33	Middle hip girth
9	Waist height (contour)	34	Thigh girth
10	Inside leg height	35	Knee girth
11	Neck girth	36	Upper knee girth
12	Neck-base girth	37	Lower knee girth
13	Head girth	38	Calf girth
14	Shoulder length	39	Minimum leg girth
15	Shoulder slope	40	Diagonal girth of heel/instep
16	Across back width	41	Cervical height—contoured
17	Back waist length (7th cervical vertebra to waist)	42	Cervical to knee hollow—contoured to hip
18	Front waist length	43	Knee height
19	Neck shoulder point to breast point	44	Tibia height
20	Breast point distance	45	Body cross section at neck level—front view
21	Trunk length	46	Shoulder width (between acromion extremities)
22	Distance of waist to hip	47	Biacromial shoulder width
23	Total crotch length	48	Armscye width
24	Crotch length front	49	Body cross section at hip level—profile
25	Center back neck to wrist length (arm length)	50	Body cross section at hip level—front view

The tables are independent of age groups and represent the age range from 18 to 87 years. The statistical evaluations show that the smaller sizes represent the younger women and the larger sizes represent the older women. As a rule, women increase in size with increasing age. The evaluations have also shown that the majority of women over 66 years of age in the sample cannot be assigned to the defined size ranges and body shapes, as they are either smaller or even narrower in the hips. As a result, they fall out of the market shares. Their age-specific body proportions therefore have no significant influence on the definition of body dimensions for the large sizes. Special age-specific size tables are necessary for the seniors (Kirchdörfer and Mahr-Erhardt, 2003).

14.7.1.2 Size charts for "plus-size men"

The men's size charts include six primary measurements and 42 secondary measurements, plus the average body weight of the men in each size. The size table is based on the SizeGERMANY size system and comprises German sizes 60–78 for five size ranges and five body shapes. This corresponds to bust girths of 120–156 cm. The size system is adapted to the SizeGERMANY size tables (Hohenstein and Human Solutions, 2008). The tables contain the five size ranges of extra-short, short, normal, long, and extra-long sizes. For each of the size ranges, the five body shapes provided are extra slim, normal, heavy, and extra heavy. This size system for men was first defined in Germany with SizeGERMANY. The following body dimensions are shown in the tables for all 25 size ranges (see Table 14.4).

The body dimensions contained in the size tables correspond to the average body dimensions of men with large sizes. However, the body measurements do not describe statistically calculated average bodies, but rather bodies that correspond to the ideal of the individual sizes on the basis of their proportions and allow the greatest market coverage.

The men's size tables are compiled independently of age groups and represent the age range of 18–81 years. The statistical evaluations show that the smaller sizes represent the younger men and the larger sizes represent the older men. As a rule, men, like women, gain in size with increasing age.

14.7.2 Morphotypes—Plus sizes

Three-dimensional body shape data are crucial for well-fitting pieces of clothing, as the body shape cannot be precisely defined using the one-dimensional body dimensions available to date. A morphotype is therefore used to identify size-specific characteristics and designates a specific body type whose definition is based on 3-D dimensions and clustered 3-D body characteristics.

The findings on the average body shapes and market shares are particularly important for the large sizes. The larger the size, the greater the inhomogeneity of the bodies. The challenge for clothing manufacturers to optimally and appropriately dress as many people as possible grows with increasing clothing sizes. Because it is impossible to serve all of them with one and the same clothing product, due to the many-body variations in the large sizes, the market shares according to morphotypes ensure the greatest possible market coverage in product development by defining the typical or average body shape.

14.7.2.1 Morphotypes—Plus-size women

The morphotypes of plus-size women are divided into three basic types with the A-shape, the H-shape, and the V-shape (see Fig. 14.14). In turn, these basic types differ in the shape of their waistline. A distinction is made between X-shape, O-shape, and no pronounced waist shape (not shown). The percentage distribution of female

Table 14.4 Body measurements of size chart "plus-size men"

Body measurements of size chart for "plus-size men"			
1	Chest girth	25	Upper arm length (acromion to elbow)
2	Hip girth	26	Biceps girth
3	Body height	27	Upper arm girth—at base of armscye
4	Waist girth	28	Elbow girth
5	Waist height (distance)	29	Forearm girth
6	Inside leg height	30	Wrist girth
7	Waistband girth	31	Center back neck to wrist length (arm length)
8	Waistband girth below the abdomen	32	Middle hip girth
9	Waistband height front	33	Thigh girth
10	Waistband height back	34	Middle thigh girth
11	Neck girth	35	Knee girth
12	Neck-base girth	36	Upper knee girth
13	Body cross section at neck level—front view	37	Lower knee girth
14	Head girth	38	Calf girth
15	Shoulder length	39	Minimum leg girth
16	Shoulder slope	40	Knee height
17	Across back width	41	Tibia height
18	Back waist length (7th cervical vertebra to waist)	42	Shoulder width (between acromion extremities)
19	Front waist length	43	Shoulder width bideltoid
20	Neck shoulder point to breast point	44	Armscye width
21	Breast point distance	45	Body cross section at waist level—profile
22	Trunk length	46	Body cross section at waist level—front view
23	Distance waist to hip	47	Body cross section at hip level—profile
24	Arm length	48	Body cross section at hip level—front view
25	Upper arm length (acromion to elbow)		

morphotypes is shown in the following table. The percentages refer to the data sample of German sizes 48–64 with 2256 data sets.

The largest share is held by H-shape, with 46.3%. With this shape the frontal view of the shoulder area is the same width as the hip area. The A-shape then follows with 29.4%. Here the shoulder area is narrower than the hip area. The A-shape is to be determined in different forms. Moderate characteristics are normal hip to the shape type; strong characteristics are to be assigned to the wide hip and extra-wide hip shape

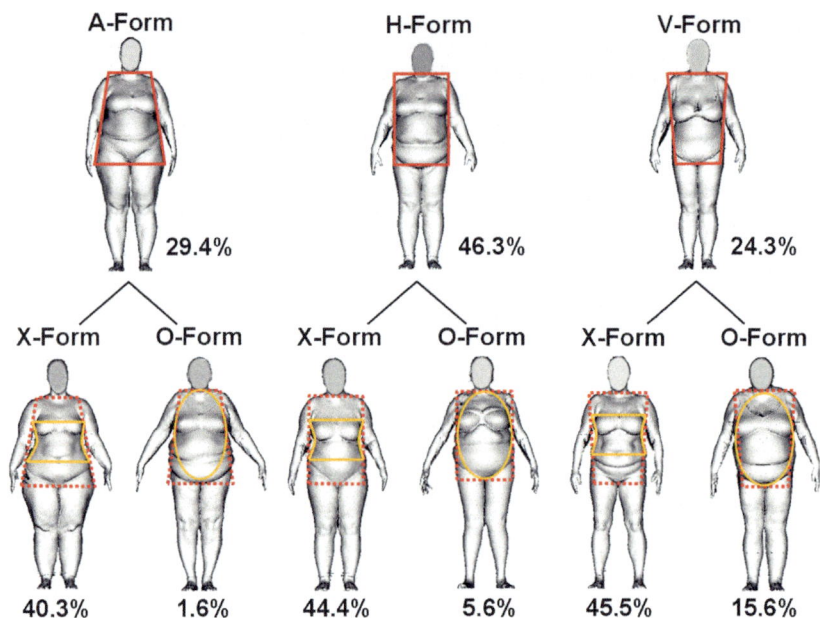

Fig. 14.14 Percentage distribution of female morphotypes.

types. The lowest share is found in the V-shape with 24.3%. This shape tends from the normal hip to the narrow hip and smaller narrow hip body shape.

The description of the morphotypes is supplemented by the description of individual body parts in detail, for example, the analysis of the correlation of shoulder width to back width. Shoulder width increases proportionally neither to the girth nor to the width of the back as the bust girth increases. If the shoulder width is larger than the back width in the smaller sizes, this ratio changes in women along with it and larger bust girth of 110 cm. From then on the back width is greater than the shoulder width for women. For clarification, in Fig. 14.15, the back and shoulder characteristics of sizes 44 (bust girth 100 cm), 48 (bust girth 110 cm), and 56 (bust girth 140 cm) are compared with each other. The comparative measurements are "shoulder width contour" and "back width." The "shoulder width contour" is the distance between the right and left acromia above the seventh cervical vertebra along the body contour. The "back width" is measured as the horizontal distance from arm base to arm base across the back, in the middle between the acromion and the lower base of the arm. In size 44 the measurement "shoulder width contour" is 9 mm larger than the back width. In size 48 the "shoulder width contour" is already 7 mm and, in size 56, even 3.5 cm smaller than the back width. These body shapes are described by the average body dimensions in the tables.

The shoulder-to-back width ratio in the larger sizes has two significant effects on the pattern design. Firstly the shoulder widths should not be too large, and secondly the armhole curve of the back should not be too round. If the specific body shape

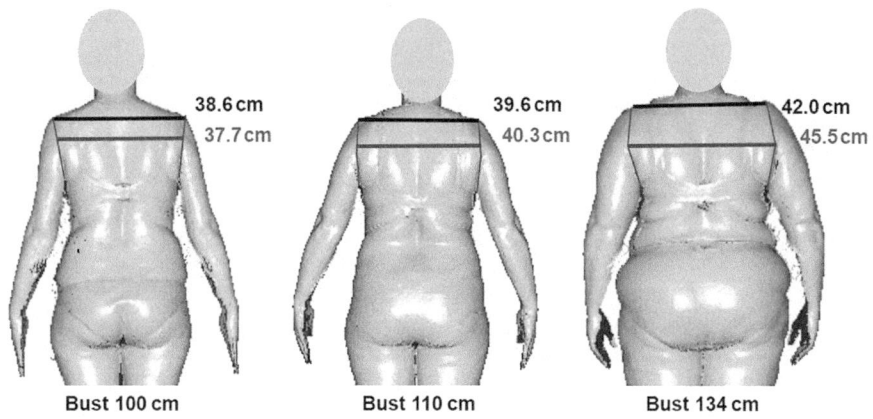

Fig. 14.15 Correlation between biacromial shoulder width and across back width.

characteristics are not taken into account during pattern development or grading, this inevitably leads to fitting problems. It becomes clear that a simple linear upgrade of the base size has to lead to fitting defects, since the actual body proportion of the respective size cannot be reproduced by deriving the parameters of the base size.

14.7.2.2 Morphotypes—Plus-size men

One of the most important men's body features determining the figure is the form of the abdomen. Accordingly the basis for the evaluation of the male morphotypes by abdominal shape was the body measurements "waist diameter side" and "hip diameter side." The statistical evaluation of the diameters was supplemented by visual inspection and clustering of the scan data.

Four abdomen types were identified: the athletic type, the abdominal type upright, the abdominal type normal, and the abdominal type hanging. Athletic types are men with large garment sizes only because of their large chest girths caused by chest muscles typical for specific athletes. Abdominal type upright describes stomachs pointing upward. Abdominal type normal characterizes stomach shapes that do not influence the hip girth. Fig. 14.16 shows hip girth marked with a line to emphasize that the hip girth may be measured underneath the belly. Abdominal type hanging specifies men with a stomach shape influencing the hip girth. This leads to a wide dispersion of the individual measurements.

The evaluation of the percentages shows that the "normal stomach type" is most frequently represented and dominates up to size 68 (chest girth 136 cm). The proportions of the "hanging stomach type" are very small in size 60 (chest girth 120 cm) and increase heavily with increasing chest girth and then determine the sizes from 70 up (chest girth 140 cm). The sporty type is mainly found in the smaller sizes but surprisingly represented up to size 68 (chest girth 136 cm). The standing stomach type is underrepresented throughout and is therefore more of an exception. In contrast to

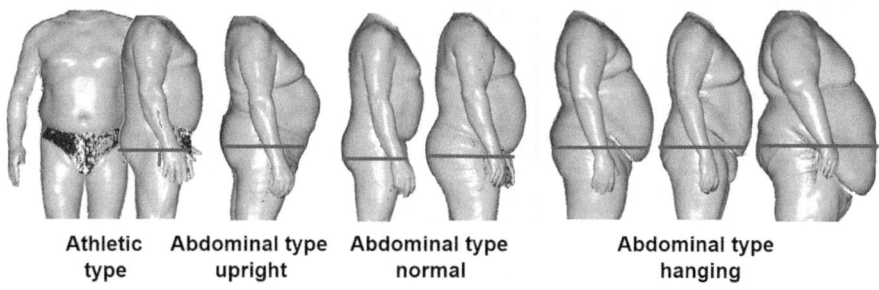

Athletic   **Abdominal type**   **Abdominal type**   **Abdominal type**
type   **upright**   **normal**   **hanging**

Fig. 14.16 Visualizing four abdomen types.

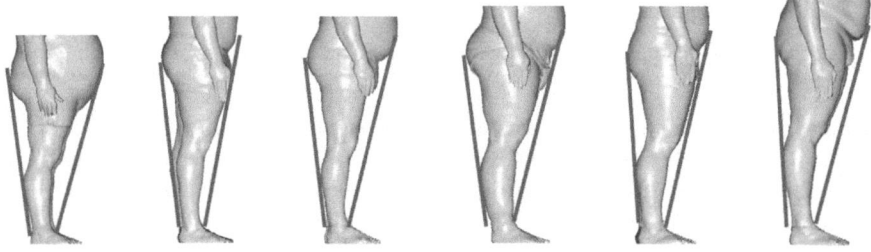

Fig. 14.17 Visualizing of typical thigh shapes based on the example size 66 (chest 132 cm).

the abdominal region, which develops significantly with increasing chest girth, the hip and thigh areas remain rather narrow. Fig. 14.17 demonstrates this.

When looking at the individual shapes, both narrow and wide thigh shapes must be identified. However, the heavily abstracted reproduction of a pants shape (shown in Fig. 14.17) shows that thigh girth plays a subordinate role in comparison with the abdominal area. A differentiation via defining different morphotypes is therefore not necessary, but there is the challenge that is placed on the pattern designer during the technical conversion of the body shapes into well-fitting pants.

14.7.3   Representative 3-D body models

Based on the statistical evaluation of body measurements and comprehensive 3-D shape analyses, average 3-D body models were developed. These visualize the typical body shapes in the different sizes and serve as a basis for fitting dummies, 3-D construction, and the simulation of clothing.

14.7.3.1   Representative 3-D body models—Women

The middle 3-D body models represent the average body shapes in women. Fig. 14.18 shows the 3-D body models of the normal sizes 50 (bust girth 116 cm), 54 (bust girth 128 cm), and 58 with a bust girth of 140 cm. For women the change in waist and hip proportions in particular must be taken into account. In the small and medium sizes,

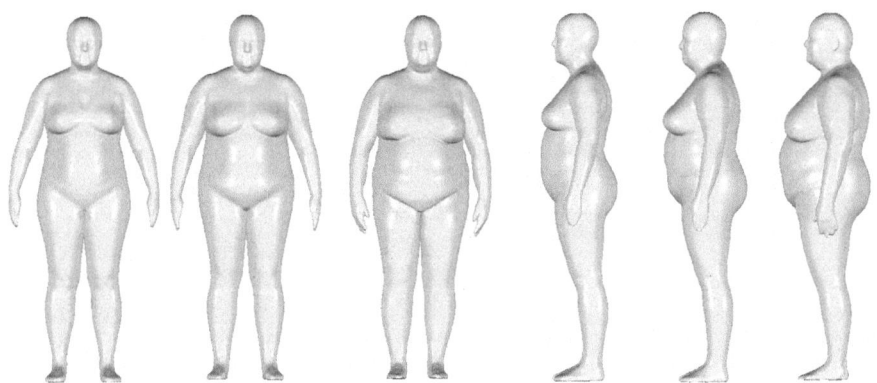

Fig. 14.18 Virtual 3-D female body models in the German sizes 50, 54, and 58.

the hip girth is very pronounced, and the waist is narrow. In the larger sizes the waist area increases significantly, and the hips remain narrow in relation to the waist. The shoulder area also remains rather narrow in relation to the total increase of the girths. The across back width increases, but less heavily than the torso in the waist area.

The 3-D body models represent the "normal" body shape, which has the largest market shares in all sizes. Separate 3-D models have to be generated for the other body shapes and morphotypes, for example, the wide hips or women with the A-silhouette.

14.7.3.2 Representative 3-D body models—Men

The average models represent the average body shape of men, which visualizes in particular the stomach shape of the individual sizes. Fig. 14.19 shows the 3-D body models of the German men's sizes 62 (chest girth 124 cm), 66 (chest girth 132 cm), and 70 (chest girth 140 cm) in the "normal" height range and the "normal" body shape. Fig. 14.20 shows the 3-D bodies in the normal size range with a chest girth of 132 cm in the five body shapes of extra heavy, heavy, normal, slim, and extra slim.

Some of the 3-D body models shown here still have individual characteristics, such as the underwear worn by the test subjects or even the abdominal characteristics, which still do not change 100% uniformly. Some body details of the 3-D shapes are even more optimized. The legs and arms in particular are difficult to convey due to different postures while scanning, and the postprocessing effort is extremely high here. Ultimately the degree of idealization of such body models depends to a large extent on their intended use.

14.7.4 Basic patterns for jackets and pants

As the chest/bust girth increases, the body proportions change, sometimes considerably as shown in some cases. The changes in the body geometry have a decisive effect on the pattern design. Designs that provide well-fitting clothing for small and medium clothing sizes result in an unsatisfactory fit for body shapes in the larger sizes. Based

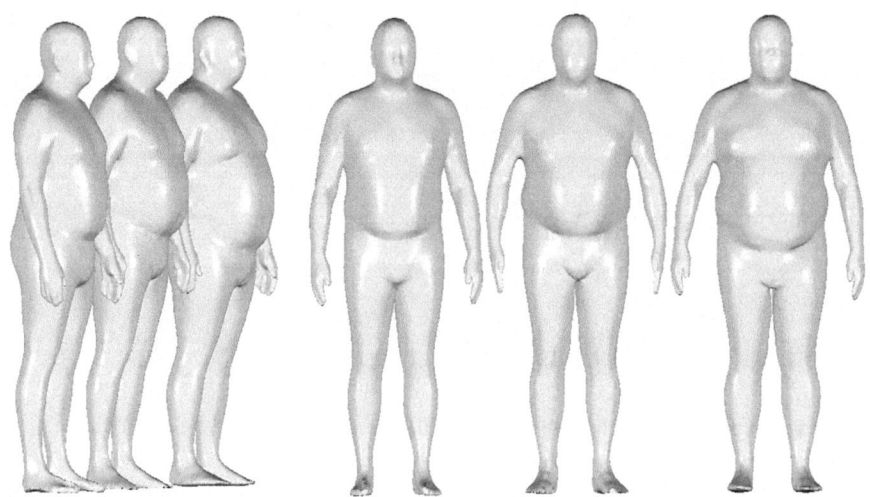

Fig. 14.19 Virtual 3-D male body models in the German sizes 62, 66, and 70.

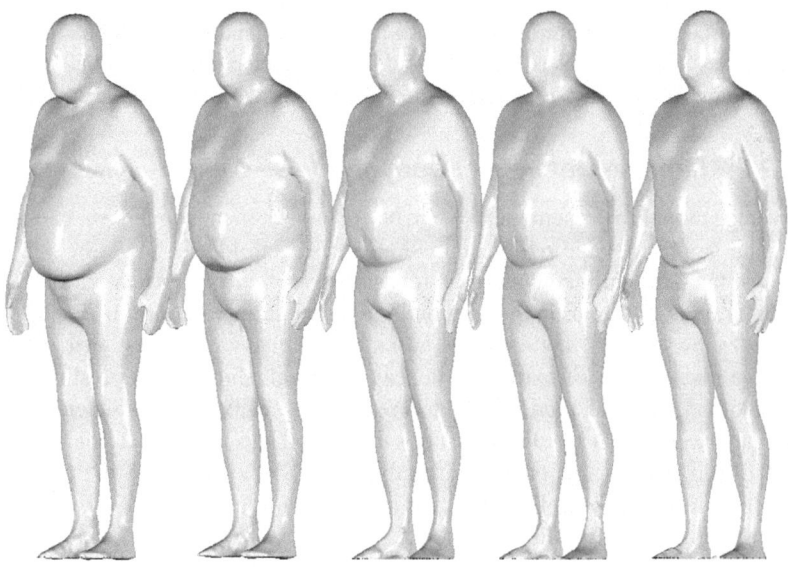

Fig. 14.20 Three-dimensional body models: chest girth of 132 cm in the five body shapes of extra heavy, heavy, normal, slim, and extra slim.

on the body measurements and 3-D body shape analyses, fit-optimized basic patterns for jackets and pants were therefore developed especially for large sizes. New solution paths have been taken here.

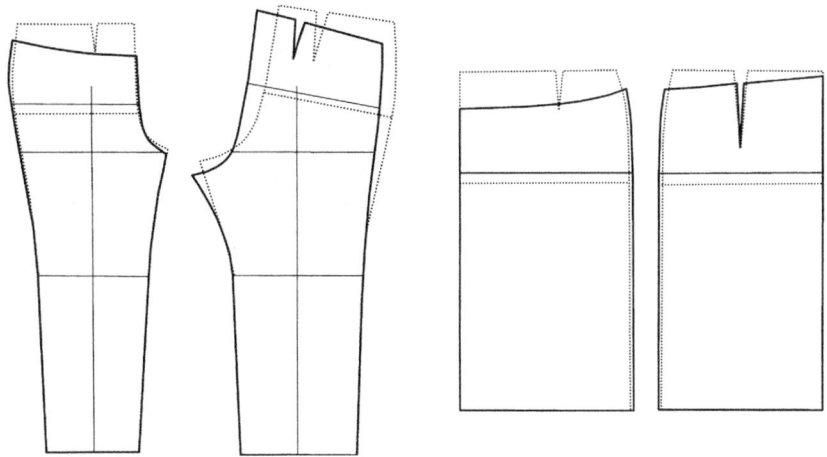

Fig. 14.21 Women's basic patterns using the examples of pants and skirt.

14.7.4.1 Women's basic patterns

For the ladies the "natural waist girth" and the "distance of waist girth-natural waist" were integrated into the basic pattern development. Both measurements listed in the size tables can be transferred 1:1 into the basic patterns for pants and skirts and thus guarantee an optimal waist shape and fit. Fig. 14.21 demonstrates the principles for plus-size patterns compared with normal sizes by means of German size 56 with bust measurement 134 cm.

Overall, in the pattern development, the following changes result for the large sizes in comparison with the small and medium sizes. The specification is to be adjusted for specific sizes.

The basic changes to the example of the skirt are as follows:

- The waistband line is to be adjusted in height and girth to the natural waist shape.
- A larger-waist size is required.
- The hip shape needs to be lower.
- The distance from waist to hip (vertical distance between waist and hips) is shortened.
- The "normal" body shape does not require a tuck in the front part (only the wide hip body shape).

The fundamental changes for the pants essentially correspond to those of the skirt. Two further parameters must be taken into account. The abdomen height must be extended, and the inner leg length must be shortened. The incline of the back trousers has to be reduced.

14.7.4.2 Men's basic patterns

In the "plus-size men" project, the traditional approach of 2-D patterning development was supplemented by a new 3-D-based solution. The idealized body surface of the average 3-D body models was transformed into a plane to derive 2-D pattern part

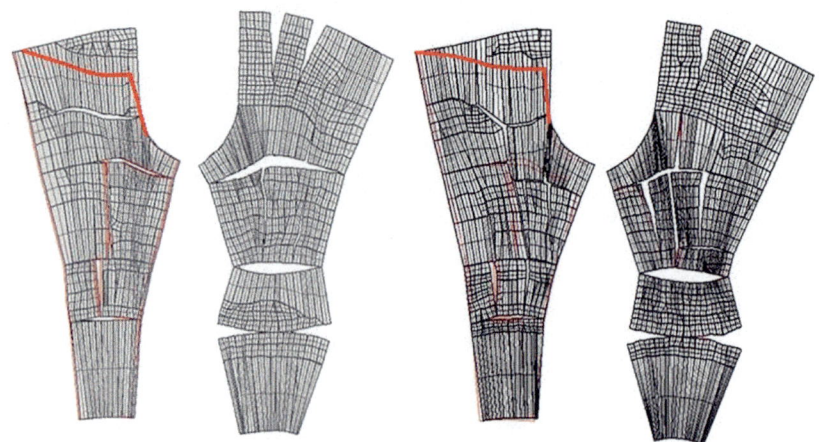

Fig. 14.22 Constructed basic trousers' pattern—size 70 "normal" and "extra heavy."

geometries based on 3-D body data. The method has already been described in detail in Chapter 6. This procedure ensures the perfect fit of basic patterns, as the 3-D body geometry is incorporated directly into the pattern pieces.

Fig. 14.22 shows the pattern piece geometries generated by unwinding and arranging the designed NURBS surfaces from the 3-D bodies. The results are basic patterning shapes in size 70 for the "normal" and "extra-heavy" body shapes.

The shape of the hips is conspicuous, as no hip curve is required in the pattern. In contrast to the conventional construction, the distribution of the seat seam length at the front part is considerably longer and correspondingly shorter in the back of the pants. This is primarily due to the determination of the vertical segmentation in the middle of the body. The differences between the unwinding with a normal waist seat and the lower abdomen unwinding are visualized. Once again, it must be pointed out that the pattern technical construction of the waistband cannot be implemented as reliably in terms of shape and height as is the case with women. Men's individual wearing preferences are too different.

Fig. 14.23 shows the 3-D body shapes, the designed NURBS surfaces on the 3-D model, and the unwound pattern piece geometries for the upper half of the body in size 70 (chest girth 140 cm) in extra-heavy body shape. Based on the pattern part geometries, a simple prototype was produced to evaluate the balance in the appropriate size on a test subject. The unwound pieces prove the necessity of large tucks in the abdominal area to optimally reproduce the body shapes. Furthermore the unwindings show an almost straight edge in the center front. Despite the reshaping of the waist, the side seams of the front parts are straight, while the back has a waist. The waist is compensated for at the front by the stomach tuck. The functional pattern depicted shows a perfect balance in size 70 that can be achieved with the aid of the large dart.

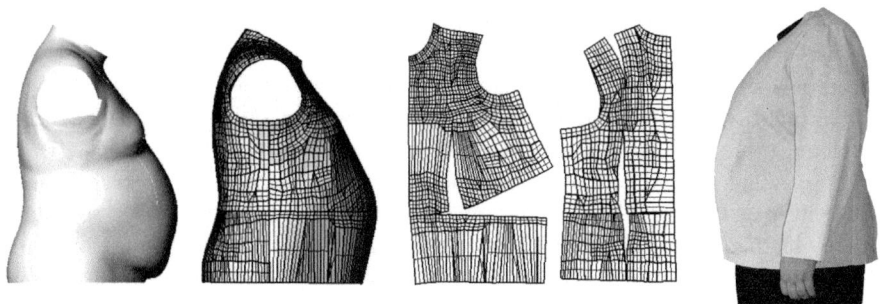

Fig. 14.23 Average 3-D body form, constructed basic jacket pattern and fitting test—German size 70 "heavy".

14.8   Conclusions

With the two presented projects, comprehensive body data for men and women of large sizes were collected and evaluated for the first time in Germany. The project results provide a valuable foundation for the production of well-fitting clothing for men and women of large sizes.

The challenge was to generate a meaningful sample that could not be realized in all previous standard series measurements. Due to the lack of data, manufacturers and retailers were forced to implement their own developments empirically. This resulted in two significant problems. Firstly the individual size interpretations of the manufacturers in retail, which have arisen in the absence of a uniform size standard, must be pointed out. This has caused an enormous variety of sizes on the market. Secondly the fitting problems in this segment also need to be pointed out. The clothing products designed for the "normal sizes" cannot be adapted to the shapes of the large sizes by linear grading or extrapolation of their dimensions. Nonobservance of the morphological characteristics of the large sizes inevitably leads to fitting defects. With the available project results, the data gap can now be closed, and target group-oriented product development can be effectively supported.

It should be noted, however, that due to the extraordinary variety of body shapes in the larger sizes, it is not possible to dress all shapes optimally with just one pattern. There cannot be one good fit for everyone. This is only conditionally possible even in small and medium sizes, but in large sizes, it is not at all feasible. The definition of the target group to be served must always be carried out carefully to achieve good fitting results. The description of the morphotypes provides valuable support in doing so.

The results presented here are based on samples collected exclusively in Germany. However, it is known that the ethnic groups of this world differ significantly in terms of their morphological body shapes. The problem of incorrectly fitting clothing is increasingly obvious with an increase in ethnic markets and international market expansion. It would therefore be preferable if comparable studies could be carried out worldwide, so that the large sizes of different population groups could be compared and contrasted with each other.

14.9   Future trends

When plus-size women and men report their problems finding well-fitting clothing, people often conclude that this is caused by inconsistent sizing. Besides the sizing problems, these difficulties mainly result from body shape and cut (more specifically grading) rather than size. The development and advancement of 3-D technologies within the past two decades allow to capture, to analyze and to interpret 3-D scan data, and to obtain meaningful anthropometric body data as basis for well-fitting clothing. The target group-specific body data—presented in this chapter—support sustainably the development of plus-size pattern and fit. Fashion designers and apparel manufacturers are supposed to utilize this measurement and shape data, to formulate adequate pattern blocks, and to create well-fitting clothing for plus-size men and women.

A further important step on the way toward well-fitting plus-size clothing is the provision of meaningful fitting tools like plus-size dress forms and plus-size virtual avatars for 3-D simulation software for fashion. One essential tool for an optimized fit approval process is the dress form to check and improve the fit of clothing. But there is still a lack of availability of plus-size dress forms with realistic body proportions. The same problem applies to the virtual avatars being used in the virtual garment simulation systems. A great challenge is the difficulty of simulating and modeling apparel and textile products. Advanced virtual 3-D garment simulation can help to enhance the communication process in the whole value chain including designers, patternmakers, merchandisers, manufacturers, suppliers, and clients by acting as a common language for everyone involved in this process. Besides the real-time simulation of mechanical properties of cloth and their influence on fit, a high-quality avatar with realistic body shapes is the prerequisite for a meaningful simulation result. So far the utilization of 3-D simulation systems for the development of plus-size clothing is still very limited because no standard plus-size avatars with realistic body shapes are available. To support the efficient advanced development of plus-size clothing, further development efforts are needed at this point. The anthropometric results presented will support the further development of useful and substantial fitting tools not only for standard sizes but also for plus-size clothing.

14.10   Source of further information and advice

The research reports including size charts, market share tables, and visualization of the target group–specific body shapes are available at Hohenstein. The contact information can be found here:

http://www.hohenstein.de/plus-size-men

http://www.hohenstein.de/plus-size-women

Further information and advice cannot be recommended due to a lack of international availability. Representative studies, standards, and reliable body data concerning this special target group are missing worldwide. This demonstrates the great need of further anthropometric research.

Acknowledgment

The IGF projects 17460N and 15144 BG under the auspices of the Research Association Forschungskuratorium Textil e.v., Reinhardtstraße 12-14, 10117 Berlin, were sponsored via the AIF as part of the program to support "Industrial Community Research and Development" (IGF), with funds from the Federal Ministry of Economics and Energy (BMWi) following an order by the German Federal Parliament.

References

Ashdown, S.P., 2007. Sizing in Clothing: Developing Effective Sizing Systems for Ready-to-Wear Clothing. Woodhead Publishing in association with The Textile Institute, Cambridge, England.

Forschungsinstitut Hohenstein, Bekleidungstechnisches Institut e.v, Forschungsgemeinschaft Bekleidungsindustrie e.v, 1980. Körpermaße und Marktanteile für Herren-Oberbekleidung 1980—Ergebnisse der HAKA-Reihenmessung 1978/79. HAKA Treuhand GmbH, Köln.

Hohenstein & Human Solution, 2008. Size Charts Women SizeGermany.

Hohenstein & Human Solutions, 2007–2009. SizeGERMANY. [Online]. Available: http://www.sizegermany.de/. Accessed 12 July 2018.

Hohenstein & Human Solutions, 2008. Size charts men SizeGermany.

Human Solutions GmbH, 2018. Anthroscan. [Online]. Available: https://www.human-solutions.com/mobility/front_content.php?idcat=257&idart=495&lang=8. Accessed 12 July 2018.

ISO—International Organization for Standardization, 1989. ISO 8559:1989: Garment Construction and Anthropometric Surveys—Body Dimensions, first ed. 1989-07-01 ed.

ISO—International Organization for Standardization, 2010. ISO 20685:2010: 3-D Scanning Methodologies for Internationally Compatible Anthropometric Database, second ed.

Kirchdörfer, E., Mahr-Erhardt, A., 2003. Oberbekleidung für Frauen über 60 Jahre—Körperdimensionen, Größenverteilung, Schnittkonstruktion. IGF 12890. Bekleidungstechnische Schriftenreihe Band 156, Hohenstein, Köln.

Kirchdörfer, E., et al., 2011. Entwicklung einer Automatischen, Körperkonformen Konstruktionssystematik für Oberbekleidung auf Basis von 3D-Körper-Informationen. IGF 15605N. Hohenstein Institut für Textilinnovation e.V., Hohenstein.

Morlock, S., Schenk, A., Harnisch, M., 2015. Passformgerechte und bekleidungsphysiologisch optimierte Bekleidungskonstruktion für Männer mit großen Größen unterschiedlicher Körpermorphologien—IGF-Vorhaben 17460N. Hohenstein Institut für Textilinnovation gGmbH, Hohenstein.

Morlock, S., Schenk, A., Klepser, A., Schmidt, A., 2016. XL plus men—new data on garment sizes. In: 7th International Conference on 3D Body Scanning Technologies, 2016 Lugano, Italien: https://doi.org/10.15221/16.255.

Morlock, S., Wendt, E., Kirchdörfer, E., Rupp, M., Krzywinski, S., Rödel, H., 2009. Grundsatzuntersuchung zur Konstruktion Passformgerechter Bekleidung für Frauen mit Starken Figuren. BPI Hohenstein, TU Dresden, Hohenstein.

Verband der Damenoberbekleidungsindustrie e.V, 1994. DOB-Größentabellen 1994: Körpermaßtabellen, Marktanteiltabellen und Konstruktionsmaße für Damenoberbekleidung. GermanFashion Modeverband Deutschland e.V., Köln.

Verband der Herrenbekleidungs-Industrie e.V. (HG.), 1960. Größentabelle für Herren- und
    Knabenoberbekleidung Nachdruck 1990, Köln.
Vitronic, 2018. Available: https://www.vitronic.de/. Accessed 11 July 2018.
Zakaria, N., Gupta, D., 2014. Apparel Sizing: Existing Sizing Systems and the Development of
    New Sizing Systems Anthropometry, Apparel Sizing and Design. Woodhead Publishing,
    Cambridge (GB), pp. 3–33.

Index

Note: Page numbers followed by *f* indicate figures and *t* indicate tables.